Single and Multicomponent Digital Optical Signal Analysis

Estimation of phase and its derivatives

Single and Multicomponent Digital Optical Signal Analysis

Estimation of phase and its derivatives

Rishikesh Kulkarni
EPFL, Lausanne, Switzerland

Pramod Rastogi
EPFL, Lausanne, Switzerland

IOP Publishing, Bristol, UK

ISBN 978-0-7503-1469-5 (ebook)
ISBN 978-0-7503-1467-1 (print)
ISBN 978-0-7503-1468-8 (mobi)

DOI 10.1088/978-0-7503-1469-5

Version: 20170901

IOP Expanding Physics
ISSN 2053-2563 (online)
ISSN 2054-7315 (print)

British Library Cataloguing-in-Publication Data: A catalogue record for this book is available from the British Library.

Published by IOP Publishing, wholly owned by The Institute of Physics, London

IOP Publishing, Temple Circus, Temple Way, Bristol, BS1 6HG, UK

US Office: IOP Publishing, Inc., 190 North Independence Mall West, Suite 601, Philadelphia, PA 19106, USA

In loving memory of my parents

O stars, in thy solitude far off
I seek thee a favor:
Keepeth twinkling in my heart
The memories of my parents,
The fuel that fires life in me
And sweetens the melody of my life
Lived every instant
In the shadows of their love,
Timeless in its moorings.

Pramod Rastogi

Dedicated to my parents

Rishikesh Kulkarni

Contents

Preface

A new era of opportunities has opened up in interferometric coherent optical imaging with the application of digital signal processing tools for the analysis of spatially displayed signals, in the form of fringe patterns that these interferometric techniques generally provide at their output. These tools aim at demystifying and extracting quantitative information from the interferometric data in line with users' interests by appropriately manipulating and handling the content rich data normally produced by the family of methods underlying these interferometric imaging techniques.

Contributing to this dawn, this book proposes a wide range of signal processing techniques to address the problem of accurate estimation of single/multiple phase and/or phase derivatives from single/multiple frame(s) of single/multi-component real/complex valued sinusoidal fringe patterns recorded in single/multi-wave digital/conventional holographic/speckle interferometry setups. The work serves as a launching pad to a more pronounced understanding of how the single/multi-component information embedded in the fringe signal at the output of these interferometric systems could be demodulated. Finally, the book also paves the way towards carving a platform that allows the closely related areas in as far as signal and image processing techniques are concerned to exploit their natural synergy in delivering reliable measurements wherever it is possible to obtain signal data in the form of interferometric images.

The book provides numerical simulations and practical examples, wherever necessary, to confirm the feasibility, effectiveness and accuracy of the panoply of methods described in the book. With much to cover in a modest space, the book's approach has been to favor a breadth of coverage at the expense of details. The book focuses on an overall view of concepts attracting current research attention by:

- Adopting a digital signal processing approach to spatial and temporal fringe demodulation;
- Offering innovative solutions for the demodulation of multicomponent signals;
- Proposing a range of ground breaking avenues for estimating simultaneously multiple phase components;
- Providing a range of methods for the simultaneous estimation of multiple phase derivatives of first order; and also the single-phase derivatives of arbitrary order p;
- Continuing a strong focus on key topics of interest such as closed fringe demodulation; and fringe denoising and phase unwrapping operations.

We hope that this book will be useful, at least, in conceptual terms, to all those who are using, or are contemplating using digital signal processing in interferometric coherent optical imaging, and in elaborate terms, to all those who are engaged, or are contemplating engaging in its sub domain of fringe analysis, the area to which this text is specifically devoted.

Rishikesh Kulkarni, PhD
EPFL, Lausanne

Professor Pramod Rastogi
EPFL, Lausanne
30 April, 2017

Acknowledgements

I would like to express the deepest of my appreciation and affection to Abhijit Patil, Rajesh Langoju, Sai Siva Gorthi, Rajshekhar Gannavarpu and Rishikesh Kulkarni, some of my past students whom it has been my pleasure and privilege to mentor for their respective PhDs in the field between 2002 and 2016 at the Swiss Federal Institute of Technology Lausanne. Abhijit and Rajesh are currently with GE Global Research Centre, Bangalore; Sai Siva is with the Indian Institute of Sciences Bangalore; Rajshekhar is with the Indian Institute of Technology Kanpur; and Rishikesh is with the Indian Institute of Technology Guwahati.

Pramod Rastogi

Author biographies

Rishikesh Kulkarni

Rishikesh Kulkarni received his MTech degree from the Indian Institute of Sciences, Bangalore, India. He received his PhD in Photonics from the Ecole Polytechnique Federale de Lausanne (EPFL). He worked as a postdoc in Applied Computing and Mechanics Laboratory at EPFL. He joined Indian Institute of Technology, Guwahati as an assistant professor in July 2017. His interest lies in digital signal processing and its applications in optical metrology.

Pramod Rastogi

Pramod Rastogi received his MTech degree from the Indian Institute of Technology Delhi, and doctorate degree from the University of Franche Comté in France. He joined the Ecole Polytechnique Federale de Lausanne (EPFL) in Switzerland 1978. He is the author or coauthor of over 170 scientific papers published in peer-reviewed archival journals. Professor Rastogi is also the author of encyclopedia articles, and has published several books in the field of optical metrology. Professor Rastogi is the 2014 recipient of the SPIE Dennis Gabor Award and is a member of the Swiss Academy of Engineering Sciences. He is a Fellow of the Society of the Photo-Optical Instrumentation Engineers (1995) and a Fellow of the Optical Society of America (1993). He is also a recipient of the Hetényi Award for the most significant research paper published in Experimental Mechanics in the year 1982. Professor Rastogi is the co-editor-in-chief of the *International Journal of Optics and Lasers in Engineering*, Elsevier.

IOP Publishing

Single and Multicomponent Digital Optical Signal Analysis

Estimation of phase and its derivatives

Rishikesh Kulkarni and Pramod Rastogi

Chapter 1

Introduction

Introduction

Measurement of variations in physical quantities is of key importance in a wide range of applications cutting across disciplines in science and engineering. The development of sophisticated measurement tools has made a great impact on scientific, engineering and technological advancements, which in turn have contributed to shaping human life. Whether it is in the validation of a hypothesis in science, in the objective evaluation of the data recorded in an experiment, or in the evaluation of the performance and control of engineering systems, high performance measurement systems are key to addressing each of these underlined tasks. Measurements provide a quantitative and standard way of evaluating and understanding systems under observation. Even in day-to-day life, quantitative information based observation is valued much more than human dependent qualitative observation.

Rapid developments in digital technology are changing the course of measurement systems from the analog to the digital domain to the point that today almost all measuring instruments are equipped with an in-built functionality of converting the measured information on the physical quantity of interest into a digital signal that can be fed into a computer. The ability of operating on digital signals using a computer has made a tremendous impact on the way the measurement data is handled. Digital signal processing is finding extensive use in dealing with the signal data in the form of modifying or rearranging it, or in realizing its reduction, conversion or transformation. This pursuit has facilitated an improvement in the interpretabilty of the data embedded in the signal. The huge computation power and the relatively low cost of measurement devices have accelerated the use of signal processing techniques in analyzing the data in an effective and efficient manner. The expression 'data is the new oil' takes all its meaning in the age of digitization that we are living in.

doi:10.1088/978-0-7503-1469-5ch1

The word 'image' is usually associated with a two-dimensional intensity map of an object (or a scene). In digital imaging, these images are generally recorded on a CCD or CMOS camera. The imaging operation in a broader sense may include light sources lying inside or even outside the visible region of the electromagnetic spectrum, such as, x-ray, infrared, or microwave. An imaging operation can be broadly classified into incoherent and coherent imaging. In incoherent imaging, the object is illuminated with spatially incoherent light and the imaging operation involves recording of the intensity reflected, emitted (self-luminous), or transmitted by the object under observation. On the other hand, in coherent imaging, the object is usually illuminated with laser light, and the imaging operation is linear in complex amplitude.

Over the years, coherent interferometric imaging has undergone intensive development, with the emergence in its fold of techniques such as synthetic aperture radar, synthetic aperture sonar, magnetic resonance imaging, digital holographic interferometry and digital speckle pattern interferometry. These techniques record both the amplitude and phase of the wavefront. The ability of recording the phase of a wavefront offers a unique opportunity to these techniques to measure physical quantities ranging from object deformation, surface topography to variations in a refractive index profile, or a magnetic field. Digital signal processing has in large part contributed to the enormous success of these techniques. In the following, we will briefly discuss some of these techniques with the importance of phase measurement clearly shown for each of them.

Interferometric synthetic aperture radar imaging (InSAR) is an important coherent interferometric imaging technique that has found use in remote sensing and high resolution terrain mapping with a resolution ranging from few centimeters to several meters. Using the coherent nature of the probing microwave signal, its working principle is based on the comparison of a coherent broadband microwave signal transmitted towards the terrain with the reflected signal recorded at the receiver after some time delay. Both the transmitter and receiver are mostly mounted on an aircraft. The processing of these two signals provides a high resolution image of the terrain under observation. In InSAR, two radar antennas are utilized to record two complex valued images of the terrain. The radar antennas are arranged such that the amplitudes of the two complex images remain the same. Accordingly, the conjugate multiplication of these two images generates a complex signal carrying the information on the height of each point in the image of the terrain in the form of phase. However, the phase map is usually obtained in a wrapped form, i.e. the phase values are mapped in the range of $(-\pi, \pi]$. Since the terrain height is proportional to the absolute phase, *phase unwrapping* is performed to obtain the absolute phase estimate. The accuracy of terrain mapping being mainly decided by the accuracy with which the phase unwrapping is performed, an extensive amount of literature is available on methods to obtain reliable phase unwrapping. An example of a wrapped phase image corresponding to a terrain map is shown in figure 1.1(a). Figure 1.1(b) shows the coherence map indicating the quality of the wrapped phase at each pixel location. Such coherence maps are helpful in achieving reliable phase unwrapping. The principle of interference of coherent waves has also been applied in

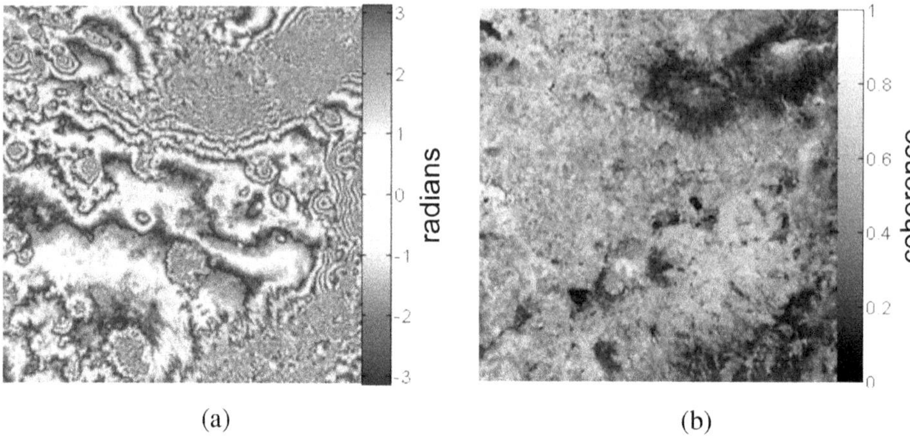

Figure 1.1. (a) Phase image corresponding to a terrain map recorded using interferometric synthetic radar interferometry; (b) coherence map indicating the quality of the phase image. Courtesy: E Cabral-Cano, Instituto de Geofisica, Universidad Nacional Autonoma de Mexico.

interferometric synthetic aperture sonar (InSAS) to image below water environments with a range independent resolution. In InSAS, similar to the InSAR, the information on the height of the seafloor is embedded in the phase that is derived from the synthetic aperture sonar images. An example of the InSAS image and the wrapped phase map corresponding to a tank at the bottom of a lake is shown in figures 1.2(a) and (b), respectively.

Magnetic resonance imaging (MRI) is a widely used diagnostic tool in medical applications. Complex valued images are recorded in MRI, with the amplitude and phase carrying the information on the body part under observation. Whereas the amplitude image carries the structural information of the body part, the phase image provides the information on the magnetic field inhomogeneities and blood flow. The phase information is particularly utilized to correct the distortions present, such as, in echo planar imaging, in measuring the local deviations from the static magnetic field, and in high-resolution phase imaging, to name a few. An example of the amplitude and phase images recorded in an MRI setup associated with a human head is shown in figures 1.3(a) and (b), respectively.

One of the important subcategories of interferometric coherent imaging techniques is optical interferometry. Although the principle of coherent optical interferometry has been understood for a long time, one has had to wait until the invention of laser, which being a coherent light source, has made it practically feasible to implement, giving rise to an array of techniques in the field. High coherence length laser allows one to probe changes in the phase of the light wavefronts. Even though light possesses both the amplitude and phase as its attributes, most of the light detecting devices, such as the CCD/CMOS camera or human eye, are sensitive only to the amplitude of the light wavefronts. Let us take the example of one such optical system, holographic interferometry, which is sensitive to the phase of light wavefronts. It uses a laser to generate object and reference wavefronts. In this technique,

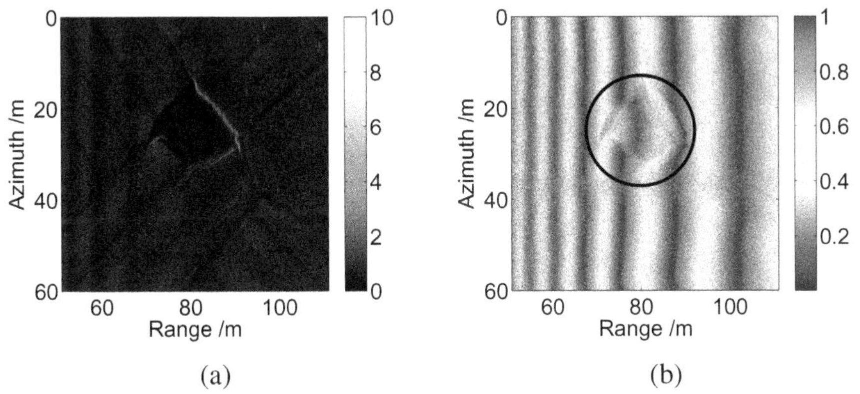

Figure 1.2. (a) Magnitude and (b) phase images recorded in interferometric synthetic aperture sonar. Courtesy: H Zhong, Naval Institute of Underwater Acoustic Technology, Naval University of Engineering, Wuhan, China.

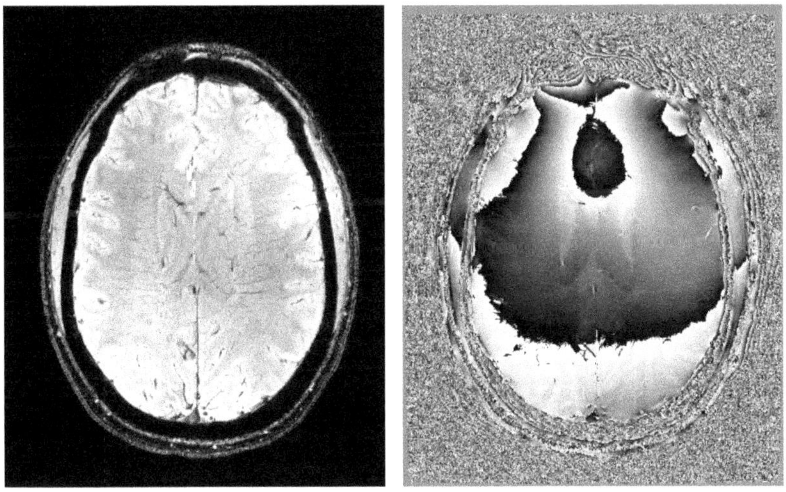

(a) (b)

Figure 1.3. (a) Magnitude and (b) phase images recorded in magnetic resonance imaging. Courtesy: S Robinson, High Field MR Center, Department of Biomedical Imaging and Image-guided Therapy, Medical University of Vienna, Austria.

the phase of the object wavefront transmitted through or scattered from an object is compared with the same wavefront but after it has undergone a change, for example, due to a force applied to it. The interference pattern carries the information on the phase difference between the two states of the object. It is interesting to note that although the interference pattern is recorded by a typically light detecting device sensitive only to the amplitude of the interference signal, the intensity of the interference pattern is a function of the phase difference. This principle of measurement is common to various other forms of optical interferometry, such as electronic speckle pattern interferometry and shearography. The change in the two states of the object appears in the form of a *fringe pattern*, which encodes the measured physical quantity. The intensity of a fringe pattern can be expressed as

$$\Gamma(x, y) = a(x, y) + b(x, y)\cos(\phi(x, y)) \tag{1.1}$$

where, $\phi(x, y)$, represents the phase carrying the information on the change in the state of the object. The fringe intensity in (1.1) is applicable to other types of interferometric setups, such as, fringe projection profilometry and moiré interferometry. Let us now consider the digital form of holographic interferometry. The compelling reason for such interferometry is that it allows for the recording of a two-dimensional complex valued fringe pattern, which can be expressed as

$$\Gamma(x, y) = A(x, y)\exp[i\phi(x, y)] \tag{1.2}$$

The expression of the fringe pattern in (1.2) is similar to the one used in the InSAR, InSAS and MRI techniques, wherein $A(x, y)$ represents the amplitude image and $\phi(x, y)$ represents the phase image (see figures 1.2 and 1.3). For the sake of clarity, we will throughout this book address equations (1.1) and (1.2) as *fringe signal* and *exponential phase field* (EPF), respectively. When the discussions are applicable to both types of signals, they will be commonly referred to as fringe signals. This should become clear from the context.

An example of the use of digital speckle interferometry in deformation analysis is shown in figure 1.4. Figure 1.4(a) shows the cross-section of a superconducting magnet employed in the Large Hadron Collider (LHC) at CERN, Geneva. The quality of the magnetic field being dependent on the deformation behavior of the magnetic coil when put under stress, the goal of this experiment is to investigate this behavior using the speckle interferometric method. The corresponding phase map is shown in figure 1.4(b). Electronic speckle pattern interferometry has been put to use to study the onset and development of heat convection from a suddenly heated thin horizontal wire in air. This is shown in figure 1.5. Another interesting example of electronic speckle pattern interferometry is in the non-destructive evaluation of art objects. The image on the left in figure 1.6 shows a 15th century fresco inside the S. Maria di Collemaggio Cathedral-L'Aquila, Italy. The rectangle in the image

(a) (b)

Figure 1.4. (a) Large Hadron Collider (LHC) superconducting dipole cross section; (b) wrapped phase corresponding to the deformation of the magnetic coil under compression obtained using digital speckle pattern interferometry.

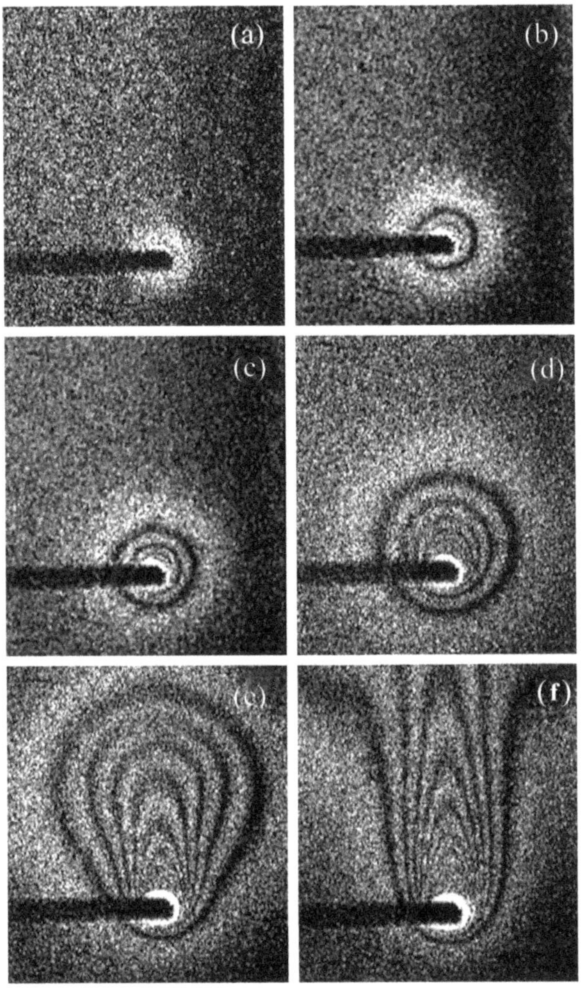

Figure 1.5. Onset and development of heat convection in air from an electrically heated tungsten wire (28.0 cm long, 0.2 mm diameter). Electronic speckle pattern interferometry images taken after (a) 0.12, (b) 0.36, (c) 0.52, (d) 0.68, (e) 1.1 and (f) 1.56 seconds, respectively, after application of current to the wire. Ambrosini D, Paoletti D, Schirripa Spagnolo G 2003 *Int J. Heat Mass Transfer* **46** 4145–55, reproduced with permission.

displays the area that is investigated. The fresco is subjected to thermal load and the resulting fringe pattern corresponding to the in-plane displacement is shown in figure 1.6(a). The discontinuities in the fringe pattern show the presence of the microcracks. That the information on the presence of these cracks can also be obtained from the correlation of the speckle images recorded prior to and post deformation is shown in figure 1.6(b). In comparision to conventional microscopy, digital holographic microscopy records both the phase and amplitude images of a biological specimen. Figure 1.7 shows an example of phase images corresponding to cells recorded using digital holographic microscopy. The wrapped and unwrapped

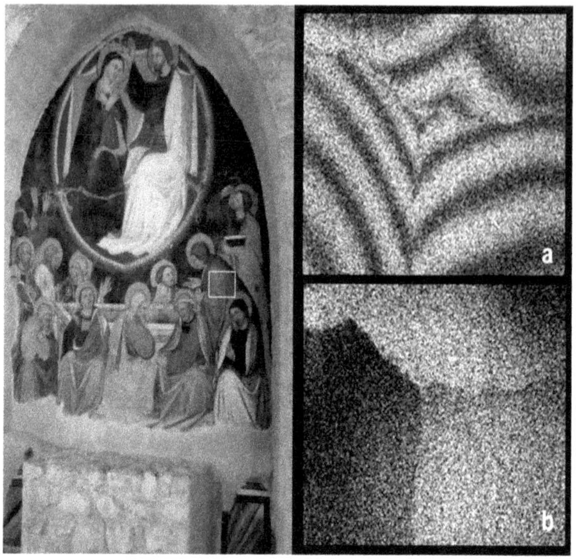

Figure 1.6. A 15th century fresco inside the S. Maria di Collemaggio Cathedral, L'Aquila, Italy (left); the yellow rectangle shows the investigated area. Detection of microcracks (a) using ESPI and (b) speckle correlation image. Reproduced with permission from Ambrosini D *et al* 1997 *Pure Appl. Opt.* **6** 557–63.

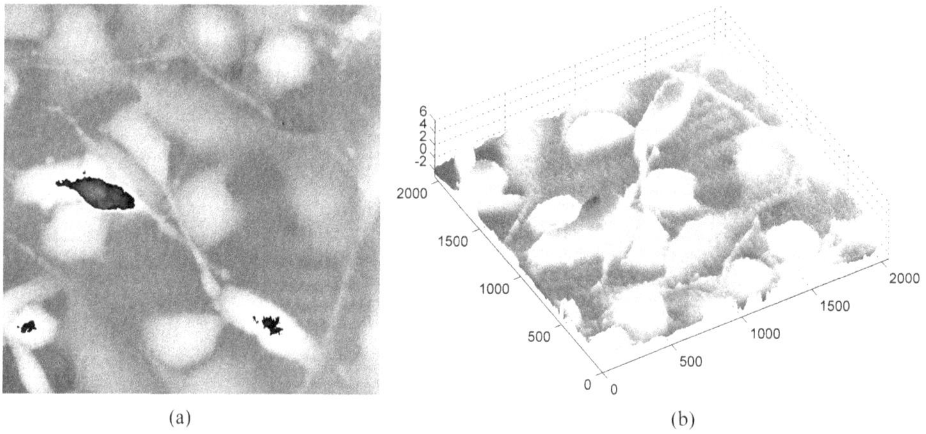

Figure 1.7. Analysis of fibroblast mouse cells by quantitative phase microcopy using a digital holography micoscope. Recorded phase images of cells displayed in their (a) wrapped and (b) unwrapped forms The phase values are in radians. Courtesy: P Ferraro, CNR-ISASI Applied Sciences and Intelligent Systems, Italy.

phase images shown in figures 1.7(a) and (b), respectively, provide the measurement of three-dimensional structural information on the cells.

Although the fringe signal and EPF carry the phase information on the physical quantity we wish to measure in the experiment at hand, this information is not directly accessible. This is further complicated by the fact that the data processing problem involved in each of these techniques is considerable. Nonetheless, the emergence of CCD/CMOS cameras has provided digital expressions and capabilities that the digital signal processing has seized upon to guide the course of these techniques to now unchartered territories, each in their own context, promising to open up new potentialities in each of the fields mentioned in the above.

Optical interferometric techniques such as holographic interferometry, electronic speckle pattern interferometry, and shearography, each involve the recording of sinusoidal fringe patterns (fringe signal) with spatially varying interference phase, or the digital holographic interferometry, which involves the recording of EPFs. Fringe analysis, which involves the estimation of the phase and phase derivatives from these fringe signals, is of utmost importance, for example, for a reliable and accurate measurement of displacements and strains undergone by a deformed object. However, the core of our understanding built in the domain over the years falls substantially short of addressing some of the issues that are relevant to practitioners, such as, the measurement of multidimensional displacement and strain components, and the pth order derivatives of displacements of diffusely scattering objects. This

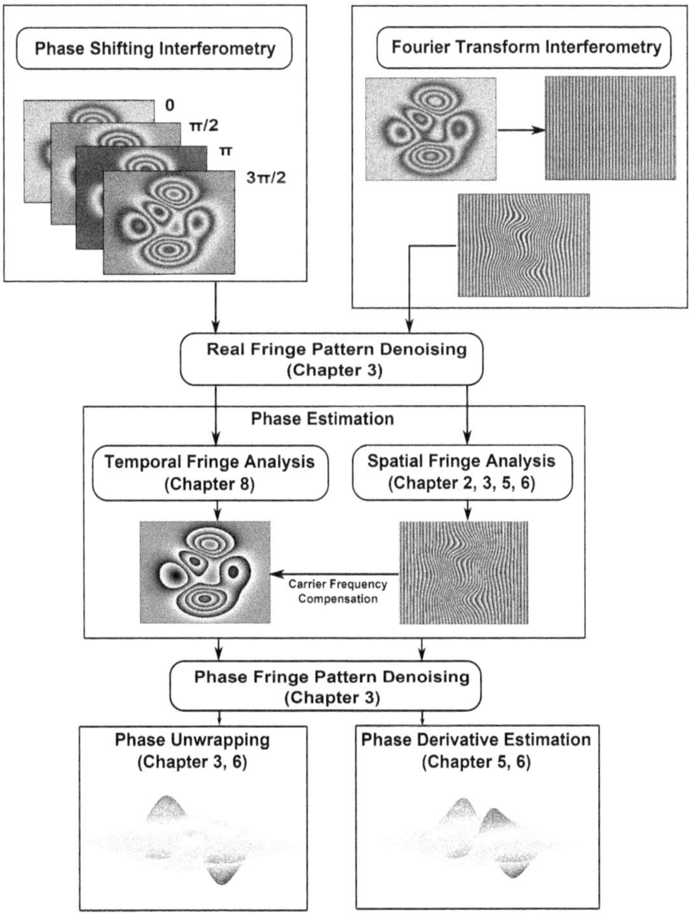

Figure 1.8. Single component fringe analysis for phase and phase derivative estimation.

book proposes a wide range of fringe analysis approaches to address the problem of accurate and simultaneous estimation of multiple phase and phase derivatives.

The work reported in this book represents an updated summary of the state of the research in the area of phase and phase derivative measurement by introducing an array of powerful methods based on digital signal processing for accurate and direct estimation of phase, and phase gradients in particular, in the optical interferometric techniques stated above, including digital holographic interferometry and fringe projection techniques. To keep the terminology simple, we have sometimes felt constrained to use the same variables more than once to represent different parameters, but due care has been assigned (as far as possible) to this task so that there is no such repeat within the same chapter. The use of these variables being context bound will help make this transition smooth for the reader as he/she navigates through the text.

The fringe analysis techniques undoubtedly hold great potential in unraveling the information carried by the fringe signal/EPF in the function of the needs of the investigator. The simplicity and the diversity of the experimental setups to which these fringe analysis techniques are applied make them particularly attractive in interferometric coherent optical imaging techniques in direct relationship to their thrust areas of applications, such as non-destructive testing, experimental

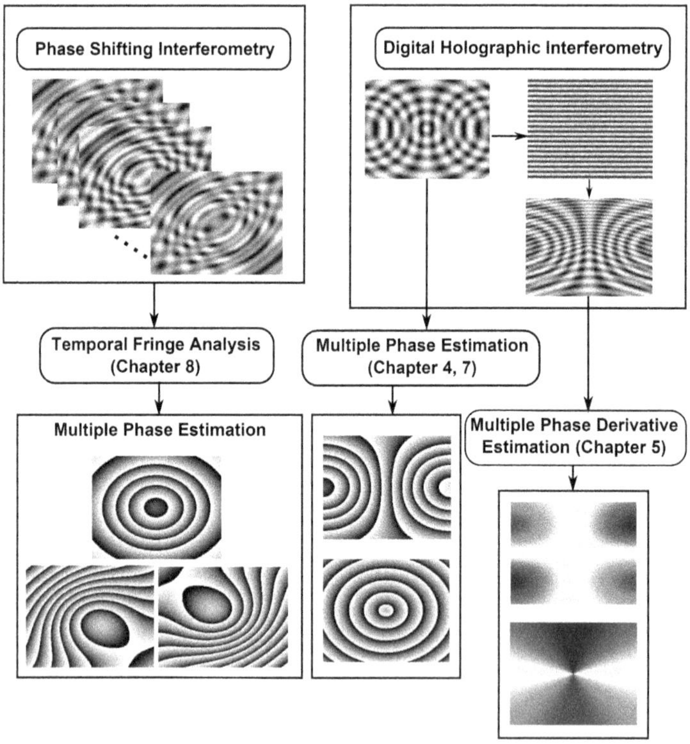

Figure 1.9. Multicomponent fringe analysis for multiple phase and phase derivative estimation. The fringe denoising and phase unwrapping steps are similar to those associated with single component fringe analysis.

mechanics, measurements in the industrial environment etc, and to conventional/digital holographic interferometry, for example.

The process of fringe analysis consists of several steps, with each step being an area of research in its own right. We will wind up this chapter by presenting flow chart descriptions of the different steps involved in fringe analysis. Chapter specific information related with various processes and algorithms designed for executing each step is also included in the flow charts. Figures 1.8 and 1.9 show the flow charts describing the different steps required for accomplishing fringe analysis in single component and multicomponent signals, respectively, thereby enabling the user in laying access to the information on phase/multiphase and/or their derivatives.

The fully fledged application of digital signal processing to retrieve the phase information and to which we are witness, has given rise to new configurations, novel concepts and approaches, and new horizons to the point of giving a new lease of life to systems that we had earlier come to recognize as being unreasonably difficult to handle. That this book is mainly focused on the application of digital signal processing techniques to fringe analysis is a coincidence based on the authors' background of activity and experience in this field. However, we believe that many of the concepts and processes considered here, translated to their proper contexts, will be applicable or of benefit to other interferometric coherent imaging techniques as well.

IOP Publishing

Single and Multicomponent Digital Optical Signal Analysis
Estimation of phase and its derivatives
Rishikesh Kulkarni and Pramod Rastogi

Chapter 2

Phase estimation: spatial fringe analysis

Fringe processing refers to the art of obtaining measurements through spatial and temporal fringe analysis techniques. Whereas in the former mode of analysis the spatial information in the fringe signal is exploited to retrieve the phase from a single fringe pattern, in the latter mode of analysis a stack of fringe patterns is processed along the temporal direction to extract the phase. Hereafter, all the chapters in this book leaving aside the last one have been assigned to have a broad range of facets in fringe processing using spatial analysis. The most notable feature of spatial analysis resides not so much in the relative simplicity of the experimental setups it may adhere to as in the possibility it offers in recording dynamic events on the time scale. This chapter is specifically devoted to fringe analysis from the viewpoint of phase estimation from a single fringe pattern. This chapter presents carrier and closed fringe patterns, the two types of patterns that have garnered major attention in any discussion of fringe processing.

2.1 Background

Typically, a two-dimensional fringe pattern can be expressed as a real sinusoidal signal:

$$\Gamma_0(x, y) = a(x, y) + b(x, y)\cos[\phi(x, y)], \tag{2.1}$$

where $a(x, y)$, $b(x, y)$, and $\phi(x, y)$ represent the background intensity, the fringe amplitude, and the phase, respectively. Depending on the optical interferometry setup, the fringe pattern can either be recorded on a photographic plate or on a solid state imaging device such as CCD or CMOS camera. The numerous advantages associated with solid state imaging devices have led to the replacement of the traditional photographic plates. In fringe patterns of size $M \times N$ recorded on a CCD or CMOS camera, the spatial coordinates x and y represent pixels along the columns and rows, respectively. In general, the only term of interest in (2.1) is the phase ϕ as it carries information on the physical quantity under measurement. Since the fringe patterns recorded in a variety of optical interferometric techniques such as

doi:10.1088/978-0-7503-1469-5ch2

holographic interferometry, electronic speckle pattern interferometry (ESPI), shearography, moiré interferometry and fringe projection profilometry, to name a few, can be described by (2.1), the fringe analysis approaches included in this chapter are applicable to all of these methods.

At a given pixel (x, y), there are three unknowns, a, b and ϕ. The assumption that there are in total K number of pixels gives rise to an underdetermined system of K equations with $3 \times K$ unknowns, each of the form given in (2.1). The main goal of any fringe analysis approach to estimate ϕ in an unambiguous manner, independent of a and b, is addressed either through optically infusing some extra information into the fringe pattern or by assuming some *a priori* knowledge on the phase to be estimated.

2.2 Carrier fringe pattern demodulation

In communication systems, the concept of carrier modulation and demodulation is of utmost importance in the reliable transmission of a signal. In frequency modulation (FM) based signal transmission, the signal that needs to be transmitted modulates the frequency of a carrier, thereby varying its instantaneous frequency. The frequency modulated carrier is demodulated at the receiver end to retrieve the signal information. A frequency modulated carrier can be represented as

$$\Gamma(x) = A \cos[\omega_x x + \phi(x)],$$

where $\phi(x)$ is the modulating signal which is synonymous with the phase term in (2.1); A represents the amplitude of a modulated carrier and ω_x is the carrier frequency. Usually, the carrier frequency is much higher compared to the maximum variation in the modulating signal. Such a signal modulated carrier is capable of transmitting the information unambiguously. In the context of fringe analysis, the introduction of carrier frequency plays an important role in obtaining sign-unambiguous estimation of phase. The carrier fringes can be introduced into the interferometric setup by simply tilting a mirror placed in the path of the reference beam. This results in a system of auxiliary fringes termed *carrier fringes*. The carrier fringe pattern in the initial unloaded state of the object can be represented as

$$\Gamma_c(x, y) = a(x, y) + b(x, y)\cos(\omega_x x + \omega_y y),$$

where ω_x and ω_y represent the carrier frequencies along the x and y directions, respectively. The object is next submitted to a load. The amount of frequency deviation the carrier frequency experiences is a measure of the change in the object caused by its deformation when put under load. The modulated carrier can be represented as

$$\Gamma(x, y) = a(x, y) + b(x, y)\cos[\omega_x x + \omega_y y + \phi(x, y)].$$

The frequency modulation introduced in the carrier by the phase ϕ is proportional to the physical quantity under measurement. The goal of spatial fringe analysis techniques therefore is to extract the phase information from the modulated carrier. The most popular of these techniques are based on the premise of the transformation of fringe signals. Transformation of signals from the time or space domain into the time–frequency or space–frequency domains simplifies their analysis. In general, Fourier transform is suitable for analyzing stationary signals (signals with constant frequencies) and wavelet transform is suitable for analyzing non-stationary signals (signals with frequency variations in function of time/space). In the context of fringe pattern analysis, the cosine fringe patterns exhibit spatial non-stationary nature due to their spatial dependence on the phase. In this chapter we discuss some of the most frequently used transform based spatial fringe analysis techniques explicitly operating in the spatial domain along with their advantages and limitations.

2.2.1 Fourier transform

Possibly, the most popular spatial fringe analysis approach is the one based on Fourier transform (FT) proposed in [1]. Let us consider a one-dimensional fringe signal along the x direction, in a given row y, in the absence of carrier fringes:

$$\Gamma_0(x) = a(x) + b(x)\cos[\phi(x)]. \tag{2.2}$$

Assuming that a carrier is introduced along the x direction, the carrier fringe signal without modulation can be represented as

$$\Gamma_c(x) = a(x) + b(x)\cos(\omega_x x).$$

Similarly, the modulated carrier fringe signal can be represented as

$$\Gamma(x) = a(x) + b(x)\cos[\omega_x x + \phi(x)]. \tag{2.3}$$

Generally, the carrier frequency is selected such that the maximum spatial variations of a, b and ϕ are much lower compared to ω_x. That is, $\partial a/\partial x \ll \omega_x$, $\partial b/\partial x \ll \omega_x$ and $\partial \phi/\partial x \ll \omega_x$. This fact can be effectively utilized in fringe analysis by transforming the signal in the frequency domain. Equations (2.2) and (2.3) can be written in the form:

$$\Gamma_0(x) = a(x) + c(x) + c^*(x), \tag{2.4}$$

$$\Gamma(x) = a(x) + c(x)\exp(i\omega_x x) + c^*(x)\exp(-i\omega_x x), \tag{2.5}$$

where $(\cdot)^*$ represents the complex conjugate of a scalar variable or conjugate transpose of a vector or matrix; $c(x) = \frac{1}{2}b(x)\exp[i\phi(x)]$. The FTs of (2.4) and (2.5) can be written as

$$\mathscr{F}_{\Gamma_0}(\omega) = \mathscr{F}_a(\omega) + \mathscr{F}_c(\omega) + \mathscr{F}_c^*(-\omega), \tag{2.6}$$

$$\mathscr{F}_{\Gamma}(\omega) = \mathscr{F}_a(\omega) + \mathscr{F}_c(\omega - \omega_x) + \mathscr{F}_c^*(-(\omega + \omega_x)). \tag{2.7}$$

We consider $\mathscr{F}(\cdot)$ as the FT operator, for example, $\mathscr{F}(c(x)) = \mathscr{F}_c(\omega)$; (2.7) uses the *frequency shift* property of the FT. Equation (2.6) shows the presence of three spectrums corresponding to the three signal components present in (2.4) in the Fourier spectrum of the fringe signal. Since these spectra overlap each other near the zero frequency region, it is not possible to separate these signal components from each other. However, a close observation of the spectrum of modulated carrier fringe signal given in (2.7) suggests that an appropriate selection of ω_x should lead to the separation of the three spectral lobes corresponding to the three signal components. As an example, a one-dimensional fringe signal of length $N = 256$ is simulated. The plots of $\phi(x)$ and $\Gamma_0(x)$ are shown in figure 2.1(a). The Fourier

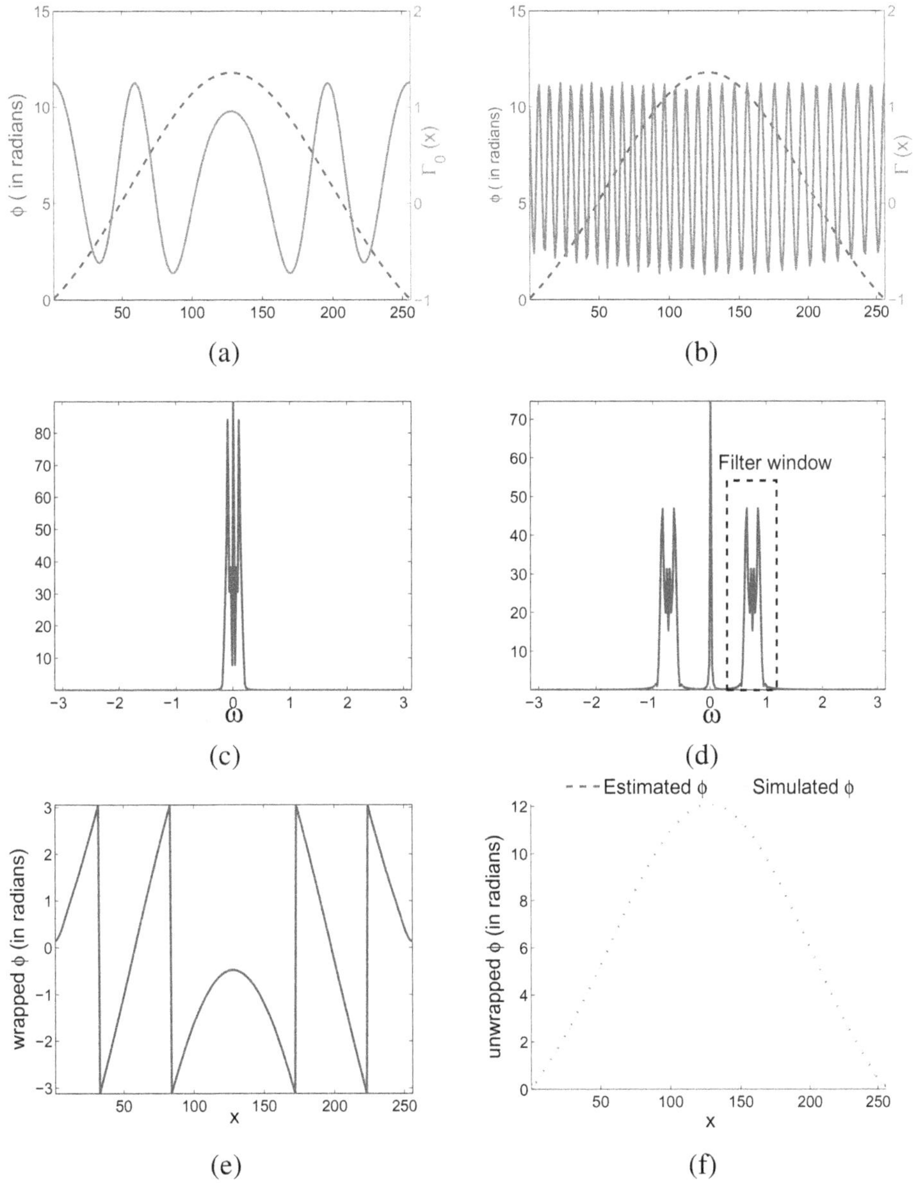

Figure 2.1. (a) One-dimensional fringe signal and phase; (b) one-dimensional modulated carrier fringe signal and phase; (c) Fourier spectrum of the fringe signal; (d) Fourier spectrum of the modulated carrier fringe signal; (e) wrapped phase estimate; (f) estimated and true unwrapped phases. The phase values are in radians.

spectrum of $\Gamma_0(x)$ in figure 2.1(c) shows the spectrum of the three signal components overlapping each other. Further, a modulated carrier is simulated with the carrier frequency of $\omega_x = 2\pi \times 30/N$ rad. The modulated carrier signal and its Fourier spectrum are shown in figures 2.1(b) and 2.1(d), respectively. The introduction of ω_x separates the spectrum of individual signal components and thereby allows one to distinguish and reconstruct the individual signal components accurately.

The phase is estimated by removing the contributions of $\mathscr{F}_a(\omega)$ and $\mathscr{F}_c^*(-(\omega + \omega_x))$ from $\mathscr{F}_\Gamma(\omega)$ by carrying out spectral filtering in the frequency plane. The spectral component $\mathscr{F}_c(\omega - \omega_x)$ carrying the required phase information can be filtered out by designing a filter window, as shown in figure 2.1(d). An inverse Fourier transformation is performed on this filtered spectrum to obtain the estimate of $c(x)\exp(i\omega_x x)$. The carrier frequency contribution is compensated from the estimated signal and the phase estimate is computed as

$$\phi_w(x) = \arctan\left\{\frac{\Im[c(x)]}{\Re[c(x)]}\right\},$$

where $\Re[\cdot]$ and $\Im[\cdot]$ represent the real and imaginary parts of their complex valued arguments. The subscript w in the estimated phase ϕ_w emphasizes its wrapped form, that is, the phase values are mapped in the range of $(-\pi, \pi]$ due to the arctan operation. Thus, the estimated phase shown in figure 2.1(e) exhibits discontinuities whenever the true phase ϕ goes beyond the $(-\pi, \pi]$ range. Such discontinuities are removed by the process of *phase unwrapping* described in the next chapter. It suffices to note here that phase unwrapping allows for removing the artificial discontinuities from the wrapped phase by appropriate addition/subtraction of the integer multiples of 2π to/from the wrapped phase. For the purpose of comparison, the plots of unwrapped estimated phase and the simulated true phase are provided in figure 2.1(f).

FT being a global operation analyzes data across the entire signal length, thereby providing high noise robustness to fringe demodulation. The immediate impact of this is that the phase estimation accuracy at any given pixel is affected by the signal quality at other pixels. Thus, the effect of a poor signal-to-noise ratio (SNR) in a certain portion of the fringe pattern is prone to permeate throughout the total estimated phase. Moreover, the discrete sampling of the fringe pattern gives rise to higher order spectral components in addition to the fundamental fringe signal. This makes it essential to carefully select the range of frequencies of the bandpass filter in the frequency plane. The criterion to be satisfied for the reliable reconstruction of the fundamental fringe signal in (2.3) [2] is:

$$\omega_x + \dot{\phi}_{\max}(x) < l[\omega_x + \dot{\phi}_{\min}(x)] \text{ for } l = 2, 3, \ldots$$
$$\omega_b < \omega_x + \dot{\phi}_{\min}(x),$$

where $\dot{\phi}_{\min}(x)$ and $\dot{\phi}_{\max}(x)$ represent the minimum and maximum instantaneous carrier-less fringe frequencies and ω_b is the cut-off frequency corresponding to the background intensity and fringe amplitude modulation; l is the order of the spectral components, with $l = 1$ representing the fundamental fringe signal. Spectral overlapping of fundamental fringe signal with the higher order terms occurs for cases such as fringe under-sampling and non-linear response of fringe recording medium, and which violates the above described criterion and results in erroneous phase estimation.

Since the fringe signal is inherently non-stationary in nature, the local fringe frequency is an important parameter that needs to be taken into account. This information is however lost in the FT based fringe demodulation. Accordingly, various space-frequency analysis methods have been developed and successfully applied for the purpose of fringe demodulation. In subsections 2.2.2 and 2.2.3 we present two of these techniques.

2.2.2 Windowed Fourier transform

The natural extension to the use of FT to analyze non-stationary signal is the windowed Fourier transform (WFT) [3]. In one-dimensional time dependent signal analysis, WFT is also termed as the short-time Fourier transform (STFT). Similar to the FT, the STFT of a non-stationary signal not only provides information on the different frequency components present in the signal, but also on the time at which they appear. Such simultaneous time and frequency information of the signal cannot be retrieved in the FT operation. The fact that the sinusoidal fringe pattern exhibits a spatially non-stationary behavior, makes the WFT a highly suitable candidate for fringe pattern demodulation. Formally, the WFT of a two-dimensional fringe pattern is defined as

$$WF_{\Gamma}(u, v, \varsigma, \nu) = \int_{-\infty}^{\infty}\int_{-\infty}^{\infty} \Gamma(x, y)h^{*}_{u,v,\varsigma,\nu}(x, y)dxdy,$$

and the inverse windowed Fourier transform (IWFT) is defined as

$$\Gamma(x, y) = \frac{1}{4\pi^2}\int_{-\infty}^{\infty}\int_{-\infty}^{\infty}\int_{-\infty}^{\infty}\int_{-\infty}^{\infty} WF_{\Gamma}(u, v, \varsigma, \nu)h_{u,v,\varsigma,\nu}(x, y)dudvd\varsigma d\nu.$$

The WFT basis function is defined as

$$h_{u,v,\varsigma,\nu}(x, y) = h(x - u, y - v)\exp[i(\varsigma x + \nu y)]. \tag{2.8}$$

Note that u and v represent the spatial coordinates as x and y, respectively. The WFT basis function is simply a modified version of the FT basis function obtained by the multiplication of the Fourier basis function with a real valued window function $h(x, y)$. The finite spatial extension of the window function limits the infinite spatial extension of the Fourier basis function. The WFT performs the signal decomposition into a number of windowed Fourier basis functions represented in (2.8). Due to the windowing operation, the FT is performed only on a selected portion of the signal at a time which allows one to identify the localized frequency content in the signal. In the context of fringe analysis, the two-dimensional WFT provides the information on the spatial frequency content in the fringe signal *at each pixel location*, thus resulting in a space-frequency representation of the fringe signal.

Although, in the WFT computation, we can use typical window functions, such as rectangular, Gaussian, Hamming, etc, the use of a Gaussian window function

$h(x, y) = \exp\left(-\frac{x^2}{2\sigma_x^2} - \frac{y^2}{2\sigma_y^2}\right)$ is generally preferred. The standard deviations σ_x and σ_y of the two-dimensional Gaussian function determine the spatial extension of the window function along the x and y directions, respectively. The WFT with a Gaussian window is also called the *Gabor transform*. The interesting point to note here is that the FT can be looked upon as a WFT computed with a rectangular window of size equal to that of the entire fringe pattern. Due to the widest possible spatial extent of the rectangular window, the FT provides the frequency measurement at the highest possible resolution while losing any information on the local presence of specific fringe frequencies. On the other hand, in the WFT based demodulation of a two-dimensional fringe signal the inherent limit set on the spatial and frequency resolutions is governed by the Heisenberg uncertainty principle as

$$\delta x \delta \varsigma \geqslant 1/2, \; \delta y \delta v \geqslant 1/2, \tag{2.9}$$

where δx, δy and $\delta \varsigma$, $\delta \nu$ represent the uncertainties in the spatial locations and spatial frequencies, respectively. The equality condition in (2.9) is satisfied for a Gaussian window. Equation (2.9) suggests that a broader spatial extent of Gaussian window results in a fine frequency resolution and poor spatial resolution and vice versa. The selection of an appropriate window size, usually dependent on the maximum frequency present in the fringe signal, is thus of crucial importance.

The above discussion is focused on a continuous version of the WFT operating on a fringe signal of an infinite length. Noting however, that the fringe analysis is performed on a digital computer on a discretized fringe pattern of finite size, the digital implementation of the WFT limits the spatial extent of the fringe pattern as $x \in [0, N-1]$ and $y \in [0, M-1]$. The discretization of the fringe pattern limits the fringe frequencies in the range of $\varsigma = (-\pi, \pi]$ and $\nu = (-\pi, \pi]$. Thus, the discrete WFT of a two-dimensional signal can be written as

$$WF_\Gamma(u, v, \varsigma, \nu) = \sum_{x=0}^{N-1} \sum_{y=0}^{M-1} \Gamma(x, y) h^*_{u,v,\varsigma,\nu}(x, y)$$

and the IWFT is computed as

$$\Gamma(x, y) = \frac{1}{4\pi^2} \sum_{\varsigma=-\pi}^{\pi} \sum_{\nu=-\pi}^{\pi} \sum_{u=0}^{N-1} \sum_{v=0}^{M-1} WF_\Gamma(u, v, \varsigma, \nu) h_{u,v,\varsigma,\nu}(x, y) \Delta u \Delta v \Delta \varsigma \Delta \nu,$$

where Δu, Δv and $\Delta \varsigma$, $\Delta \nu$ represent the spatial and frequency resolutions. Typically, we set $\Delta u = 1$ and $\Delta v = 1$. Whereas the range of frequencies can be adjusted depending on the knowledge of the approximate values of the maximum and minimum of fringe frequencies, the frequency resolutions are set based on the variations of the fringe frequency over the fringe pattern. The availability of *a priori* information on the fringe frequencies thus helps in the appropriate selection of the parameters in the implementation of the WFT based fringe analysis. The frequency and window size parameters are usually set by the user on an ad hoc basis.

Windowed Fourier filtering
In the Fourier spectrum in figure 2.1(d), a bandpass filter is used to select one of the complex signal components in the FT based fringe demodulation. The spectral contribution of noise lying outside the bandpass region is removed during the filtering operation. However, the spectral contribution of noise lying within the bandpass region is retained and is reproduced during the complex fringe signal ($c(x)$) reconstruction. A similar operation of fringe filtering and complex fringe signal reconstruction can be performed using windowed Fourier filtering (WFF), a method based on the WFT.

A fringe pattern is usually corrupted by the white Gaussian noise. Due to the *whiteness* of the noise, the spectral coefficients, albeit of small values, associated with it are spread throughout the frequency plane compared to the spectral coefficients associated with the noiseless fringe signal. This fact is effectively utilized in the WFF based fringe denoising and phase estimation. The frequency bands are selected as $\varsigma = [\varsigma_{\min}, \varsigma_{\max}]$ and $\nu = [\nu_{\min}, \nu_{\max}]$ depending on the carrier frequency and an *a priori* knowledge of the approximate values of maximum fringe frequencies. It is quite obvious that fringe patterns with a wide range of fringe frequencies require broader frequency bands and fringe patterns with a small range of fringe frequencies require narrow frequency bands around the carrier frequency.

Considering that the carrier fringes are added only along the x direction, the unwanted spectral lobe can be removed from the fringe signal spectrum by computing the WFT with the frequency bands selected as $\varsigma = [\omega_x - (\dot{\phi}_x)_{\max}, \omega_x + (\dot{\phi}_x)_{\max}]$ and $\nu = [-(\dot{\phi}_y)_{\max}, (\dot{\phi}_y)_{\max}]$. Here, $(\dot{\phi}_x)_{\max}$ and $(\dot{\phi}_y)_{\max}$ represent the absolute maximum values of instantaneous fringe frequencies along the x and y directions, respectively. To remove the noise contribution from the fringe pattern, the WFT spectrum is *thresholded* such that the spectral coefficients below a pre-defined threshold *Thr* are set to zero. The thresholded WFT spectrum is represented as

$$\overline{WF_\Gamma}(u, v, \varsigma, \nu) = \begin{cases} WF_\Gamma(u, v, \varsigma, \nu), & \text{if } |WF_\Gamma(u, v, \varsigma, \nu)| \geqslant Thr \\ 0, & \text{otherwise} \end{cases}.$$

The inverse WFT applied on this thresholded WFT spectrum provides the noise filtered complex fringe signal $\bar{\Gamma}(x, y)$. The argument of this complex signal provides the phase estimate in the wrapped form as

$$\phi_w(x, y) = \arctan\left\{\frac{\Im[\bar{\Gamma}(x, y)]}{\Re[\bar{\Gamma}(x, y)]}\right\}.$$

The resulting phase map contains a carrier fringe contribution that has to be compensated to obtain the phase estimate associated with the measurand.

Windowed Fourier ridges
Another approach for fringe demodulation based on the WFT is the windowed Fourier ridges (WFRs). In this method, the fundamental assumption is that within a small window around each pixel the fringe pattern contains a single fringe frequency. Accordingly, the windowed fringe signal around each pixel is compared

with a number of windowed Fourier basis functions of different frequencies ς and ν. The Fourier basis function displaying the best match with the windowed fringe signal is termed the *windowed Fourier ridge*. The respective values of ς and ν provide the estimates of local fringe frequencies, termed the *instantaneous frequencies*. It should be noted that in the present case the word 'instantaneous' is referred to in the spatial and not in the temporal context. Based on the assumption of a single frequency windowed fringe signal around the pixel (u, v), the phase within the window can be represented as

$$\phi(x, y) = \phi(u, v) + \dot{\phi}_x(u, v)(x - u) + \dot{\phi}_y(u, v)(y - v),$$

where $\dot{\phi}_x(u, v)$ and $\dot{\phi}_y(u, v)$ represent the fringe frequencies at (u,v) along x and y directions, respectively. For simplicity of analysis, consider a one-dimensional fringe signal given in (2.3). The one-dimensional phase within a window around u is

$$\phi(x) = \phi(u) + \dot{\phi}_x(u)(x - u). \tag{2.10}$$

Let us consider that within the analysis window the second term in (2.5) can be written as

$$\Gamma_1(x) = \frac{1}{2} b(x) \exp[i(\omega_x x + \phi(x))]. \tag{2.11}$$

Substituting the phase representation in (2.10) into (2.11), we have,

$$\Gamma_1(x) = \frac{1}{2} b(u) \exp[i(\phi(u) + \dot{\varphi}_x(u)(x - u))],$$

where $\dot{\varphi}_x(u) = \omega_x + \dot{\phi}_x(u)$ and within the Gaussian window the fringe amplitude is considered to be constant ($b(x) = b(u)$). Basically, the WFT of this one-dimensional fringe signal results in a two-dimensional spectrum $WF_\Gamma(u, \varsigma)$ in function of u and ς. Locating the maximum amplitude of $WF_{\Gamma_1}(u, \varsigma)$, the spectrum corresponding to $\Gamma_1(x)$, with respect to ς, we have

$$\dot{\varphi}_x(u) = \arg\max_{\varsigma} WF_{\Gamma_1}(u, \varsigma).$$

Finally, we obtain the instantaneous fringe frequency as $\dot{\phi}_x(u) = \dot{\varphi}_x(u) - \omega_x$. At $\dot{\varphi}_x(u)$, the spectral contribution due the remaining two terms in (2.7) in the spectrum $\mathscr{F}_\Gamma(u, \varsigma)$ is small and hence can be neglected [4]. Consequently, the phase estimate at u can be computed as

$$\phi(u) = \arctan\left\{\frac{\Im[WF_{\Gamma_1}(u, \dot{\varphi}_x(u))]}{\Re[WF_{\Gamma_1}(u, \dot{\varphi}_x(u))]}\right\}.$$

Note that a similar analysis is applicable while considering the third term in (2.5) and which provides the estimates of $-\dot{\varphi}_x(u)$ and $-\phi(u)$. The two-dimensional extension of the above analysis is straightforward. The WFT of a two-dimensional fringe signal results in a four dimensional spectrum in function of u, v, ς and ν.

In the WFR based method, the instantaneous frequencies are computed independently of the phase estimation. The phase estimation can also be performed by integrating the instantaneous frequency. Thus, the phase at u can be computed as

$$\phi(u) = \phi(0) + \int_0^u \dot{\phi}_x(x)dx.$$

The estimated phase obtained in this manner is in an *unwrapped* form. Thus, no further phase unwrapping is required.

The WFT based analysis of the modulated carrier fringe signal in figure 2.1(b) is shown in figure 2.2. The WFT spectrum plotted in figure 2.2(a) allows for easily identifying the spectral contribution of the three fringe signal components in (2.7). In the same figure, the fringe frequency $\dot{\varphi}(x)$ corresponding to the complex fringe signal $\Gamma_1(x)$, plotted in red, is seen to overlap the maximum amplitude of $\mathscr{F}_{\Gamma_1}(x, \varsigma)$. In fact, the trace of this maximum amplitude represents the WFR. The phase of $\mathscr{F}_{\Gamma_1}(x, \varsigma)$ corresponding to the WFR shown in figure 2.2(b) provides the estimate of the phase ϕ after compensating for the carrier frequency contribution.

The performance of the WFT based fringe analysis depends on a number of user defined parameters such as the window size, the frequency band, the frequency resolution and the threshold selection. Selection of the values of these parameters is made depending on the SNR of the fringe pattern and the fringe density. It is these values that will finally determine the accuracy, the noise robustness and the computational cost in the phase estimation. Typically, a large window size provides high noise robustness at high computational cost, while a small window size provides higher spatial resolution but at the expense of noise robustness. Reference [3] has proposed the size of the Gaussian window to be taken as $(4\sigma_x + 1) \times (4\sigma_y + 1)$. Whereas the implementation of the WFF speaks in favor of the selection of coarse increment values of $\Delta\varsigma$ and $\Delta\nu$, the implementation of the WFR supports the selection of finer increment values of $\Delta\varsigma$ and $\Delta\nu$ in order to estimate the instantaneous frequencies accurately. As a result, the computation time

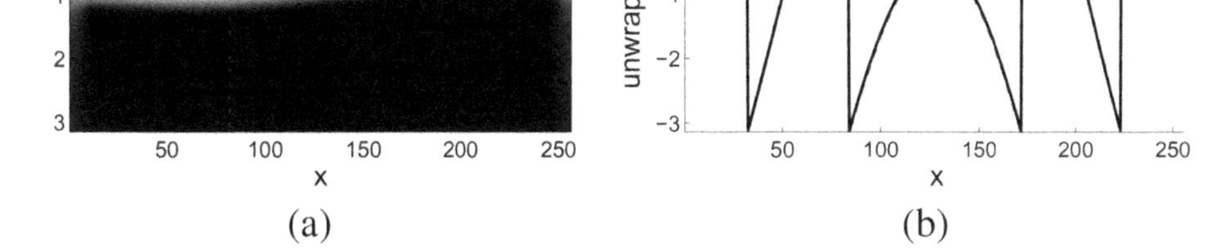

Figure 2.2. (a) Space-frequency representation of the one-dimensional fringe signal; (b) estimated phase using the WFT based fringe analysis.

required for implementing the WFF algorithm is less as compared to that for the WFR algorithm. A literature survey shows that for the purposes of phase estimation, the WFF algorithm is to be preferred over the WFR algorithm. However, when it comes to the instantaneous fringe frequency determination, the WFR algorithm is to be preferred.

A detailed study on the windowed Fourier fringe analysis and its implementation has been presented in [3,4]. The ease of implementation and its effectiveness makes the WFT an important tool for fringe analysis.

2.2.3 Continuous wavelet transform

Wavelet transform (WVT) has proved to be a versatile tool for analyzing non-stationary signals. A detailed discussion on the WVT and its properties can be found in [5,6]. This subsection will describe the WVT within the context of fringe analysis.

In the case of time dependent signals, the wavelet trasform provides a time-frequency representation of a signal where the information on the frequency content is available at each instant of time. Since the fringe signals also exhibit a spatial non-stationary behavior, the WVT can be effectively utilized for fringe demodulation. To recall, the wavelet transform was mainly introduced with a view to overcoming the limitations inherent to the STFT for analysing non-stationary signals. In the STFT based time-frequency analysis, the window size remains fixed at each instant of time. Accordingly, the signal analysis is performed with a fixed time-frequency resolution. On the other hand, in the WVT of the signal, the window size is chosen depending on the local frequency. This allows for the signal analysis to be performed with an optimal time-frequency resolution in a manner such that the low and high frequency signals are represented with finer frequency and time resolution, respectively. Although several types of WVT have been developed and are being utilized in many image and signal processing applications, the continuous wavelet transform (CWT) has been by far the one that has found most extensive use in fringe analysis [7–9].

A brief introduction to wavelets is included before formally moving on to the concept of continuous wavelet transform. Wavelets are continuously oscillating functions of limited temporal (or spatial) extent. A function $\psi(x)$ qualifies to be a wavelet function only if it is zero mean and centered at $x = 0$ with a finite extent. Mathematically, it has to satisfy the following *admissibility condition*:

$$\int_{-\infty}^{\infty} \psi(x)dx = 0.$$

The above wavelet function can be scaled and translated to form a family of wavelets. Such a family of wavelets associated with their respective wavelet coefficients is capable of faithfully representing non-stationary signals. The family of wavelets can be represented as

$$\psi_{u,s}(x) = \frac{1}{\sqrt{s}}\psi\left(\frac{x-u}{s}\right),$$

where s and u represent the scale and translation (spatial) parameters, respectively. It can be observed that the family of wavelets $\psi_{u,s}(x)$ is generated as the translated and dilated versions of $\psi(x)$, appropriately termed the 'mother wavelet'. The real valued parameter s performs the frequency scaling of wavelets in inverse proportionality to frequency. Thus, a high value of s corresponds to a low frequency and a low value of s corresponds to a high frequency wavelet. In general, the translation parameter u takes integer values corresponding to pixel locations in the digital implementation of the wavelet transform. At this point, let us formally introduce the one-dimensional CWT as:

$$W_\Gamma(u, s) = \langle \Gamma, \psi_{u,s} \rangle = \int_{-\infty}^{\infty} \Gamma(x) \frac{1}{\sqrt{s}} \psi^*\left(\frac{x-u}{s}\right) dx,$$

where $\langle \cdot,\cdot \rangle$ represents the inner product operation with complex valued arguments. Note that the one-dimensional CWT of the fringe signal results in a two-dimensional map of the CWT coefficients (W_Γ) in function of s and u.

Several types of mother wavelets have been reported in the literature [10], such as Morlet, Paul, and the Mexican hat. However, we need to select an analytic wavelet function for the purpose of fringe demodulation. A function $f(x)$ is said to be analytic if its FT $\mathscr{F}_f(\omega) = 0$, for $\omega < 0$. As an example, the one-dimensional complex Morlet wavelet is defined as

$$\psi_{\text{Morlet}}(x) = \frac{1}{\sqrt[4]{\sigma^2\pi}} \exp(i\varsigma_0 x) \exp\left(\frac{-x^2}{2\sigma^2}\right),$$

where ς_0 represents the frequency of oscillation of the Morlet wavelet; and the width of the Gaussian envelope is controlled by the parameter σ. In order to satisfy the admissibility condition, the frequency of oscillation of the Morlet mother wavelet should be set such that $\varsigma_0 \geqslant 5.34$ rad/pixel [5]. Further, the family of wavelets derived using the complex Morlet wavelet can be represented as

$$\psi_{u,s}(x) = \frac{1}{\sqrt{s}} \frac{1}{\sqrt[4]{\sigma^2\pi}} \exp\left[i\varsigma_0\left(\frac{x-u}{s}\right)\right] \exp\left[-\frac{1}{2}\left(\frac{x-u}{\sigma s}\right)^2\right]. \tag{2.12}$$

From the comparison of the CWT basis function given in (2.12) with the WFT basis with the Gaussian window in (2.8), it can be observed that whereas the Gaussian window size of the WFT basis is *fixed* by the parameter σ at all pixel locations, the width of the Gaussian envelope in the CWT computation is controlled by σ *and* the frequency scaling parameter s. This difference provides the CWT with the unique opportunity of analyzing the non-stationary fringe signals with optimal space-frequency resolutions.

The plots of real parts of complex Morlet wavelets computed at two pixel locations for two different values of frequency scaling parameter s are provided in figure 2.3. The wavelet overlapping on a modulated carrier fringe signal gives a good indication of the wavelet best matched to the fringe signal at the selected pixel

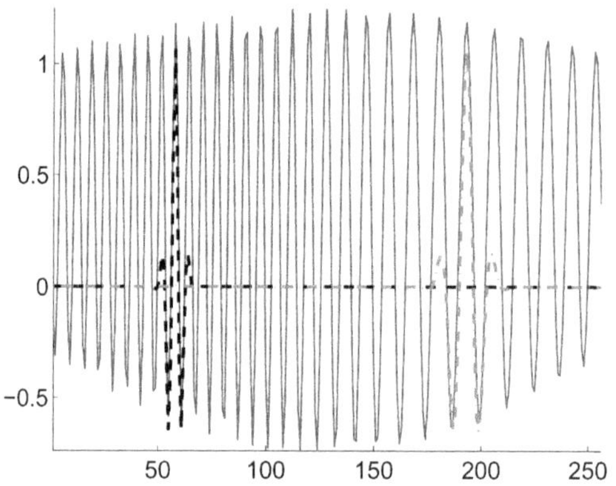

Figure 2.3. The plots of real parts of the complex Morlet wavelets at two pixel locations serve to indicate the best match to the local fringe signal.

location. It is now left to see how the fringe signal at each pixel location can be represented using the family of wavelets. To do so, we resort to an analysis similar to the one that was carried out while discussing the WFR algorithm. Therein, we considered a linear phase representation within a Gaussian window around a pixel location u as given in (2.10). Considering the complex fringe signal component given in (2.11), its 1D-CWT can be represented as

$$\begin{aligned} W_{\Gamma_1}(u, s) &= \int_{-\infty}^{\infty} \Gamma_1(x)\frac{1}{\sqrt{s}}\psi^*\left(\frac{x-u}{s}\right)dx \\ &= \frac{1}{2}\int_{-\infty}^{\infty} b(x)\exp[i(\phi(u) + \dot{\varphi}_x(u)(x-u))]\frac{1}{\sqrt{s}}\psi^*\left(\frac{x-u}{s}\right)dx. \end{aligned}$$

Substituting the complex Morlet wavelet as the mother wavelet, we obtain

$$\begin{aligned} W_{\Gamma_1}(u, s) = \frac{1}{2\sqrt{s}\sqrt[4]{\sigma^2\pi}}b(u)\exp[i\phi(u)]\int_{-\infty}^{\infty} \exp[i(\dot{\varphi}_x(u)(x-u))] \\ \times \exp\left[-i\varsigma_0\left(\frac{x-u}{s}\right)\right]\exp\left[-\frac{1}{2\sigma^2}\left(\frac{x-u}{s}\right)^2\right]dx. \end{aligned}$$

Substituting $t = \frac{x-u}{s}$ in the above equation gives

$$W_{\Gamma_1}(u, s) = \frac{\sqrt{s}}{2\sqrt[4]{\sigma^2\pi}}b(u)\exp[i\phi(u)]\int_{-\infty}^{\infty} \exp\left[-\frac{t^2}{2\sigma^2}\right]\exp[-i(\varsigma_0 - \dot{\varphi}_x(u)s)t]dt. \quad (2.13)$$

Now, let us represent the Gaussian window as $h(t) = \exp\left[-\frac{t^2}{2\sigma^2}\right]$. If the FT of $h(t)$ is represented as $\mathscr{F}_h(\varsigma)$, then from (2.13) we obtain,

$$W_{\Gamma_1}(u, s) = \frac{\sqrt{s}}{2\sqrt[4]{\sigma^2\pi}}b(u)\exp[i\phi(u)]\mathscr{F}_h(\varsigma_0 - \dot{\varphi}_x(u)s). \quad (2.14)$$

The above result is important from the viewpoint of fringe demodulation. The maximum value of $|W_{\Gamma_1}(u, s)|$ is achieved when $\varsigma_0 = \dot{\varphi}_x(u)s$. Thus, at a given pixel location u, a specific scale parameter s_{ridge} corresponding to the maximum value of $|W_{\Gamma_1}(u, s)|$ is identified. The instantaneous frequency of the fringe signal at this location can then be computed as

$$\dot{\varphi}_x(u) = \frac{\varsigma_0}{s_{\text{ridge}}}. \tag{2.15}$$

The path traced by the maximum value of $|W_{\Gamma_1}(u, s)|$ in the space-frequency plane is termed the *wavelet ridge*. Once the wavelet ridge is detected, equations (2.14) and (2.15) based on the 1D-CWT of the fringe signal lead to two methods of phase estimation: *phase method* and *phase gradient method*. Note that with the selected wavelet frequencies, the remaining two fringe signal components in (2.5) do not contribute in the computation of the CWT.

Phase gradient method
The specific value of the scale parameter s associated with the pixel location u provides the instantaneous frequency, i.e., the phase gradient estimate, in accordance with (2.15). The integration of this phase gradient provides us with the required phase estimate. This method of phase estimation is termed the *phase gradient method*. Since the phase estimated using this method is in an unwrapped form, no further phase unwrapping is required. However, this method is capable of obtaining accurate phase estimation only as long as the phase gradients have been estimated correctly.

The discretization of scale parameter s decides the accuracy with which the phase gradients are estimated. To achieve a high accuracy phase gradient estimate, the range of s has to be divided into intervals of small sizes. The fine segmentation of the range of s naturally increases the overall computation time. The phase gradient estimation errors, if present, get accumulated during the integration process which essentially results in an inaccurate phase estimation. This fact makes the phase gradient method a less interesting choice for phase estimation.

Phase method
On the other hand, the *phase method* does not depend on the phase gradient for phase estimation. Instead, the phase of the complex wavelet coefficient belonging to the wavelet ridge provides the phase estimate in accordance with (2.14). Thus, the phase estimate is computed as

$$\hat{\phi}(x) = \arctan\left\{\frac{\Im\left[W_\Gamma(x, s_{\text{ridge}})\right]}{\Re\left[W_\Gamma(x, s_{\text{ridge}})\right]}\right\}.$$

Contrary to the phase gradient method, the phase method does not require fine discretization of s. Accurate phase estimation can thus be performed with coarse discretization of s with relatively high computational efficiency. Since due to the arctan function involved in the computation the estimated phase is in a wrapped form, the method still requires going through the additional step of phase unwrapping. Note that the carrier fringe contribution has also to be compensated for from the estimated phase.

The CWT of the one-dimensional fringe signal given in figure 2.1(b) is computed using the complex Morlet wavelet with $\varsigma_0 = 6$ rad/pixel and $\sigma = 0.5$. The frequency scaling is performed with $s = [1, 20]$ and the step size of 0.01. The plot of the modulus of CWT coefficients ($W_\Gamma(x, s)$) is provided in figure 2.4(a). The maximum value of the modulus along the columns, i.e. the wavelet ridge, is represented by the red color. The phases of the CWT coefficients computed as

$$\phi(x, s) = \arctan\left\{\frac{\Im[W_\Gamma(x, s)]}{\Re[W_\Gamma(x, s)]}\right\},$$

are displayed in figure 2.4(b). The phase gradients are obtained, by using the frequency scaling parameter s_{ridge} corresponding to the wavelet ridge, and the relationship given in (2.15), as

$$\dot{\phi}_x(x) = \frac{\varsigma_0}{s_{\text{ridge}}} - \omega_x.$$

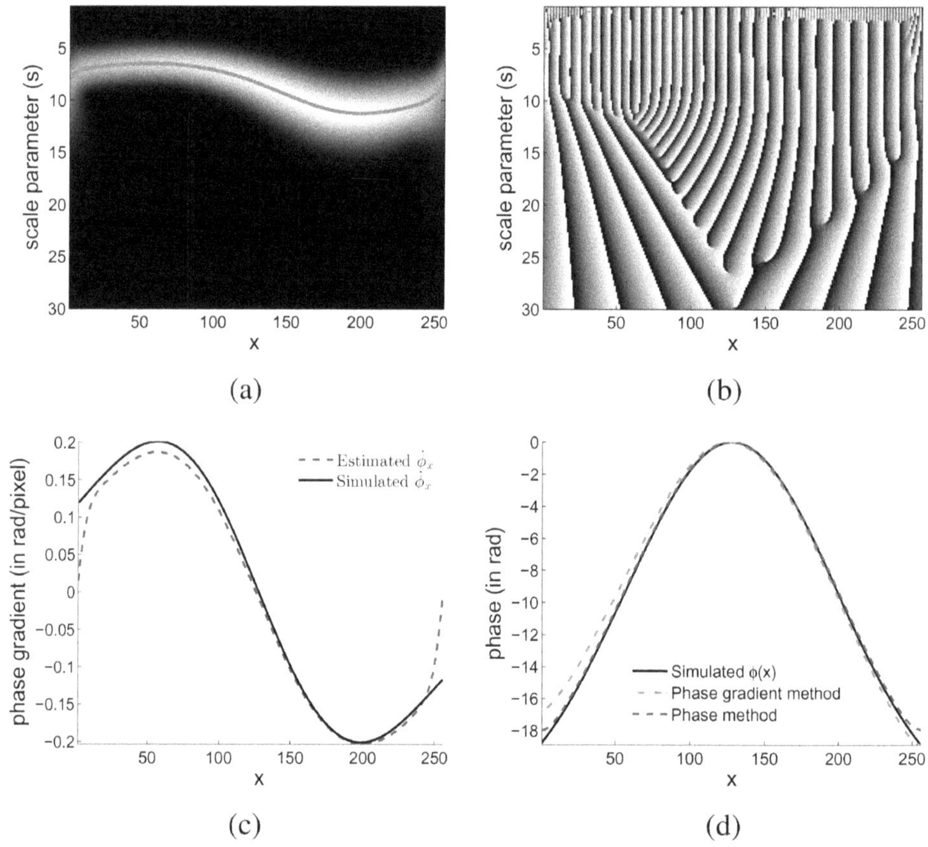

Figure 2.4. (a) $|W_\Gamma(x, s)|$, modulus of wavelet coefficients; the red colored trace indicates the wavelet ridge. (b) Phase of wavelet coefficients; (c) simulated and estimated phase gradients; (d) unwrapped phase estimates using the phase gradient and phase methods.

The simulated and estimated phase gradients are shown in figure 2.4(c). The phase estimates are obtained by using the phase gradient and the phase methods. The estimated and simulated phases are provided in figure 2.4(d). Figure 2.4(c) indicates that the phase derivative is estimated with a certain amount of error, with the error at the borders being significantly higher due to the abrupt termination of the fringe data. The errors in the estimation of phase and phase derivatives at the fringe borders can be significantly reduced or even avoided by extrapolating the fringe signal beyond the borders using one of the linear predictive techniques. Alternatively, the estimates near the borders can simply be ignored.

In the phase gradient based phase estimation, the phase derivative estimation errors get propagated during the integration of the estimated phase derivative. Consequently, large errors appear in the estimated phase as shown in figure 2.4(d). On the other hand, the phase estimate computed using the phase method is relatively more accurate, thereby making the latter a promising method for phase estimation. The phase gradient method can be used if the estimation of instantaneous fringe frequency is of interest.

The use of two-dimensional CWT based fringe analysis proposed in [11] has been shown to provide more robust phase estimation than one-dimensional CWT based fringe analysis.

2.2.4 Spatial phase shifting technique

Apart from the transform based fringe demodulation techniques, the phase shifting algorithms developed in the context of temporal fringe analysis have also seen their adaptation in the spatial fringe analysis. Consider the one-dimensional modulated carrier fringe signal given in (2.3). The fringe intensities at three consecutive pixels centered around the pixel x can be represented as

$$\begin{aligned}\Gamma_{-1} &= a(x-1) + b(x-1)\cos[\omega_x x - \omega_x + \phi(x-1)]\\ \Gamma_0 &= a(x) + b(x)\cos[\omega_x x + \phi(x)]\\ \Gamma_1 &= a(x+1) + b(x+1)\cos[\omega_x x + \omega_x + \phi(x+1)].\end{aligned} \tag{2.16}$$

If we assume that the background intensity $a(x)$, the fringe amplitude $b(x)$ and the phase $\phi(x)$ remain constant over the three pixels, that is,

$$\begin{aligned}a(x-1) &\approx a(x) \approx a(x+1)\\ b(x-1) &\approx b(x) \approx b(x+1)\\ \phi(x-1) &\approx \phi(x) \approx \phi(x+1),\end{aligned}$$

then (2.16) can be represented as

$$\begin{aligned}\Gamma_{-1} &\approx a(x) + b(x)\cos[\omega_x x - \omega_x + \phi(x)]\\ \Gamma_0 &\approx a(x) + b(x)\cos[\omega_x x + \phi(x)]\\ \Gamma_1 &\approx a(x) + b(x)\cos[\omega_x x + \omega_x + \phi(x)].\end{aligned}$$

These three equations are equivalent to the ones obtained in three-step phase shifting interferometry [12]. Since the intensities at three consecutive pixels Γ_{-1}, Γ_0 and Γ_1 are phase shifted each with phase steps of ω_x, the phase estimate at pixel x can be computed as,

$$\omega_x x + \phi(x) = \arctan\left\{\frac{1 - \cos(\omega_x)}{\sin(\omega_x)} \frac{\Gamma_{-1} - \Gamma_1}{2\Gamma_0 - \Gamma_{-1} - \Gamma_1}\right\}. \tag{2.17}$$

The above procedure describes the principle of *spatial phase shifting (SPS)* for phase estimation from a modulated carrier fringe pattern. The methods described in [13] and [14] consider $\omega_x = \pi/2$ and $2\pi/3$, respectively. In order to introduce these high values of phase steps between the consecutive pixels, a large amount of carrier fringes have to be introduced. However, in line with the sampling criterion, the spatial frequency of carrier fringe pattern is limited by the pixel size of the CCD. The SPS method also assumes the phase to be constant over the three consecutive pixels. This assumption restricts the use of this method to only those applications of phase estimation which exhibit smooth spatial variations of phase, that is, low values of phase slopes. The noise robustness offered by the method depends on the chosen value of ω_x. The analyses for the minimization of phase estimation errors in [15,16] have shown that the phase step of $\omega_x = 2\pi/3$ provides the highest noise robustness.

The SPS method of phase estimation proposed in [15] overcomes the limitative assumption of phase constancy in the implementation of (2.17) by instead calling for the first order derivative of phase to be constant over the three consecutive pixels. The equations in (2.16) can thus be rewritten as

$$\begin{aligned}
\Gamma_{-1} &\approx a(x) + b(x)\cos\left[\omega_x x - \omega_x + \phi(x) - \frac{\partial\phi(x)}{\partial x}\right] \\
\Gamma_0 &\approx a(x) + b(x)\cos[\omega_x x + \phi(x)] \\
\Gamma_1 &\approx a(x) + b(x)\cos\left[\omega_x x + \omega_x + \phi(x) + \frac{\partial\phi(x+1)}{\partial x}\right].
\end{aligned} \tag{2.18}$$

Here, the partial derivatives of phase are considered as functions of both x and y. The first order Taylor series expansion of the phase considered in (2.18) can be written as

$$\begin{aligned}
\frac{\partial\phi(x)}{\partial x} &= \phi(x) - \phi(x-1) \\
\frac{\partial\phi(x+1)}{\partial x} &= \phi(x+1) - \phi(x).
\end{aligned}$$

Under the assumption of the constancy of phase derivatives, i.e. $\frac{\partial\phi(x)}{\partial x} = \frac{\partial\phi(x+1)}{\partial x}$, we obtain

$$\begin{aligned}
\Gamma_{-1} &\approx a(x) + b(x)\cos\left[\omega_x x + \phi(x) - \omega_x - \frac{\partial\phi(x)}{\partial x}\right] \\
\Gamma_0 &\approx a(x) + b(x)\cos[\omega_x x + \phi(x)] \\
\Gamma_1 &\approx a(x) + b(x)\cos\left[\omega_x x + \phi(x) + \omega_x + \frac{\partial\phi(x)}{\partial x}\right].
\end{aligned} \tag{2.19}$$

Equation (2.19) indicates that the effective phase step is basically equal to the instantaneous phase step, $\omega_x + \frac{\partial \phi(x)}{\partial x}$. Since there are three equations in (2.19) with each containing four unknowns, at least one unknown has to be eliminated. Accordingly, the background intensity $a(x)$ is removed by high pass filtering of the fringe signal. The high pass filtered fringe signals can be represented as

$$\Gamma_{-1} \approx b(x)\cos\left(\omega_x x + \phi(x) - \omega_x - \frac{\partial \phi(x)}{\partial x}\right)$$
$$\Gamma_0 \approx b(x)\cos(\omega_x x + \phi(x))$$
$$\Gamma_1 \approx b(x)\cos\left(\omega_x x + \phi(x) + \omega_x + \frac{\partial \phi(x)}{\partial x}\right).$$

Using these filtered fringe signals, the phase estimate is computed as

$$\omega_x x + \phi(x) = \arctan\left\{\frac{\Gamma_{-1} - \Gamma_1}{sgn(\Gamma_0)[4\Gamma_0^2 - (\Gamma_{-1} - \Gamma_1)^2]^{1/2}}\right\}. \tag{2.20}$$

where $sgn(\cdot)$ is a function with value -1 if $\Gamma_0 < 0$ and 1 otherwise.

In another SPS method for phase estimation proposed in [16], the condition on constancy of background intensity and fringe amplitude is relaxed. Let $O(x)$ and $R(x)$ represent the intensities associated with the object and reference waves, respectively. Assuming that the reference beam intensity R is constant over the CCD surface, the modified formula for phase estimation is given as

$$\omega_x x + \phi(x) = \arctan\left\{\frac{O_0^{1/2}(\Gamma_{-1} - \Gamma_1) - \cos(\omega_x)\left[O_{-1}^{1/2}(\Gamma_0 - \Gamma_1) + O_1^{1/2}(\Gamma_{-1} - \Gamma_0)\right]}{\sin(\omega_x)\left[O_{-1}^{1/2}(\Gamma_0 - \Gamma_1) + O_1^{1/2}(\Gamma_{-1} - \Gamma_0)\right]}\right\}$$

where O_{-1}, O_0 and O_1 represent the object wave intensities at $(x - 1)$, x and $(x + 1)$, respectively; and, $\Gamma_{-1} = \Gamma(x - 1)$, $\Gamma_0 = \Gamma(x)$ and $\Gamma_1 = \Gamma(x + 1)$. The SPS method of phase estimation given in (2.20) is applied to the one-dimensional fringe signal with the phase shown in figure 2.1(b). The carrier frequency is selected as $\omega_x = 2\pi/3$. The error in phase estimation based on the assumption of phase constancy and phase derivative constancy is shown in figure 2.5(a). The improvement in phase estimation accuracy under the assumption of phase derivative constancy is apparent. The phase estimation is performed in the presence of noise with SNR = 10 dB. The phase estimates are computed, using (2.20), for two values of carrier frequencies, $\omega_x = \pi/2$ and $\omega_x = 2\pi/3$. The error plots in figure 2.5(b) are indicative of the noise robustness offered by the method in function of the carrier frequency, here $\omega_x = 2\pi/3$.

2.2.5 Robust quadrature filtering

In the FT based fringe analysis, a band-pass filter is used to select one of the spectral lobes in the spectrum of the fringe signal to compute the complex fringe signal. The phase of this complex fringe signal provides the phase estimate. The FT operates on the entire data length and thus is a global operation. However, the spatial extent of

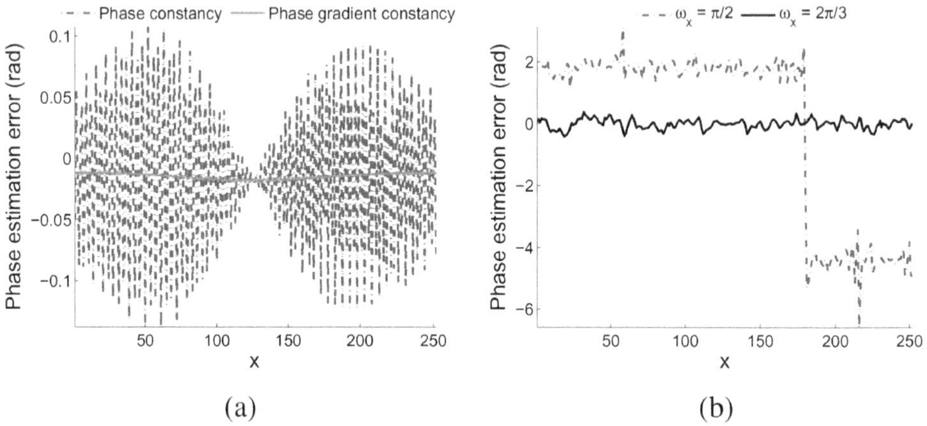

Figure 2.5. Plots of (a) phase estimation errors in the SPS method under the assumption of phase and phase derivative constancies; (b) phase estimation error in the SPS method in function of the carrier frequency.

the fringe pattern is limited by the size of the CCD frame. Additionally, certain pixels may contain invalid data on account of noise or fringe losses in some areas due to the presence of shadows or holes on it. As a result, using method in sections 2.2.1–2.2.4, we observe a deterioration in the accuracy of the phase estimate, especially at the border of valid and invalid fringe areas. Therefore, special care has to be taken while dealing with the invalid fringe areas and the border pixels. The robust quadrature filtering (RQF) based phase estimation method proposed in [17], explicitly deals with these issues. This filtering technique developed for the computation of complex fringe signal in a *pixelwise* manner has the capability of adjusting the filter performance at the border pixels.

Considering the one-dimensional fringe signal given in (2.4), the RQF is designed to retrieve the complex fringe signal $\Gamma_1(x) = \exp[i(\omega_x x + \phi(x))]$ in a way which minimizes the following cost function

$$U(\Gamma_1) = \sum_{x\in L} |\Gamma_1(x) - 2\Gamma(x)|^2 + \lambda \sum_{x=2}^{N} |\Gamma_1(x) - \Gamma_1(x-1)\exp(i\omega_x)|^2, \tag{2.21}$$

where $x \in L$ refers to the pixel locations in the interior of the fringe signal where valid fringe data is available; the value of parameter λ usually termed *regularization parameter* depends on the noise strength. While the first term attempts to match the estimated signal with the observed fringe signal, the second term called the *regularization term* imparts an extra condition of spatial continuity to the phase. Due to the use of the *first order regularization operator* in (2.21), the RQF acts as the first order regularization filter. The minimization of $U(\Gamma_1)$ with respect to Γ_1 is achieved by setting its spatial derivatives with respect to the real and imaginary parts of Γ_1 equal to zero. Accordingly, we obtain a system of linear equations for $x \in L$ as follows [17]:

$$\Gamma_1(x) - 2\Gamma(x) + \lambda[-\Gamma_1(x-1)\exp(i\omega_x) + 2\Gamma_1(x) - \Gamma_1(x+1)\exp(-i\omega_x)] = 0. \tag{2.22}$$

Note that the above equation for the extraction of complex fringe signal is applicable only for the valid pixels belonging to the interior of the fringe signal ($x \in L$). In the case of invalid pixels, that is, those lying in the invalid fringe regions ($x \notin L$), the linear equations take the form

$$-\Gamma_1(x-1)\exp(i\omega_x) + 2\Gamma_1(x) - \Gamma_1(x+1)\exp(-i\omega_x) = 0, \tag{2.23}$$

and, at the border pixels, $x = 0$ and $x = N - 1$, we have

$$\Gamma_1(0) - 2\Gamma(0) + \lambda[\Gamma_1(0) - \Gamma_1(1)\exp(-i\omega_x)] = 0 \tag{2.24}$$

$$\Gamma_1(N-1) - 2\Gamma(N-1) + \lambda[\Gamma_1(N-1) - \Gamma_1(N-2)\exp(i\omega_x)] = 0. \tag{2.25}$$

Solving this linear system of equations, defined for each x, involves the computation of the inverses of matrices. Since for typical size of fringe patterns, this operation would require a large amount of memory and computation power, a more efficient method of iterative gradient descent is generally used for obtaining the estimate of the complex fringe signal.

Considering the infinite extension of the fringe signal for the purpose of analysis, the response of the first order RQF can be represented by (2.22) alone. Accordingly, the frequency response of the first order RQF can be computed as

$$H(\omega) = \frac{1}{1 + 2\lambda[1 - \cos(\omega - \omega_x)]}. \tag{2.26}$$

The frequency response indicates the band-pass filtering characteristic of the RQF with ω_x as its center frequency and the filter width is controlled by λ. For a reliable filtering of the unwanted spectral lobes in (2.7), it is necessary that the quadrature conditions in the RQF frequency response, that is, $H(0) = 0$ and $H(-\omega_x) = 0$, are satisfied. These conditions ensure that the contributions of the background intensity and the unwanted complex fringe signal (lying around $-\omega_x$) in (2.7) are removed from the retrieved complex fringe signal. However, in accordance with (2.22)–(2.25), the response of the RQF *varies* pixel-wise depending on whether the pixel belongs to the valid or invalid region of the fringe signal. Thus, RQF operates as a linear filter with a spatially varying filter response. This key feature of RQF allows one to obtain the complex fringe signal from a real fringe signal containing invalid regions while the phase estimation at the valid pixels remains unaffected.

Equation (2.26) indicates that the frequency response of the first order RQF depends on the carrier frequency ω_x and the regularization parameter λ. Thus, it is not always guaranteed that the required quadrature conditions are met. This problem becomes even more evident for lower values of ω_x, in which case the contribution due to the background intensity causes the accuracy of the phase estimation to deteriorate. These limitations associated with the first order RQF, represented by (2.22)–(2.26), can be overcome by the design of the second order RQF with an additional background intensity filtering capability. The design criterion and implementation details of these filters have been discussed in [17,18]. The important precept behind the design of these different RQFs is to retrieve the complex fringe signal carrying the phase information from the modulated carrier fringe signal by rejecting the unwanted spectral components as effectively as possible.

2.2.6 Phase locked loop

The phase estimate is obtained in wrapped form in the transform and SPS based methods of carrier fringe pattern demodulation. The ensuing use of an unwrapping algorithm is essential for obtaining the absolute phase estimate. A novel approach based on the synchronous demodulating system has been proposed in [19] to directly compute the unwrapped phase estimates from a single recording of the modulated carrier fringe pattern. This technique termed *phase locked loop* (PLL) is well-established in communications engineering. The basic principle of the PLL system is described in figure 2.6. The phase changes in the input signal are tracked by varying the voltage applied to the voltage controlled oscillator (VCO) such that the phase difference between the input signal and the output of VCO is minimized. The synchronous demodulation property of the PLL is then used to obtain the accurate estimation of phase from a noisy fringe pattern.

Let us consider the one-dimensional fringe signal

$$\Gamma(x) = \cos[\omega_x x + \phi(x)], \tag{2.27}$$

where for simplicity of analysis we consider $a(x) = 0$ and $b(x) = 1$. The working principle of a PLL based phase estimation is governed by the following differential equation

$$\frac{d\hat{\phi}(x)}{dx} = -\tau\Gamma(x)\sin(\omega_x x + \hat{\phi}(x)), \tag{2.28}$$

where τ is a low valued constant on which depends the frequency response of the PLL; and $\hat{\phi}(x)$ represents the demodulated phase. Substituting (2.27) into (2.28) and simplifying we obtain

$$\frac{d\hat{\phi}(x)}{dx} = -\frac{\tau}{2}[\sin[\phi(x) - \hat{\phi}(x)] + \sin(2\omega_x x + \phi(x) + \hat{\phi}(x))]. \tag{2.29}$$

Under the assumption of continuous and smoothly varying phase, we can consider the first term in (2.29) as slowly varying, and the second term as varying at twice the rate of the fringe signal. In the case of a small value of τ, the first order system

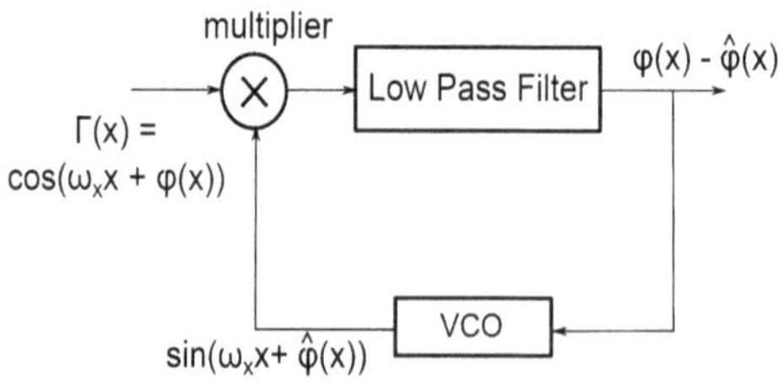

Figure 2.6. Phase locked loop system for fringe demodulation.

described in (2.29) operates as a low pass filter. In addition, since the true phase $\phi(x)$ is followed closely by the estimated phase $\hat{\phi}(x)$, we can write

$$\frac{d\hat{\phi}(x)}{dx} = -\tau[\sin(\phi(x) - \hat{\phi}(x))],$$
$$\approx \tau(\hat{\phi}(x) - \phi(x)).$$

In the digital implementation of the PLL based phase estimation, the derivative operation is approximated with the first order difference equation. Thus, from (2.28), we obtain

$$\hat{\phi}(x+1) = \hat{\phi}(x) + \tau\Gamma(x)\sin(\omega_x x + \hat{\phi}(x)).$$

The above equation indicates that the phase estimate at $(x+1)$ is computed as the *sum* of the already estimated phase at x and the driving term containing the quadrature of the fringe signal. As a consequence, the resulting phase is in an *unwrapped* form.

A simulation example of fringe pattern demodulation is provided in figure 2.7. A fringe pattern of size $M \times N = 256 \times 256$, shown in figure 2.7(d), is simulated at

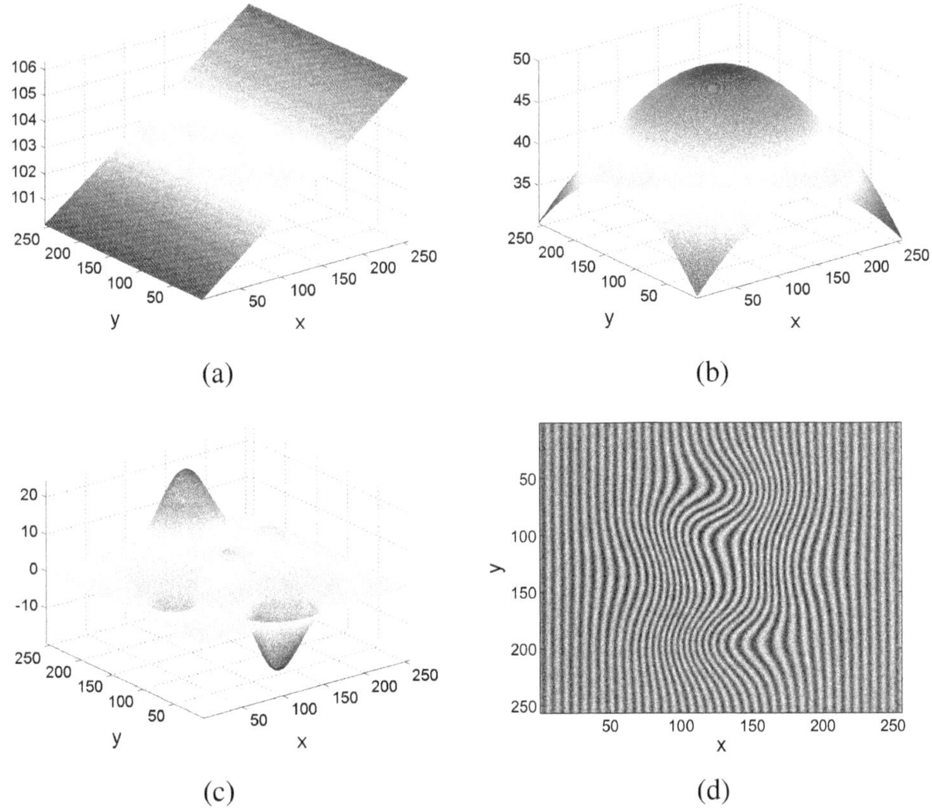

Figure 2.7. (a) Background intensity $a(x, y)$; (b) fringe amplitude $b(x, y)$; (c) phase $\phi(x, y)$; (d) modulated carrier fringe pattern simulated at SNR = 20 dB with $\omega_x = 2\pi \times 40/N$.

SNR = 20 dB with the plots of its background intensity, fringe amplitude and phase shown in figures 2.7(a), (b) and (c), respectively. For an FT based fringe demodulation example, let us choose the carrier frequency of $\omega_x = 2\pi \times 40/N$ rad/pixel and a filter window, as shown in figure 2.8 (b). The Fourier spectra of the fringe pattern without and with the carrier fringes added are shown in figures 2.8(a) and (b), respectively. Figure 2.8(c) shows the Fourier spectrum of the filtered fringe signal. The inverse Fourier transformation of the filtered fringe signal allows for computing the phase. The estimated wrapped phase obtained post the carrier fringe compensation is shown in figure 2.8(d).

For a WFF based fringe demodulation example, equally applied to the fringe pattern in figure 2.7(d), let us choose the frequency bands (i) $\varsigma = [\omega_x - 0.5, \omega_x + 0.5]$; $v = [-0.5, 0.5]$ and (ii) $\varsigma = [\omega_x - 1, \omega_x + 1]$; $v = [-1, 1]$, the other parameters, i.e. the window size and threshold being kept the same in both the cases. The aim of this example is also to highlight the effect the selection of frequency bands has on the phase estimates. The estimated phase maps post the

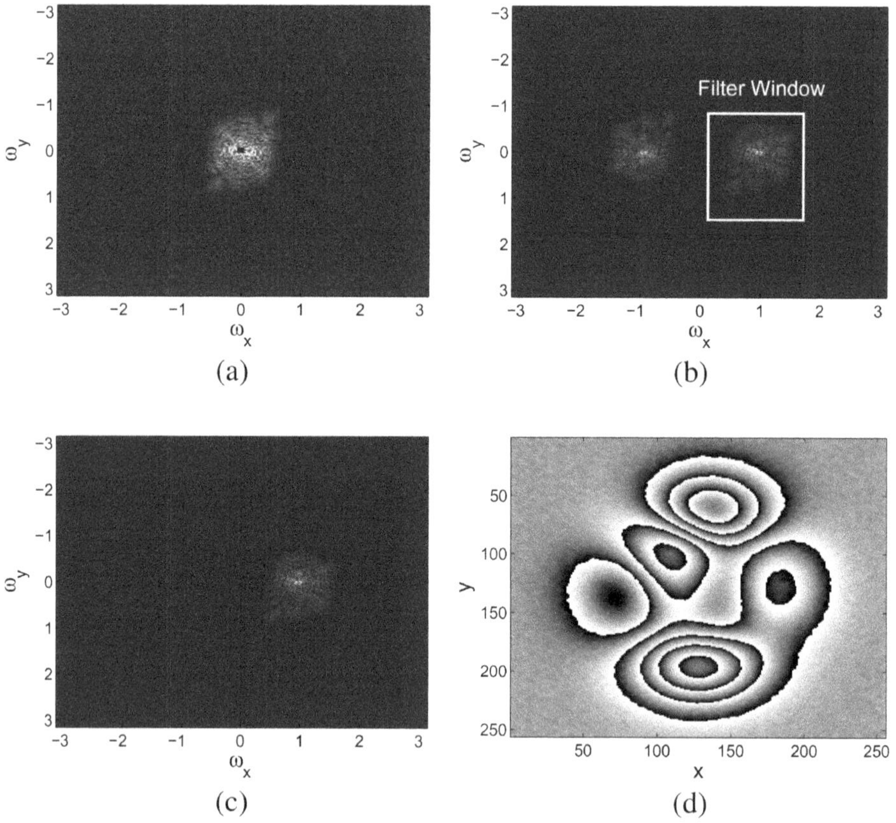

Figure 2.8. (a) Fourier spectrum of the closed fringe pattern; (b) Fourier spectrum of the modulated carrier fringe pattern described in figure 2.7(d); The selected filter window is shown as the white rectangle. (c) Fourier spectrum of the filtered fringe pattern and (d) wrapped phase estimate obtained post the carrier frequency compensation.

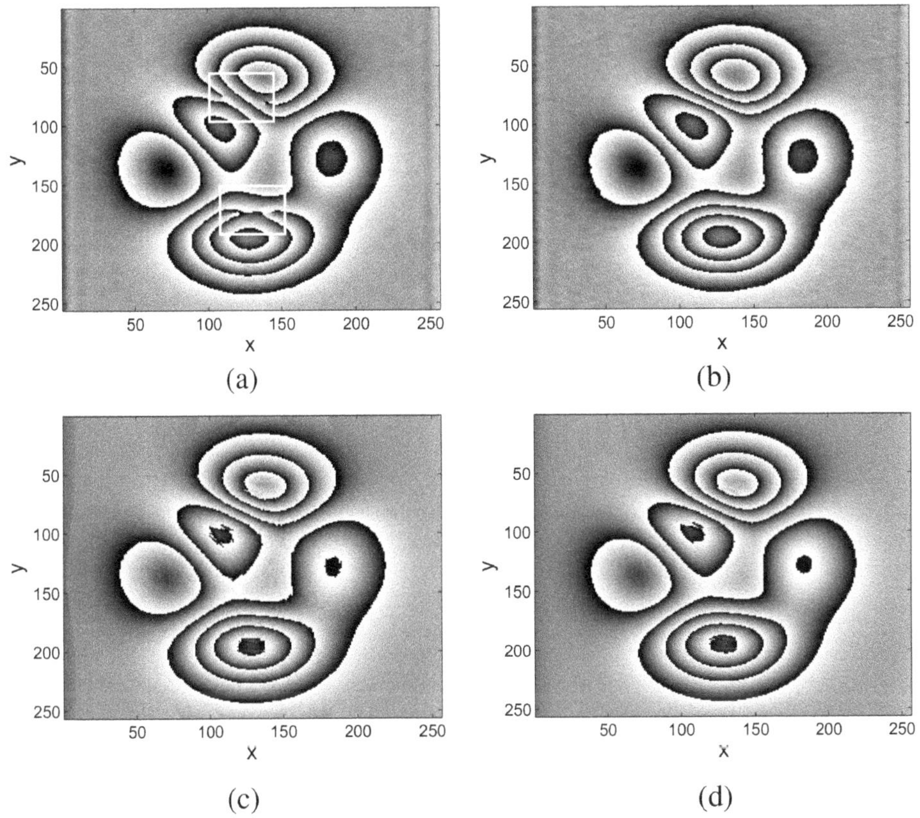

Figure 2.9. Wrapped phase estimates computed using the WFF method with (a) inappropriate and (b) appropriate selection of frequency bands. Wrapped phase estimate computed using the WFR method with (c) coarse and (b) fine increments in the frequencies.

carrier frequency compensation are shown in figures 2.9 (a) and (b), respectively. The rectangular regions shown in figure 2.9(a) indicate the locations in the fringe pattern where the phase has not been estimated accurately. This is due to an inappropriate selection of the frequency range in case (i). On the other hand, the appropriate selection of the frequency range in case (ii) gives rise to an accurate phase estimate, as shown in figure 2.9(b). For the sake of comparison, the example of the WFR algorithm implemented with an appropriate selection of the frequency range is shown in figures 2.9(c) and (d), with the particularity that (c) and (d) are obtained, respectively, with coarse and fine increments in the frequencies. The results of phase estimation that are shown in figures 2.9(c) and (d), respectively, show that the quality of phase estimates obtained using the WFR algorithm is inferior to those obtained using the WFF algorithm. This example also illustrates that some information on the fringe density is essential to an appropriate selection of the frequency range, a parameter of paramount interest in achieving a successful implementation of the WFF and WFR algorithms.

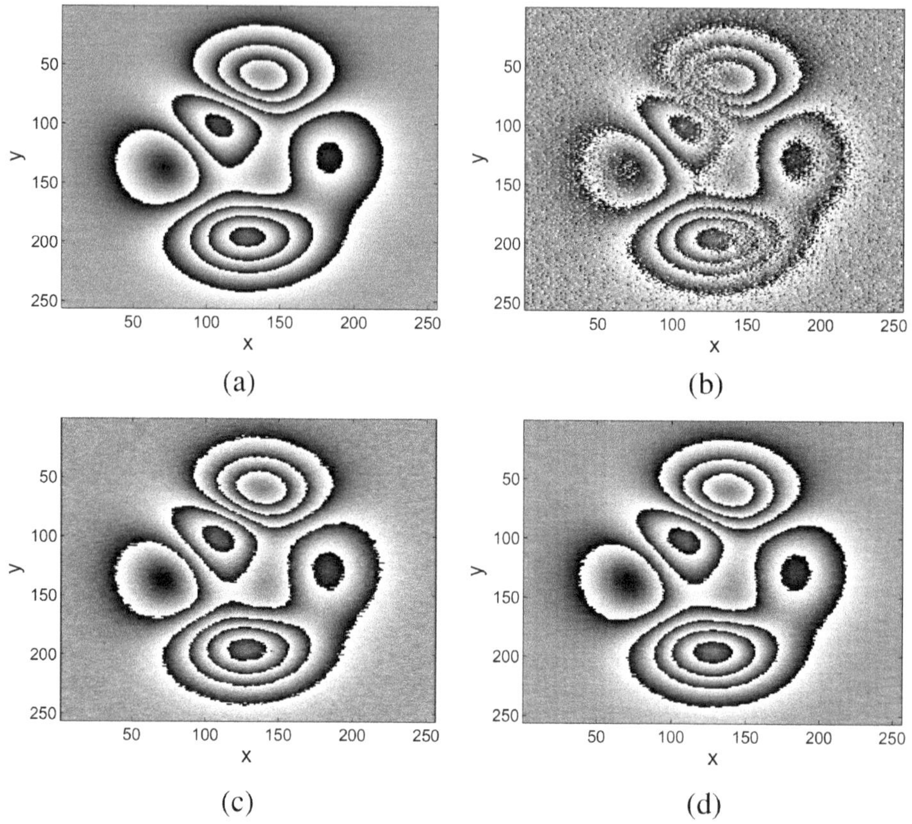

Figure 2.10. Wrapped phase estimates computed using (a) CWT, (b) SPS with the specified carrier frequency (c) SPS with carrier frequency of $\omega_x = 2\pi/3$ radians/pixel and (d) RQF methods.

The demodulation of the fringe pattern in figure 2.7(d) using the CWT, the SPS, the SPS with higher carrier frequency and the RQF methods is shown in figures 2.10(a), (b), (c) and (d), respectively. While the CWT and RQF methods are found to provide reliable phase estimation, the SPS method fails to provide an accurate phase estimate, as can be observed from figure 2.10(b). The failure of the SPS method can be traced to its inability to satisfy the criterion for the required phase shift between the two consecutive pixels. However, if this phase shift is set to $\omega_x = 2\pi/3$, a reliable phase estimate is obtained as shown in figure 2.10(c).

The example of fringe pattern shown in figure 2.11(a) is recorded in fringe projection profilometry. The modulated carrier fringe pattern carries the information on the three-dimensional shape of an object. The demodulation of this fringe pattern using the FT, WFF, CWT and RQF methods is shown in figures 2.11(b), (c), (d) and (e), respectively.

2.3 Closed fringe pattern demodulation

Phase estimation based on carrier fringe demodulation, as described in section 2.2, and on the phase shifting principle has certain limitations. The phase shifting based

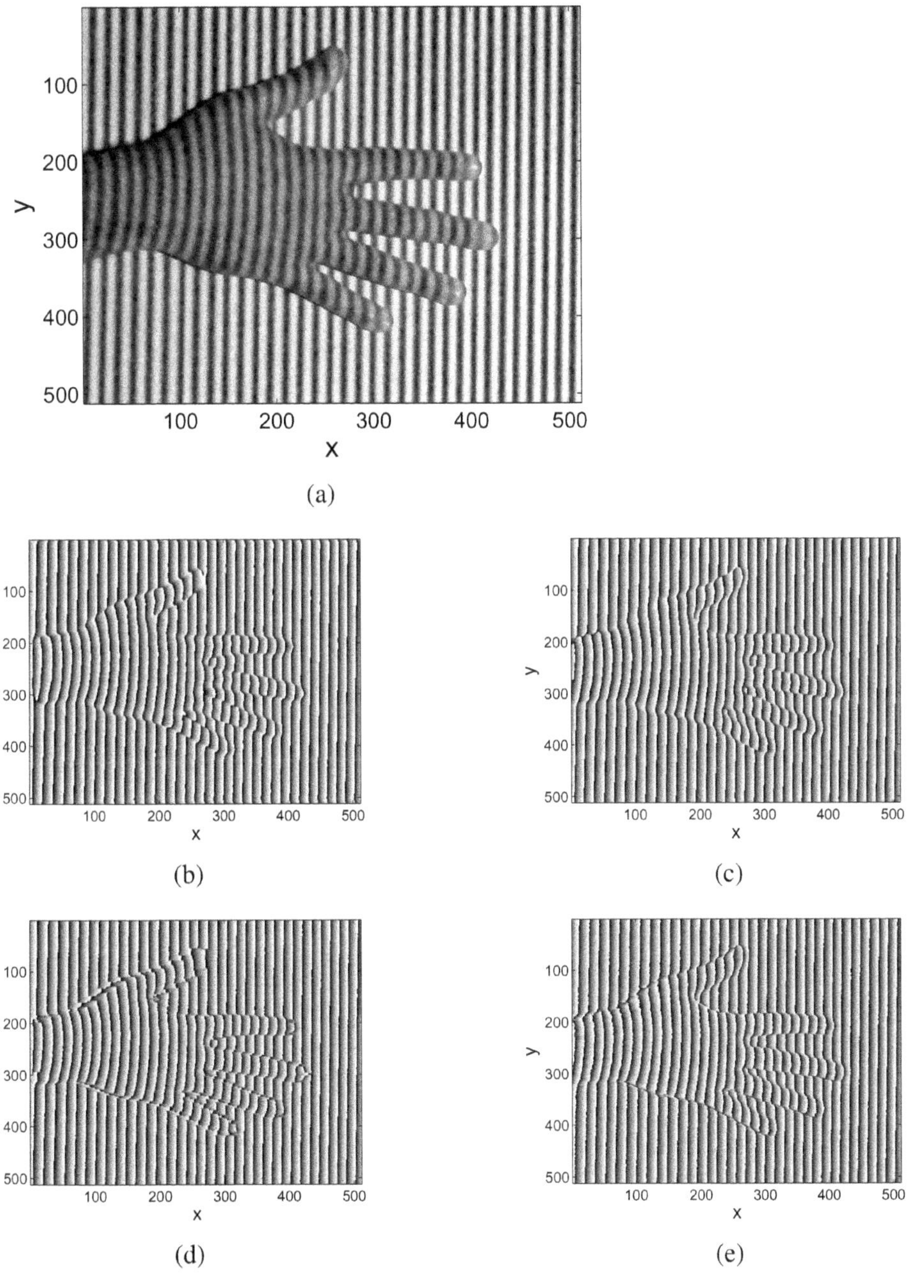

Figure 2.11. (a) Fringe pattern recorded in a fringe projection profilometry setup. Wrapped phase estimates obtained using (b) FT (c) WFT (d) WVT and (e) RQF methods.

methods require a number of fringe patterns, albeit phase shifted, to be recorded while assuming the phase to be constant in all of these frames. This assumption makes the phase shifting technique look suitable only for applications pertaining to static measurements. On the other hand, the sole requirement of a single recording of a fringe pattern in the carrier fringe based phase estimation offers promising avenues of applications to dynamic measurements. Two limitative factors that however restrict the use of this latter technique is that first, the spatial resolution of the recording device, that is, the CCD, limits the amount of carrier fringes that can be introduced, and second, the underlying condition that the maximum spatial variation of the phase associated with the measurand has to be much smaller than the phase associated with the carrier has to be satisfied. These two factors limit the dynamic range of the phase that can be reliably estimated.

To overcome these limitations, a fringe analysis technique should be able to demodulate a fringe pattern from a single recording and that too without the addition of carrier fringes. The phase embedded in such a fringe pattern can no longer be assumed to be a monotonically increasing function and the corresponding fringe is termed to be in a closed form. The closed fringe pattern in (2.1) contains three unknown parameters a, b and ϕ. We proceed by applying successively high pass filtering and fringe amplitude normalization techniques [20–22] for eliminating the parameters a and b, respectively. The thus obtained fringe pattern is given by

$$\Gamma(x, y) = \cos[\phi(x, y)].$$

The above equation suggests that the problem of phase estimation from a closed fringe pattern is ill-posed due to the existence of more than one possible solution. The following subsections discuss some of the closed fringe demodulation techniques that have been proposed in the literature. These techniques based on the assumption that the phase is spatially smoothly varying limit the possible solutions of phase estimates to the continuous function space.

2.3.1 Spiral phase quadrature transform

One of the important methods for closed fringe demodulation is based on spiral phase quadrature transform [23,24]. A critical step while estimating phase from a closed fringe pattern is to obtain the quadrature component, that is, $\sin[\phi(x, y)]$, of the fringe pattern. The spiral phase quadrature transform is promising as a method for reliably computing this quadrature component from a noisy closed fringe pattern. In the case of a one-dimensional modulated carrier fringe signal, its Hilbert transform provides the estimate of the quadrature fringe signal. Let us first consider the one-dimensional fringe signal

$$\Gamma(x) = \cos[\phi(x)].$$

The Hilbert transform of this signal is

$$\sin[\phi(x)] = \mathscr{H}\{\Gamma(x)\} = \mathscr{H}\{\cos[\phi(x)]\}.$$

Using these, the equivalent complex fringe signal follows as

$$\exp(i\phi(x)) = \Gamma(x) + i\mathscr{H}\{\Gamma(x)\}.$$

The FT of the Hilbert transform of the given signal is

$$\mathscr{F}[\mathscr{H}\{\Gamma(x)\}] = -i\ \mathrm{sgn}(\varsigma)\mathscr{F}[\Gamma(x)],$$

where sgn[ς] is the one-dimensional *signum* function defined in the frequency domain as

$$\mathrm{sgn}(\varsigma) = \frac{\varsigma}{|\varsigma|} = \begin{cases} -1, \varsigma < 0 \\ 0, \ \varsigma = 0. \\ +1 \ \varsigma > 0 \end{cases}$$

A two-dimensional extension of the one-dimensional signum function which keeps intact the latter's point discontinuity property has been proposed [23] to compute the two-dimensional Hilbert transform. The two-dimensional signum function in the spatial frequency domain is defined as

$$\mathrm{sgn}(\varsigma, v) = \frac{\varsigma + iv}{\sqrt{\varsigma^2 + v^2}} = \exp[i\theta(\varsigma, v)]. \tag{2.30}$$

The signum function with its spiral phase property is the crux of spiral phase transform based closed fringe demodulation. The unit magnitude and odd symmetry properties of the 1D-signum function are carried forth into the 2D-signum function in (2.30) with equal response along all directions.

The important steps involved in the spiral quadrature transform based closed fringe demodulation are as follows:

1. The spiral phase function given in (2.30) is multiplied with the FT of the closed fringe pattern. The product is inverse Fourier transformed:

$$\begin{aligned} SP[\Gamma(x, y)] &= \mathscr{F}^{-1}\{\exp[i\theta(\varsigma, v)]\mathscr{F}[\Gamma(x, y)]\} \\ &= \exp(i\beta(x, y))\sin(\phi(x, y)), \end{aligned}$$

   where $\beta(x, y)$ represents the unique fringe orientation angle computed at each pixel location. It is estimated using a fringe orientation algorithm.
2. Since the fringe orientation angle map, $\beta(x, y)$, is computed in a wrapped form, its unwrapped form is estimated by applying an unwrapping algorithm [23].
3. Once the unwrapped fringe orientation angle map has been estimated, the quadrature signal estimate of the fringe pattern is computed as

$$\sin(\phi(x, y)) = \Im\{\exp[-i\beta(x, y)]SP[\Gamma(x, y)]\}.$$

The accurate estimation of the fringe orientation is essential to obtaining a reliable estimate of the quadrature fringe signal. Larkin has suggested two fringe orientation estimation algorithms [25] based on energy operators. It should be noted that although this method has been explained using a normalized fringe pattern, the use of the fringe normalization procedure is not essential to the successful accomplishment of the method. However, the background intensity has to be removed before this method can be applied to fringe demodulation. The spiral phase quadrature transform based phase estimation has been shown in [23] to be inherently robust to the errors in the estimation of fringe orientation in the presence of noise.

2.3.2 Regularized phase tracking

In the case of carrier fringe demodulation, the phase locked loop technique (2.2.5) has been seen to provide the phase measurement directly in an unwrapped form. An idea of phase-tracking similar to the one used in PLL based phase estimation, named regularized phase tracking (RPT), is proposed [26] for the demodulation of closed fringe patterns. In addition to phase tracking, the fringe frequency tracking is also performed in the RPT based fringe demodulation. An important feature of the RPT based method lies in its introduction of an additional constraint to the problem of cost function minimization by the estimated phase. This constraint ensures that the estimated phase is a smoothly varying function of the spatial coordinates x and y. The introduction of this additional constraint is termed *regularization.*

In the RPT technique the closed fringe pattern is first normalized, and the phase estimation is next performed by analyzing the fringe pattern within a small window around each pixel at a time. The fringe intensity in this window is assumed to be a pure sinusoid of a single frequency. In other words, the phase modulating the fringe intensity within the window is considered to be linear. Such a linear phase approximation around a pixel (x,y) can be expressed as

$$\phi_L(x, y, u, v) = \hat{\phi}(x, y) + \hat{\omega}_x(x, y)(x - u) + \hat{\omega}_y(x, y)(y - v), \tag{2.31}$$

where u and v indicate the pixel locations along column and row, respectively, within the analysis window; $\hat{\omega}_x$ and $\hat{\omega}_y$ indicate slopes, that is, the spatial fringe frequency estimates at (x, y); $\hat{\phi}(x, y)$ indicates the phase estimate. The phase estimate is obtained by minimizing the cost function $U(x, y)$

$$\begin{aligned} U_{x,y}(\hat{\phi}, \omega_x, \omega_y) = \sum_{u,v\in L} \Big\{ & \big[\Gamma(u, v) - \cos(\phi_L(x, y, u, v))\big]^2 \\ & + \lambda\big[\hat{\phi}(u, v) - \phi_L(x, y, u, v)\big]^2 \mathbb{1}(u, v)\Big\}, \end{aligned} \tag{2.32}$$

with respect to $(\hat{\phi}, \omega_x, \omega_y)$. In (2.32), $u, v \in L$ represent the neighborhood of the current pixel; λ represents the regularization parameter. The role of the first term in (2.32) is to minimize the difference between the fringe model, $\cos[\phi_L(x, y, u, v)]$, and the fringe intensity within the analysis window L in a least squares sense. However, minimization of this term alone is not sufficient to obtain the required phase estimate since there will be more than one possible solution. Therefore, an additional constraint is imposed in the form of the second term in (2.32). This term ensures the spatial smoothness and continuity of the sought phase. Note that this additional constraint is applied using only those pixels within the analysis window for which the phase estimates have already been computed. Such pixels are indicated with an indicator function $\mathbb{1}(u, v)$, which has a value 1 for the already processed pixels and 0 for the unprocessed pixels.

Now, let us represent the spatial co-ordinates as $\mathbf{r} = (x, y)$. The cost function given in (2.32) is a function of three variables, $\hat{\phi}(\mathbf{r})$, $\hat{\omega}_x(\mathbf{r})$ and $\hat{\omega}_y(\mathbf{r})$. The phase estimate is obtained by minimizing the cost function with respect to these variables. A fixed-step gradient descent method can be used to obtain the optimal values of these three variables as

$$\begin{aligned}
\hat{\phi}(\mathbf{r})^{k+1} &= \hat{\phi}(\mathbf{r})^{k} - \tau\frac{\partial U}{\partial\hat{\phi}(\mathbf{r})}\\
\hat{\omega}_x(\mathbf{r})^{k+1} &= \hat{\omega}_x(\mathbf{r})^{k} - \tau\frac{\partial U}{\partial\hat{\omega}_x(\mathbf{r})}\\
\hat{\omega}_y(\mathbf{r})^{k+1} &= \hat{\omega}_y(\mathbf{r})^{k} - \tau\frac{\partial U}{\partial\hat{\omega}_y(\mathbf{r})},
\end{aligned}$$

where k indicates the iteration step; τ is the step size. The initial conditions of the iterative process are set as $[\hat{\phi}(\mathbf{r})^0, \hat{\omega}_x(\mathbf{r})^0, \hat{\omega}_y(\mathbf{r})^0]$.

Initialization of the iterative process plays an important role for the values of $\hat{\phi}(\mathbf{r})$, $\hat{\omega}_x(\mathbf{r})$ and $\hat{\omega}_y(\mathbf{r})$ to converge reliably towards their true values. A crystal-growing algorithm has been proposed [26] to ensure an appropriate initialization of these variables. Accordingly, the initial conditions at the current pixel $\mathbf{r}$ are set equal to the values of $(\hat{\phi}, \omega_x, \omega_y)$ estimated at a pixel $\mathbf{r}' = (x', y')$ lying around its neighborhood and which has previously been processed. That is,

$$\begin{aligned}
\hat{\phi}(\mathbf{r})^0 &= \hat{\phi}(\mathbf{r}'),\\
\hat{\omega}_x(\mathbf{r})^0 &= \hat{\omega}_x(\mathbf{r}'),\\
\hat{\omega}_y(\mathbf{r})^0 &= \hat{\omega}_y(\mathbf{r}').
\end{aligned}$$

A notable feature of this initialization procedure is that it ensures that the estimated phase is in an unwrapped form. In the case of a convoluted fringe pattern, a relatively robust close fringe demodulation approach based on a fringe following strategy has been proposed in [27]. The fringe demodulation is achieved by following a specific pixel scanning strategy instead of a conventional row-by-row or column-by-column scanning.

2.3.3 Local adaptable quadrature filter

The concept of the robust quadrature filter (RQF) has been effectively used in the demodulation of a modulated carrier fringe pattern. In the RQF based fringe analysis, the filtering is applied to the entire modulated carrier fringe pattern. The same procedure when applied to a closed fringe pattern gives rise to sign ambiguities in the estimated phase. Consequently, in the local adaptable quadrature filter (LAQF) based phase estimation, the quadrature filtering is applied *locally* around the current pixel (x,y) within the analysis window L [28]. Similar to the RPT method, here too we assume a linear phase approximation, as in (2.31), inside the analysis window. Accordingly, the complex fringe signal $\Gamma_1(x, y) = \exp[i\phi(x, y)]$ is computed in a way which minimizes the cost function

$$U(\Gamma_l) = \sum_{(u,v)\in L} |\Gamma_{lx}(u, v) - 2\Gamma_x(u, v)|^2 + |\Gamma_{ly}(u, v) - 2\Gamma_y(u, v)|^2$$
$$+ \lambda \sum_{(u,v)\in L} |\Gamma_l(u, v) - \Gamma_l(u - 1, v)\exp(i\omega_x)|^2 + |\Gamma_l(u, v) - \Gamma_l(u, v - 1)\exp(i\omega_y)|^2,$$

where $\Gamma_{lx}(x, y)$ and $\Gamma_x(x, y)$ represent the finite differences of $\Gamma_l(x, y)$ and $\Gamma(x, y)$, respectively, along the x direction. This is similar for $\Gamma_{ly}(x, y)$ and $\Gamma_y(x, y)$. In digital implementation, the finite difference is computed as $\Gamma_x(x, y) = \Gamma(x, y) - \Gamma(x - 1, y)$. With a view to achieving computational efficiency, we retrieve the complex fringe signal in an iterative manner using the steepest descent method similar to that employed in the RQF.

The fundamental difference between the RQF and LAQF is that, whereas the former is applied to the entire modulated carrier fringe pattern, where the carrier frequency is known *a priori*, the latter is applied locally to the closed fringe signal within a small area around each pixel, where the carrier frequency information (ω_x and ω_y) needs to be computed. The phase estimate within the analysis window is computed as

$$\hat{\phi}(u, v) = \arctan\left\{\frac{\Im[\Gamma_l(u, v)]}{\Re[\Gamma_l(u, v)]}\right\}.$$

The local frequency estimates computed at the current pixel location are used as the tuning frequencies in the phase estimation at neighboring pixel locations. Once the phase is obtained at the current pixel (x,y), the estimates of local frequencies are computed as

$$\omega_x = \arctan\left\{\frac{\sin[(\hat{\phi}(x, y) - \hat{\phi}(x', y))(x - x')]}{\cos[(\hat{\phi}(x, y) - \hat{\phi}(x', y))(x - x')]}\right\},$$
$$\omega_y = \arctan\left\{\frac{\sin[(\hat{\phi}(x, y) - \hat{\phi}(x, y'))(y - y')]}{\cos[(\hat{\phi}(x, y) - \hat{\phi}(x, y'))(y - y')]}\right\}$$

where x' and y' represent the location of pixels for which the phase has already been estimated previously. In order to obtain a continuous and sign unambiguous phase estimation, each pixel is visited based on a fringe following strategy similar to the one adopted in the RPT based method.

References

[1] Takeda M, Ina H and Kobayashi S 1982 Fourier-transform method of fringe-pattern analysis for computer-based topography and interferometry *J. Opt. Soc. Am.* **72** 156–60

[2] Takeda M and Mutoh K 1983 Fourier transform profilometry for the automatic measurement of 3-D object shapes *Appl. Opt.* **22** 3977–82

[3] Kemao Q 2007 Two-dimensional windowed Fourier transform for fringe pattern analysis: principles, applications and implementations *Opt. Lasers Eng.* **45** 304–17

[4] Kemao Q 2004 Windowed Fourier transform for fringe pattern analysis *Appl. Opt.* **43** 2695–702

[5] Daubechies I 1990 The wavelet transform, time-frequency localization and signal analysis *IEEE Trans. Inf. Theory* **36** 961–1005

[6] Mallat S 2009 *A Wavelet Tour of Signal Processing* 3rd edn (Burlington: Academic)

[7] Watkins L R 2012 Review of fringe pattern phase recovery using the 1-D and 2-D continuous wavelet transforms *Opt. Lasers Eng.* **50** 1015–22

[8] Gdeisat M A, Abid A, Burton D R, Lalor M J, Lilley F and Moore C 2009 Spatial and temporal carrier fringe pattern demodulation using the one-dimensional continuous wavelet transform: recent progress, challenges, and suggested developments *Opt. Lasers Eng.* **47** 1348–61

[9] Li S, Su X and Chen W 2010 Wavelet ridge techniques in optical fringe pattern analysis *J. Opt. Soc. Am.* A **27** 1245–54

[10] Watkins L R 2014 *Phase Estimation in Optical Interferometry* ed Rastogi (Boca Raton, FL: CRC Press) pp 69–120

[11] Gdeisat M A, Burton D R and Lalor M J 2006 Spatial carrier fringe pattern demodulation by use of a two-dimensional continuous wavelet transform *Appl. Opt.* **45** 8722–32

[12] Creath K 1988 V phase-measurement interferometry techniques *Progress in Optics* vol 26 (Amsterdam: Elsevier) pp 349–93

[13] Shough D M, Kwon O Y and Leary D F 1990 High-speed interferometric measurement of aerodynamic phenomena *Proc. SPIE* **1221** 394–403

[14] Mertz L 1983 Real-time fringe-pattern analysis *Appl. Opt.* **22** 1535–9

[15] Servin M and Cuevas F J 1995 A novel technique for spatial phase-shifting interferometry *J. Mod. Opt.* **42** 1853–62

[16] Bothe T, Burke J and Helmers H 1997 Spatial phase shifting in electronic speckle pattern interferometry: minimization of phase reconstruction errors *Appl. Opt.* **36** 5310–6

[17] Marroquin J L, Figueroa J E and Servin M 1997 Robust quadrature filters *J. Opt. Soc. Am.* A **14** 779–91

[18] Servin M, Quiroga J A and Padilla M 2014 *Fringe Pattern Analysis for Optical Metrology: Theory Algorithms, and Applications* (Weinheim: Wiley) pp 1–327

[19] Servin M and Rodriguez-Vera R 1993 Two-dimensional phase locked loop demodulation of interferograms *J. Mod. Opt.* **40** 2087–94

[20] Quiroga J A, Antonio Gómez-Pedrero J and García-Botella A 2001 Algorithm for fringe pattern normalization *Opt. Commun.* **197** 43–51

[21] Ochoa N A and Silva-Moreno A A 2007 Normalization and noise-reduction algorithm for fringe patterns *Opt. Commun.* **270** 161–8

[22] Bernini M B, Federico A and Kaufmann G H 2009 Normalization of fringe patterns using the bidimensional empirical mode decomposition and the Hilbert transform *Appl. Opt.* **48** 6862–9

[23] Larkin K G, Bone D J and Oldfield M A 2001 Natural demodulation of two-dimensional fringe patterns. I. General background of the spiral phase quadrature transform *J. Opt. Soc. Am.* A **18** 1862–70

[24] Larkin K G 2001 Natural demodulation of two-dimensional fringe patterns. II. Stationary phase analysis of the spiral phase quadrature transform *J. Opt. Soc. Am.* A **18** 1871–81

[25] Larkin K G 2005 Uniform estimation of orientation using local and nonlocal 2-D energy operators *Opt. Exp.* **13** 8097–121

[26] Servin M, Cuevas F J, Malacara D, Marroquin J L and Rodriguez-Vera R 1999 Phase unwrapping through demodulation by use of the regularized phase-tracking technique *Appl. Opt.* **38** 1934–41
[27] Servin M, Marroquin J L and Cuevas F J 2001 Fringe-follower regularized phase tracker for demodulation of closed-fringe interferograms *J. Opt. Soc. Am.* A **18** 689–95
[28] Estrada J C, Servin M and Marroquin J L 2007 Local adaptable quadrature filters to demodulate single fringe patterns with closed fringes *Opt. Exp.* **15** 2288–98

Chapter 3

Fringe denoising and phase unwrapping

In the previous chapter we have described spatial fringe analysis techniques for phase estimation from a single fringe pattern. The performance of spatial fringe analysis techniques greatly depends on their ability to handle noisy fringe patterns. Whereas some techniques are inherently noise robust, others require fringe denoising as a pre-processing operation. Denoising of fringe patterns reduces the operational load on subsequent fringe analysis. Especially in the case of closed fringe demodulation, denoising of the fringe pattern is highly recommended for obtaining a reliable phase estimation. In this chapter we will discuss a number of representative fringe denoising techniques that are applicable to both real sinusoidal fringe patterns such as those recorded in an ESPI setup, and wrapped phase fringe patterns such as those recorded in digital holographic interferometry.

In general, the phase is a spatially smoothly varying function and the sinusoidal fringe encoding this phase is a low frequency signal. In coherent interferometry, this signal is corrupted by the *speckle noise*. The speckle noise appears mainly because of the coherent nature of the light source and the scattering of this light from a rough object surface. The fringe pattern is also affected by the electronic noise generated into the imaging device, that is, the CCD camera. Whereas the speckle noise is multiplicative, the electronic noise is additive in nature. If such a fringe pattern is operated upon with a linear filter, the noise in the output fringe signal tends to additive white Gaussian in keeping with the central limit theorem.

3.1 Real sinusoidal fringe pattern denoising

A number of fringe denoising algorithms based on different signal and image processing techniques [1–4] have been developed for denoising the real sinusoidal fringe patterns. A comparative study of the performances of a few selected fringe denoising methods can be found in [5,6]. In the design of a fringe denoising algorithm, a knowledge of features that distinguish a noiseless fringe signal from noise is essential. To start with, we have two important observations to make: (1) the

doi:10.1088/978-0-7503-1469-5ch3

true fringe signal is of non-stationary sinusoidal nature. (2) The noise is of additive white Gaussian type. Whereas, a true fringe signal typically exhibits low-frequency characteristics, the noise contribution covers the entire frequency range. This distinctive feature is used by most of the fringe denoising algorithms. Space related limitations have restricted us to include only a few representative fringe denoising algorithms.

3.1.1 Moving average filter

Probably the most commonly used filter in image and signal processing is the moving average filter. Although simple in operation and implementation, the moving average filtering is found to produce satisfactory results in most situations. It effectively performs the averaging of the image intensities over a window. Since averaging is a low pass filtering operation, the size of the averaging window is an important parameter in denoising wideband fringe patterns. Typically, a symmetric window of size $(2L + 1) \times (2L + 1)$ can be defined around each pixel, where L is a window size parameter to be defined by the user. A denoised fringe pattern can then be represented as

$$\Gamma_f(x, y) = \frac{1}{(2L+1)^2} \sum_{u=-L}^{L} \sum_{v=-L}^{L} \Gamma(x-u, y-v). \tag{3.1}$$

In principle, the moving average operation can be represented as a convolution of the fringe signal with a function $g_L(x, y)$,

$$g_L(x, y) = \frac{1}{(2L+1)^2} \sum_{u=-L}^{L} \sum_{v=-L}^{L} \delta(x-u, y-v),$$

where $\delta(\cdot)$ represents the delta. The subscript L indicates the dependency of the filter response on L. The moving average filter in (3.1) is basically a non-causal 2D finite impulse response filter. Accordingly, the frequency response of the filter can be written as

$$\mathcal{F}g(\varsigma, \nu) = \frac{1}{(2L+1)^2}\left(1 + \sum_{k=1}^{L} 2\cos(k\varsigma)\right)\left(1 + \sum_{k=1}^{L} 2\cos(k\nu)\right).$$

Specifically, a typical 3×3 average filter applied to a fringe pattern produces the denoised fringe pattern as

$$\Gamma_f(x, y) = \frac{1}{9} \sum_{u=-1}^{1} \sum_{v=-1}^{1} \Gamma(x-u, y-v).$$

The magnitude of the frequency responses of 3×3 and 5×5 filters are shown in figures 3.1 (a) and 3.1(c), respectively. The figures indicate that the moving average filter acts as a low pass filter with its cut-off frequency defined by the parameter L. A repeated application of the moving average filter has been observed to produce a better denoising effect by effectively shifting the frequency response towards the low

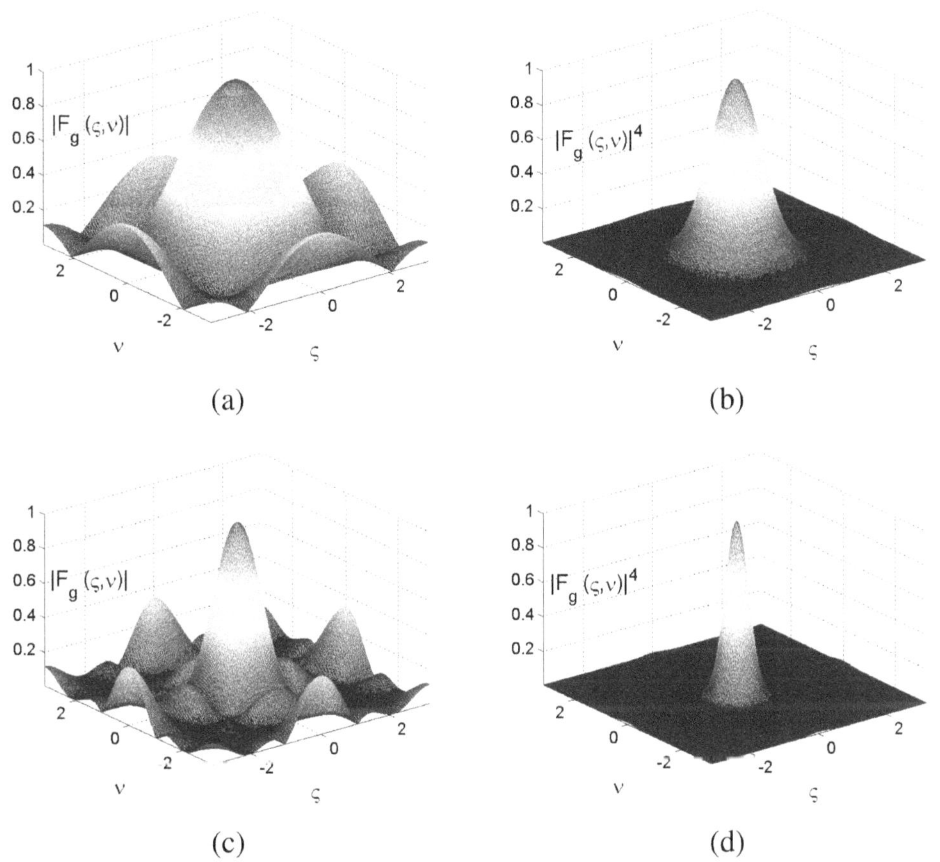

Figure 3.1. Frequency responses of the moving average filters with (a) $L = 1$ and (c) $L = 2$. Idem after four repeated applications of moving average filters with (b) $L = 1$ and (d) $L = 2$.

pass characteristic of the Gaussian filter. To illustrate, the frequency responses obtained after four repeated applications of 3×3 and 5×5 moving average filters are shown in figures 3.1(b) and (d), respectively.

An example of moving average fringe denoising is demonstrated in figure 3.2. Figure 3.2(a) shows the noise corrupted fringe pattern simulated at SNR=10 dB. The results of fringe denoising performed with moving average filters with $L = 1$ and $L = 2$ are shown in figures 3.2(b) and (c), respectively. The improvement in fringe denoising is much more apparent using a moving average filter with $L = 2$ than with $L = 1$. Figure 3.2(d) shows the denoised fringe pattern obtained for $L = 2$ with filter application repeated four times. The effect of the narrow pass-band of the effective filter frequency response shown in figure 3.1(d) is visible in figure 3.2(d). Whereas the low density fringe region is effectively filtered, the high fringe density region away from the center of the fringe pattern appears to be blurred out.

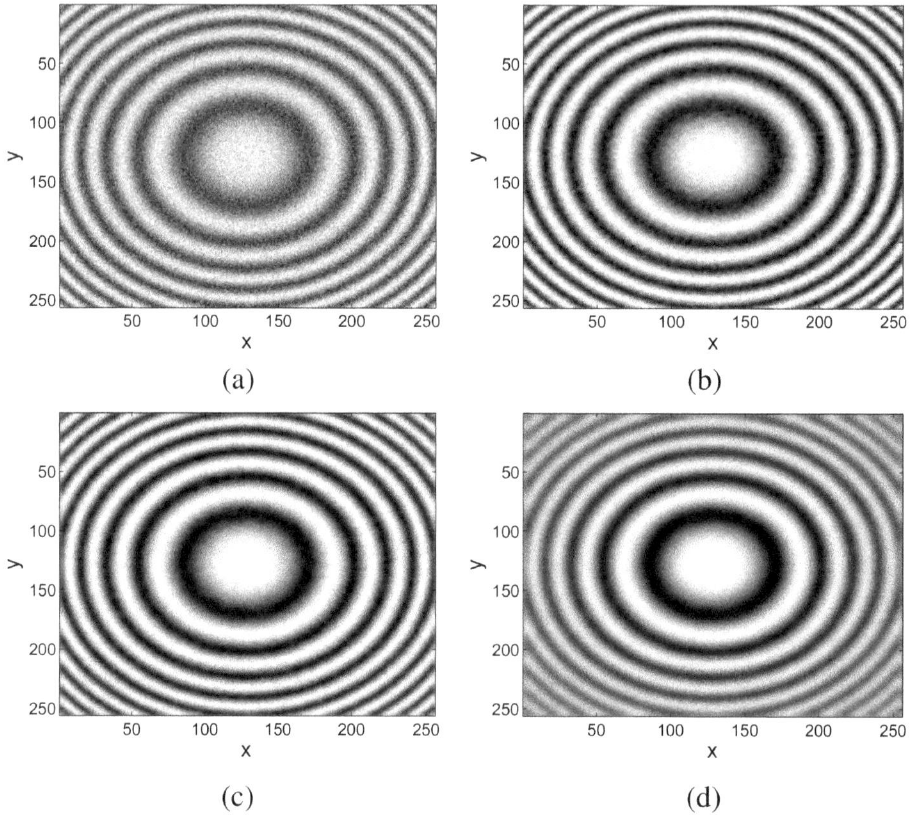

Figure 3.2. (a) Simulated fringe pattern of size 256 × 256 at SNR = 10 dB. Filtered fringe patterns obtained using moving average filters with (b) $L = 1$ (c) $L = 2$ and (d) $L = 2$ with four repeated applications of the moving average filter.

Since, in practice, only a finite size fringe data is available depending on the pixel array of the CCD camera, there is a lack of sufficient information at the border pixels to perform the averaging operation. Typically, the fringe pattern is either padded with zeros or the fringe intensities are extrapolated beyond the border pixels. As there is no explicit provision of handling the border pixels in the moving average filter based fringe denoising, the deterioration of denoised fringe data occurs at these pixels.

3.1.2 Windowed Fourier filtering

Apart from phase estimation, the windowed Fourier filtering (WFF) described in the last chapter can also be used for the purposes of denoising fringe patterns. To recall, the WFT of a fringe signal is computed as

$$WF_{\Gamma}(u, v, \varsigma, \nu) = \sum_{x=0}^{N-1} \sum_{y=0}^{M-1} \Gamma(x, y) h^*_{u,v,\varsigma,\nu}(x, y).$$

In the WFF based fringe analysis, the WFT spectrum is thresholded such that

$$\overline{WF_{\Gamma}}(u, v, \varsigma, \nu) = \begin{cases} WF_{\Gamma}(u, v, \varsigma, \nu), & \text{if } |WF_{\Gamma}(u, v, \varsigma, \nu)| \geqslant Thr \\ 0, & \text{otherwise} \end{cases}.$$

Subsequently, the inverse WFT is applied to the so thresholded WFT spectrum $\overline{WF_{\Gamma}}(u, v, \varsigma, \nu)$ to reconstruct the modified fringe signal.

The important distinction between the WFF based phase estimation and fringe denoising is noted as follows. In the case of the WFF based phase estimation from a carrier modulated fringe pattern, the frequency bands in the computation of the WFT are set such that the windowed Fourier filter acts as a localized band-pass filter to enable retrieving the complex fringe signal associated with the real modulated carrier fringe signal. By comparison, the fringe denoising described in the present discussion is mainly applied to the carrier-less fringe patterns. Since in the WFF based fringe denoising such a fringe signal is low-pass in nature, the frequency bands in the computation of the WFT are set such that the windowed Fourier filter acts as a localized low-pass filter. The retrieved fringe signal is thus real sinusoidal in character similar to the original, albeit in its denoised form.

The applicability of the WFF based fringe denoising is illustrated using the fringe pattern in figure 3.2(a). In the WFT computation, the frequency bands are selected as $\varsigma = [-0.5, 0.5]$ and $\nu = [-0.5, 0.5]$. Based on the WFF algorithm, the denoised fringe patterns obtained with the window size parameters set as $\sigma_x = 3$, $\sigma_y = 3$ and $\sigma_x = 10$, $\sigma_y = 10$ are shown in figures 3.3(a) and 3.3(b), respectively. The improvement in the results with the increase in the window size shows the robust fringe denoising capability of the WFF based method. Since WFT is a convolution operation, the WFF similar to the moving average filter based fringe denoising also suffers from the deblurring of fringes near the border pixels. For example, the fringe blurring effect evident in figure 3.3(b) is due to the large window size used in the WFF based fringe denoising algorithm. Basically, all the convolution based

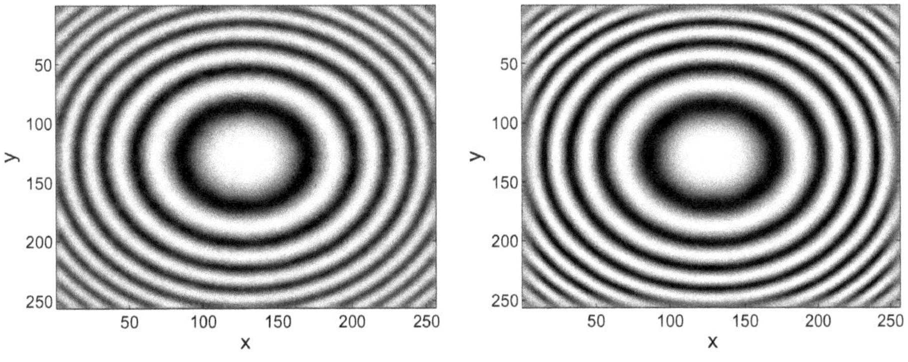

(a) (b)

Figure 3.3. Denoised fringe patterns obtained using the WFF algorithm with (a) $\sigma_x = 3$, $\sigma_y = 3$, $Thr = 1$ and (b) $\sigma_x = 10$, $\sigma_y = 10$, $Thr = 6$.

filters, including the wavelet transform based fringe filtering [1], suffer from this limitation.

3.1.3 Regularized low pass filter

The advantages of the WFF based technique can be extended to the fringe denoising operation by calling upon some of the features of robust quadrature filtering. Indeed, the RQF renders itself well to the application of fringe denoising by its ability to perform localized low-pass filtering instead of band-pass filtering. In principle, the low-pass filter can be seen as a special case of the band-pass filter with its center frequency shifted to zero. Accordingly, the frequency response of a regularized low-pass filter (RLPF) can be written in accordance with (2.26) as

$$H(\omega) = \frac{1}{1 + 2\lambda[1 - \cos(\omega)]}. \tag{3.2}$$

The center frequency ω_x given in (2.26) is set at $\omega_x = 0$. In effect, the above equation is similar to the frequency response of a low pass filter. The parameter λ decides the filter performance.

The extension of 1D-RLPF to 2D-RLPF is straightforward when it comes to handling the two-dimensional fringe patterns. The important point to note here is that since the RLPF is entirely implemented in the spatial domain alone, it can be explicitly tuned to handle the finite sized fringe patterns, including even those that may contain invalid data regions due to the presence of shadows, obstruction or noise.

The 2D extension of cost function minimization considered in (2.21)–(2.25) is simple. The denoised fringe pattern $\Gamma_f(x, y)$ is computed by minimizing the cost function

$$\begin{aligned} U(\Gamma_f) = \sum_{(x,y)\in L} & |\Gamma_f(x, y) - \Gamma(x, y)|^2 + \lambda\, |\Gamma_f(x, y) - \Gamma_f(x - 1, y)|^2 \\ & + \lambda\, |\Gamma_f(x, y) - \Gamma_f(x, y - 1)|^2, \end{aligned} \tag{3.3}$$

where $(x, y) \in L$ refers to the pixel locations in the interior of the fringe signal where valid fringe data is available; the value of parameter λ depends on the noise strength. In order to handle fringe patterns containing invalid fringe regions, the 2D-RLPF cost function, defined in (3.3), is modified as

$$\begin{aligned} U(\Gamma_f) = \sum_{x=0}^{N-1}\sum_{y=0}^{M-1} & |\Gamma_f(x, y) - \Gamma(x, y)|^2 \mathbb{1}(x, y) \\ & + \lambda\, |\Gamma_f(x, y) - \Gamma_f(x - 1, y)|^2 \mathbb{1}(x, y)\mathbb{1}(x - 1, y) \\ & + \lambda\, |\Gamma_f(x, y) - \Gamma_f(x, y - 1)|^2 \mathbb{1}(x, y)\mathbb{1}(x, y - 1), \end{aligned}$$

where $\mathbb{1}(x, y)$ is an indicator function which has a value of 1 for the already processed pixels and 0 for the unprocessed pixels. The denoised fringe signal $\Gamma_f(x, y)$

is computed iteratively by using a gradient descent method. Considering the infinite spatial extent of the fringe signal, the two-dimensional frequency response of the 2D-RLPF can be obtained in the same way as the frequency response of the 1D-RLPF given in (3.2). This frequency response is also found to depend on the selection of the regularization parameter λ. Note that (3.3) has accounted for the first order regularization in cost function minimization. Similarly, the second order regularization can also be used in the definition of the cost function.

In the example of fringe denoising in figure 3.2, the fringe data in the lower right quadrant of the fringe pattern is set to zero by using a binary mask, as shown in figure 3.4(a). The so masked fringe patterns filtered using a 5×5 moving average filter and a WFF based filter are shown in figures 3.4(b) and (c), respectively. The effect of convolution operation involved in these denoising methods is evident at the mask borders, with the blurring of fringe intensities at the border pixels. By comparison, the 2D-RLPF filter applied to the masked fringe pattern accurately restores the denoised fringe pattern without any blurring of the fringe intensities at the border. This is shown in figure 3.4(d). However, it should be noted that the

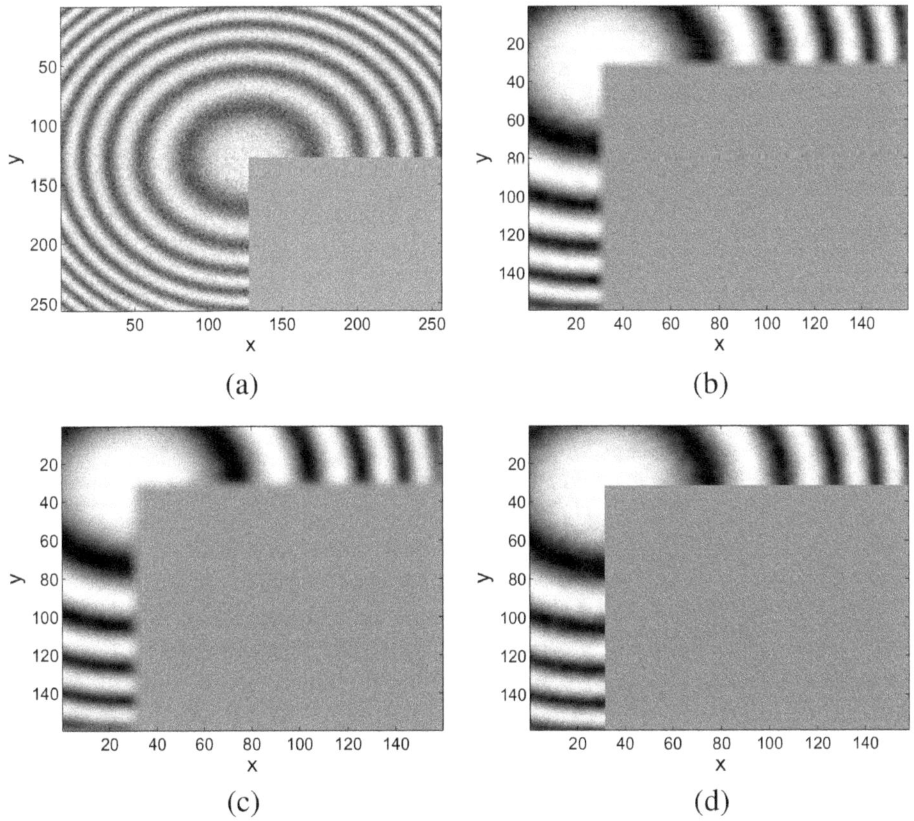

Figure 3.4. (a) Masked fringe pattern. Denoised fringe patterns computed with (b) moving average ($L = 2$) (c) windowed Fourier and (d) regularized low pass filters.

indicator function defining the valid and invalid regions of the fringe data has to be provided as an input to the RLPF based fringe denoising algorithm.

3.1.4 Adaptive monogenic filter

In the WFT based fringe analysis, the fringe pattern is locally approximated as a single frequency sinusoidal signal. The basic principle is to obtain within a window, around each pixel, the best match for the recorded fringe signal in terms of the WFT basis. In order to find the best match, a large number of the WFT basis functions might be required and which, in turn, could affect the total computational time consumed in the fringe denoising operation.

A novel method of fringe filtering based on the design of an adaptive monogenic filter is proposed in [7]. The ingenuity of this method lies in the design of a ring-shaped isotropic bandpass filter in the frequency plane. Such a design of filter allows one to perform fringe denoising with a relatively smaller number of isotropic bandpass filters. This is especially helpful if there are wide spatial variations in the fringe frequency throughout the fringe pattern. The additional capability of adaptive monogenic filtering is in obtaining the normalized fringe signal. This capability proves to be of great help in demodulating closed fringe patterns.

The polar representation of a two-dimensional frequency vector (ς_0, ν_0) in the Cartesian frequency plane can be given as $(r\cos\theta, r\sin\theta)$. Accordingly, three filters are designed in the polar separable form as

$$G_{0q}(r, \theta) = G(r - r_q) \tag{3.4}$$

$$G_{1q}(r, \theta) = iG(r - r_q)\sin(\theta) \tag{3.5}$$

$$G_{2q}(r, \theta) = iG(r - r_q)\cos(\theta). \tag{3.6}$$

In equations (3.4)–(3.6), $G(r)$ represents a 2D bell-shaped function; r_q represents the tuning frequency of the qth filter; G_{1q} and G_{2q} represent the two companion filters with odd symmetric frequency response. These filters are essential to computing the local amplitude of the sinusoidal fringe signal which further enables the normalization of the fringe pattern. Thus, the output of the qth monogenic filter is a vector valued function $[G_{0q}, G_{1q}, G_{2q}]$.

As the local fringe frequency information is not known *a priori*, the fringe signal is passed through a set of Q monogenic filters and the denoised fringe signal is obtained as the weighted linear combination of the outputs of the individual filters. The Q monogenic filters are designed such that their frequency responses overlap each other. The weights of their linear combination depend on the magnitudes of the output of the monogenic filters. Let $A_q(x, y)$ represent the magnitude of the output of the qth monogenic filter. The normalized filtered fringe pattern is then obtained as

$$\Gamma_f(x, y) = \frac{\sum_{q=1}^{Q} w_q(x, y)\Gamma_{0q}(x, y)}{\sum_{q=1}^{Q} w_q(x, y)A_q(x, y)},$$

where, Γ_{0q} represents the output of the qth isotropic filter G_{0q}. The weights w_q are computed as

$$w_q(x, y) = \left(\frac{A_q(x, y)}{A_{\max}(x, y)}\right),$$

with $A_{\max}(x, y) = \max\{A_q(x, y)\}$. The selection of the Q monogenic filter set depends on the dynamic range of the fringe frequency. The minimum ($r_{\min}$) and maximum ($r_{\max}$) fringe frequencies decide the maximum and minimum widths of the monogenic filter, respectively. The selection of the filter width also depends on the desired level of noise robustness. Therefore, an appropriate selection of the filter width is of great importance in order to obtain the best compromise between the two aforesaid parameters. A 2D bell-shaped function proposed in [7] is given as

$$G(r) = \frac{1}{2}\left[1 + \sin\left(\frac{(h + 2r)\pi}{2h}\right)\right],$$

where $h = (r_{\max} - r_{\min})/(Q - 1)$; the qth monogenic filter is designed with $r_q = r_{\min} + h(q - 1)$ for $q = 1, 2, \ldots Q$. The results of fringe denoising provided in [7] strongly encourage the use of the adaptive monogenic filter for the denoising and normalization of a closed fringe pattern in a computationally efficient manner.

An example of the adaptive monogenic filter based denoising of the fringe pattern in figure 3.2(a) is shown in figure 3.5. The selected parameters associated with the filters are $r_{\min} = 0.1$, $r_{\max} = 0.5$, $Q = 15$. A 2D Gaussian bell shaped function is used as $G(r)$. Figure 3.5(a) shows the denoised fringe pattern. The plots of noisy and denoised fringe signals at $y = 128$ are provided in figure 3.5(b). That the adaptive monogenic filter effectively provides both the fringe amplitude normalization and fringe denoising is shown by the plots in figure 3.5(b).

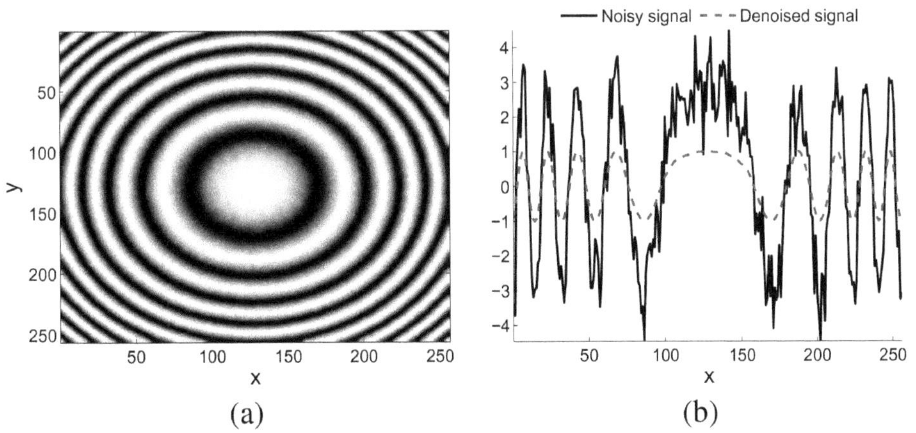

Figure 3.5. (a) Filtered fringe pattern computed using the adaptive monogenic filter with $r_{\min} = 0.1$, $r_{\max} = 0.5$ and $Q = 15$ using 2D Gaussian $G(r)$. (b) Noisy and denoised fringe signals along row $y = 128$. Note that the filtering process provided both the fringe amplitude normalization and fringe denoising.

The methods described above treat the problem of fringe denoising from the viewpoint of frequency analysis. The additive nature of the white Gaussian noise establishes the presence of noise throughout the frequency plane. Under the assumption of the low frequency nature of the original fringe signal, the denoising approaches described herein have attempted to separate the low frequency signal from the high frequency noise. The non-stationary nature of the fringe signal requires that the band-pass filtering of the fringe pattern is performed in an adaptive manner. The adaptiveness in this context is related to selecting the center of the band-pass filter in accordance with the local frequency of the fringe signal. By its definition the local fringe signal is a sinusoid of constant frequency. The matching of the local fringe signal with a number of pre-defined sinusoidal signals is performed and the best match is selected as the denoised local fringe signal.

3.1.5 Dimensionality reduction technique

Recent advancements in digital optical imaging techniques have allowed the user to record a huge amount of information in a relatively easier and cost effective manner. The realization that all of this information or data may not be useful for the analysis, led to a significant amount of research towards the development of data reduction techniques [8]. The aim being to extract a meaningful reduced representation of a data set from a collection of a much larger acquired data set.

In the context of fringe pattern analysis, the fringe data is acquired as a set of single or multiple two-dimensional images. Any *a priori* information on the underlying structure of the data is welcome in the process of data reduction. For example, information such as on the local frequency of the fringe signal is potentially helpful for the purpose of fringe denoising. This has motivated researchers to perform image analysis of fringe signal from the viewpoint of data reduction. An example of the *a priori* information that has found use in this context is the lower dimensionality nature of a noiseless two-dimensional fringe signal. The dimensionality of any image basically refers to the number of independent rows or columns in the image. The dimensionality of an image can be expressed as the rank of the image matrix. For an image $\Gamma(x, y)$ of size $M \times N$ pixels, the $rank(\Gamma) \leqslant \min(M, N)$. In a strict sense when $rank(\Gamma) < \min(M, N)$, the information content in the image can be represented accurately by the lower rank, that is, the lower dimensional image. This fact has been effectively explored for the purpose of fringe denoising.

The sinusoidal fringe signal being deterministic in nature, the pixel intensities at different locations in the fringe image are not completely independent of each other. In effect, the rank of a two-dimensional fringe image is usually lower than $\min(M, N)$ in noiseless condition. In the presence of random noise, the rank of the fringe image is enhanced to $\min(M, N)$ due to the introduction of an independent random component at each pixel. It is, however, still possible to reconstruct a fringe image closely similar to the noiseless fringe signal in a least squares sense. To do so, the fringe image is first decomposed into a number of image components. The approximation to the noiseless fringe image is then reconstructed by utilizing a reduced number of the most meaningful image components such that the image contribution due to the noise is eliminated. Singular value decomposition (SVD) of the fringe image provides a unique decomposition of the image matrix as follows:

$$\boldsymbol{\Gamma} = \boldsymbol{U}\boldsymbol{\Lambda}\boldsymbol{V}^T,$$

where $\boldsymbol{U}$ and $\boldsymbol{V}$ are the $M \times M$ and $N \times N$ sized matrices, respectively, containing $\boldsymbol{u}_k$ and $\boldsymbol{v}_k$ as the column vectors usually termed left and right singular vectors corresponding to the kth singular value λ_k, respectively; $\boldsymbol{\Lambda}$ represents $M \times N$ diagonal matrix containing singular values as its diagonal elements. The kth image component is represented as

$$\Gamma_k(x, y) = \boldsymbol{u}_k \lambda_k \boldsymbol{v}_k^T.$$

The fringe signal can be expressed as the combination of image components as

$$\Gamma(x, y) = \sum_{k=1}^{\min(M,N)} \Gamma_k(x, y).$$

In a singular matrix $\boldsymbol{\Lambda}$, the singular values are arranged in a descending order. The highest singular value and its associated singular vectors together represent the most meaningful component $\Gamma_1(x, y)$ of the image data. Similarly, $\Gamma_2(x, y)$, $\Gamma_3(x, y)$, ... represent the image components in decreasing order of meaningfulness. In the noiseless condition, where $rank(\Gamma) \leqslant \min(M, N)$, the combination of the first $\mathscr{R}_\Gamma$ image components accurately represents the fringe image. Since in the presence of noise too, the first $\mathscr{R}_\Gamma$ singular values are dominant, and the first $\mathscr{R}_\Gamma$ numbers of image components are able to reconstruct the approximation of the noiseless fringe image in a least squares sense. Thus, the filtering of noise from the fringe image is achieved by discarding the contributions of insignificant image components. The denoised fringe signal is computed as

$$\Gamma_f(x, y) = \sum_{k=1}^{\mathscr{R}_\Gamma} \Gamma_k(x, y).$$

This algorithm is illustrated with the help of the simulated fringe pattern shown in figure 3.2(a). First, consider the fringe pattern in noiseless condition. The SVD of the fringe pattern is computed and the diagonal of the singular value matrix is plotted in figure 3.6(a). The plot shows three dominant singular values with the remaining singular values keeping very low magnitudes. Note that only the first 20 singular values have been plotted for the purpose of display. Figure 3.6(c) shows the reconstructed fringe signal as a combination of image components associated with the first three dominant singular values. As expected, a perfect reconstruction is achieved. Now, consider the noisy fringe pattern simulated at SNR = 10 dB. The SVD of this noisy fringe pattern is computed and a plot of the diagonal of the singular value matrix associated with this noisy fringe image is shown in figure 3.6(b). The plot shows that apart from the first three singular values, the remaining singular values are also of non-negligible magnitudes. However, the first three

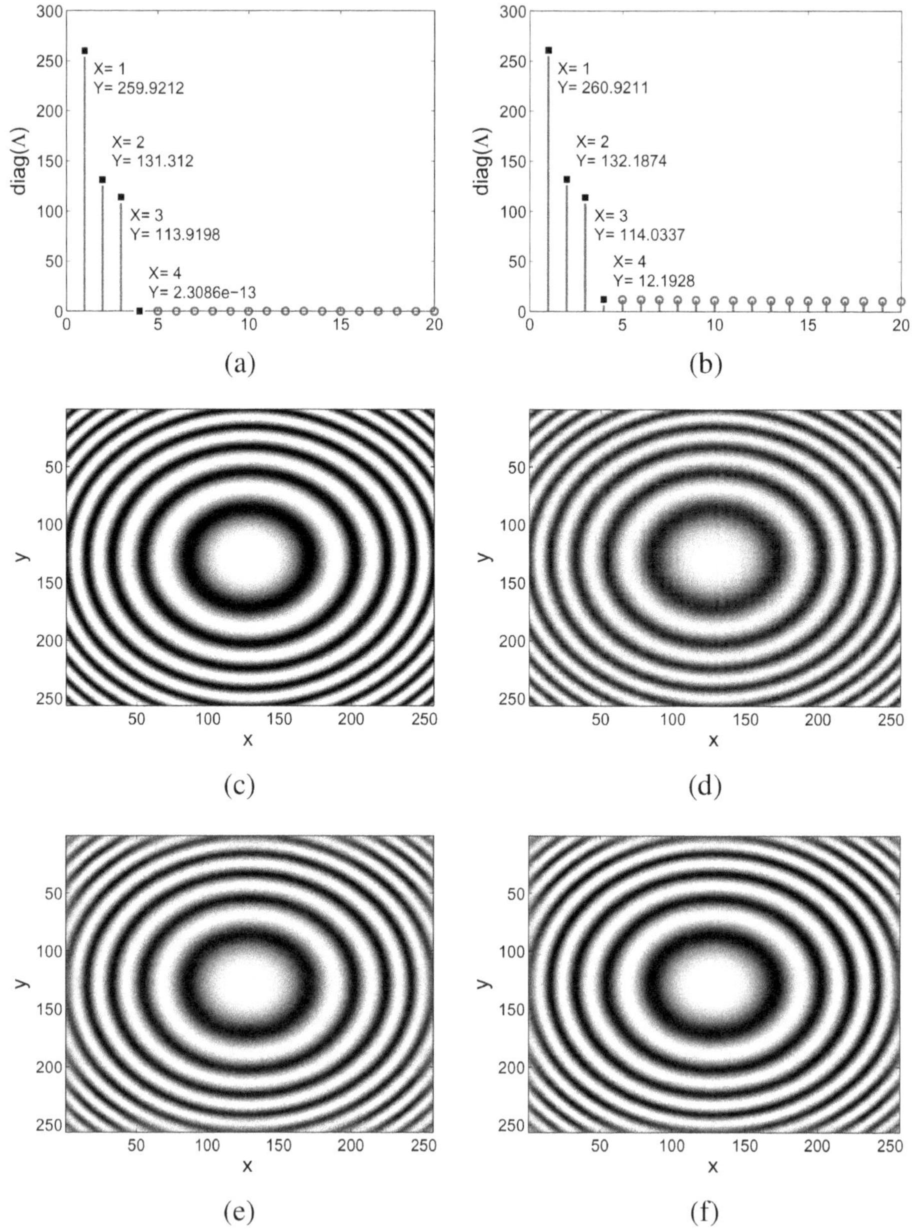

Figure 3.6. Display of the first 20 diagonal elements of singular matrix $\mathbf{\Lambda}$ at (a) SNR = ∞ dB and (b) SNR = 10 dB. Denoised fringe patterns at (c) SNR = ∞ dB and (d) SNR = 10 dB. Denoised fringe pattern obtained for (e) $J = 5$ and (f) $J = 15$. The rank $\mathscr{R}_\Gamma$ is approximated to be $\mathscr{R}_\Gamma = 10$.

singular values still dominate. The reconstruction of the fringe image obtained with these three singular values, shown in 3.6(d), is seen to give the closest similarity to the noiseless fringe signal. In the SVD based fringe denoising, the artifacts appear especially along the x and y directions in the reconstructed fringe signal (see figure 3.6(d)) on account of the directionality of the SVD operation. A scheme to overcome these artifacts is proposed in [9]. Accordingly, the noisy fringe image is rotated J times with each increment in the angle of rotation equal to $2\pi/J$. The fringe denoising is performed on each rotated image. The final denoised fringe image is computed as

$$\Gamma_f(x, y) = \sum_{j=1}^{J} \overline{\lambda}_j \Gamma^{(j)}(x, y) \Bigg/ \sum_{j=1}^{J} \overline{\lambda}_j. \tag{3.7}$$

In (3.7), $\Gamma^{(j)}(x, y)$ represents the denoised fringe image corresponding to the jth rotated noisy image; $\overline{\lambda}_j = \sum_{k=1}^{\mathscr{R}_\Gamma} \lambda_k^{(j)}$ with $\lambda_k^{(j)}$ representing the kth diagonal element of the singular matrix corresponding to the jth rotated image. Typically, the rank $\mathscr{R}_\Gamma$ of the true fringe image is not known *a priori*. Therefore, it is a user defined parameter which varies depending on the fringe pattern at hand. For the example considered in figure 3.6, the denoised fringe patterns shown in figures 3.6(e) and (f) are obtained with $J = 5$ and $J = 15$ number of rotated images, respectively.

A practical example of fringe denoising is considered in figure 3.7(a). The fringe pattern is recorded in an ESPI setup corresponding to the in-plane displacement of a circular aluminum plate. Since the fringe data is available only on the circular region of the plate, a binary mask is defined to identify the circular object region within the recorded rectangular image shown in figure 3.7(b). The moving average filter and the WFF based algorithms are applied to obtain the denoised fringe patterns shown in figures 3.8(a) and (b), respectively. In fringe denoising performed by these convolution based methods, the pixels outside this circular region unnecessarily and undesirably contribute to the denoising operation. On the other hand, the RLPF based fringe denoising, however, allows one to accommodate the indicator function

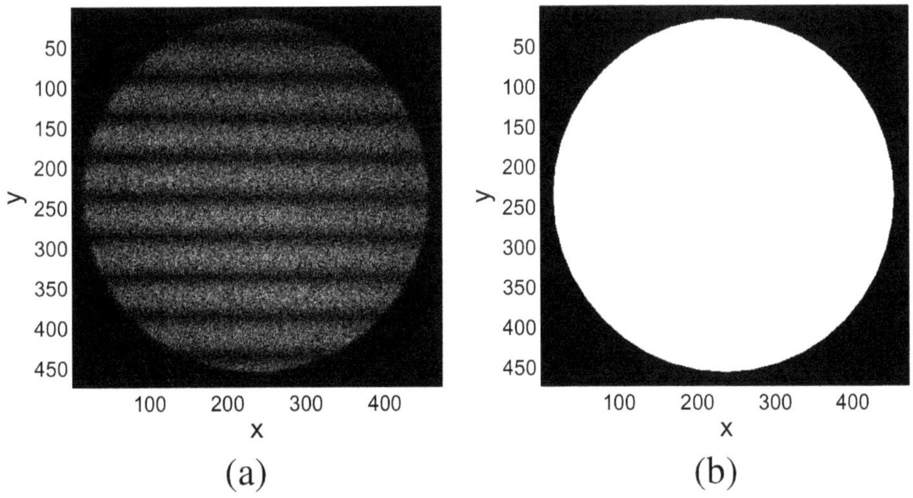

Figure 3.7. (a) An example of a circular aluminum plate recorded in an ESPI setup. (b) The display of the binary mask defining the circular object region within the recorded rectangular image.

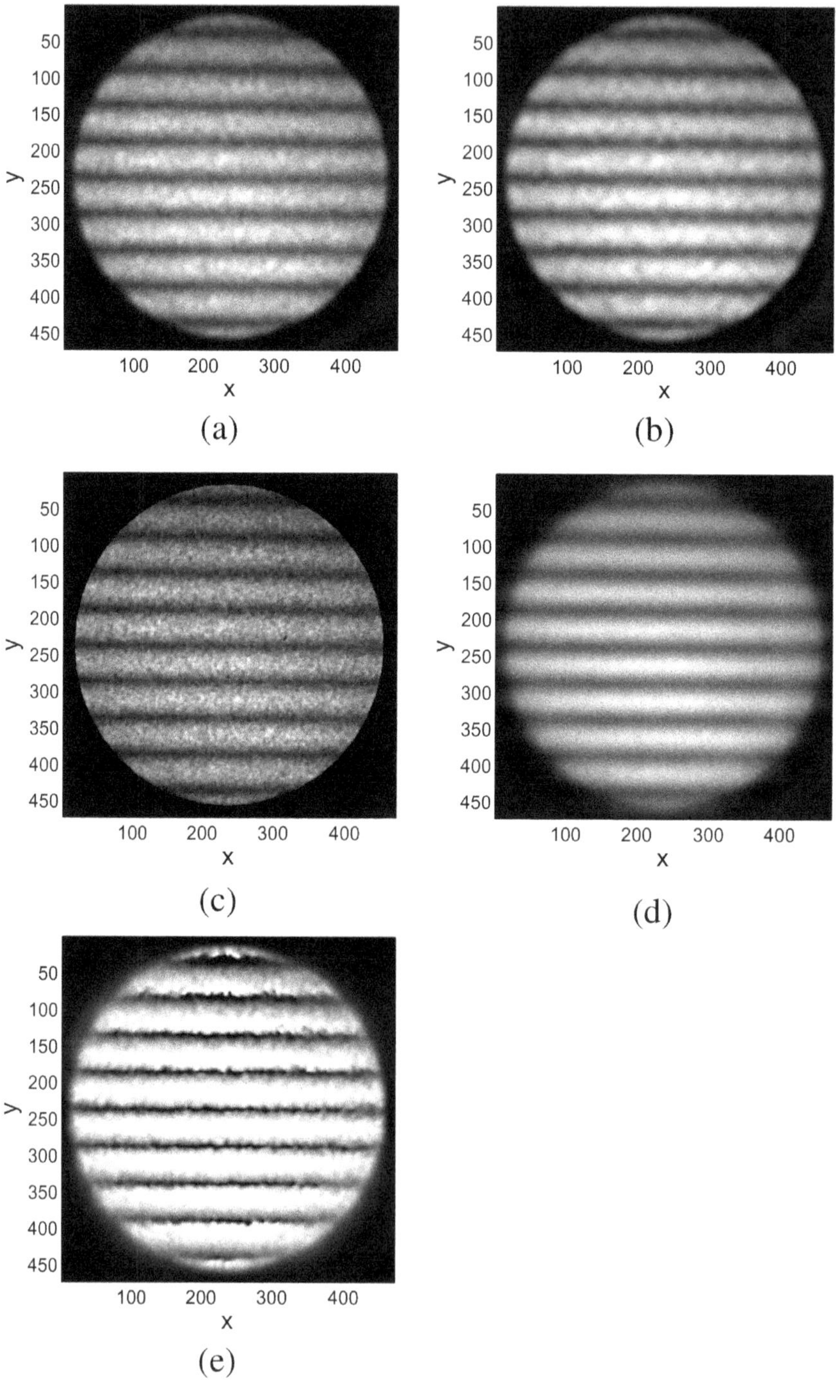

Figure 3.8. Denoised fringe patterns obtained on filtering the fringe pattern in figure 3.7(a) using (a) moving average filter, (b) windowed Fourier filter (c) regularized low-pass filter (d) dimensionality reduction technique, and (e) adaptive monogenic filter.

in order to discriminate the valid and invalid fringe data. The mask shown in figure 3.7(b) acts as an indicator function and the result of the RLPF based fringe denoising is shown in figure 3.8(c). A comparison of this result with those shown in figures 3.8(a) and 3.8(b) allows one to identify the difference in the performance of these methods in the context of fringe denoising, especially at the border pixels. Similar to the moving average filter, repeated applications of the RLPF also allows for achieving better denoising results. The result of fringe denoising obtained with the dimensionality reduction technique is shown in figure 3.8(d). Although this method seems to provide a comparatively better denoising capability in the valid fringe region, the lower value of rank approximation results in the blurring of fringe intensities near the mask border. The application of adaptive monogenic filter provides not only the filtered but the normalized amplitude fringe pattern as well. This is shown in figure 3.8(e).

We have described a wide range of fringe denoising methods, each of which possess their own distinct advantages and limitations. Apart from the methods we have covered, various fringe denoising techniques and their variants include techniques based on wavelet thresholding [5], partial differential equations [3,10,11], and empirical mode decomposition [4,12].

3.2 Phase fringe pattern denoising

There are broad categories of methods, such as, the spatial and temporal fringe analysis techniques or even system configurations, such as, digital holography, which provide the phase estimates in a wrapped form with the true phase values mapped in the range of $[-\pi, \pi)$. The wrapped phase map is generally termed phase fringe pattern. The phase fringe pattern carries the required phase information albeit with a 2π phase ambiguity. If the real fringe pattern is not filtered prior to its submission to fringe analysis, the resulting phase fringe pattern is found to carry over the noise associated with the fringe recording process. Naturally, the subsequent operation of phase unwrapping for removing the 2π phase ambiguity gets affected by the presence of noise in the phase fringe pattern. Hence, the denoising of the phase fringe pattern is helpful and often an essential step prior to the phase unwrapping operation. An important point to note here is that the 2π phase discontinuities present in the phase fringe patterns have to be retained during the denoising process. It is therefore recommended that the denoising should be performed on the EPF carrying the phase information rather than on the phase fringe pattern itself. The EPF associated with the phase fringe pattern is represented as

$$\Gamma(x, y) = \exp\left[i\phi_w(x, y)\right] = \exp[i\phi(x, y)].$$

3.2.1 Sine–cosine average filter

The moving average filter described in section 3.1.1 can be extended to filter the phase fringe patterns. Termed in its new avatar a sine–cosine average filter (SCAF), it operates on the real (cosine) and imaginary (sine) parts of the EPF. The denoised sine and cosine fringe patterns are combined to obtain the denoised phase fringe pattern:

$$\phi_w^{(f)}(x, y) = \arctan\left\{\frac{\sin^{(f)}[\phi_w(x, y)]}{\cos^{(f)}[\phi_w(x, y)]}\right\},$$

where the superscript (f) indicates the filtered quantities. Similar to the moving average filter, the phase fringe patterns can be effectively denoised by repeated applications of the SCAF. This approach works especially well in the case of slow spatial variations of the phase. However, the fringe deblurring effect of the moving average filter appearing in the regions of rapidly varying phase, that is, in the high density fringe areas, adversely affects the denoised phase fringe pattern quality. This limitation can be effectively circumvented by using an anisotropic SCAF [13] instead of the conventional isotropic SCAF (for example, the 3×3 average filter). Depending on the knowledge of the direction of high fringe density, we choose the SCAF filter window to be anisotropic. Whereas a smaller window dimension is selected along the direction of higher fringe density, a larger window dimension is selected along the direction of lower fringe density. While observing that this window size selection strategy is user dependent, it is nonetheless important to highlight the method's underying assumption that the direction of higher density fringe lies only along one direction. As exemplified in [13], this method is especially useful in denoising phase fringe patterns that have been recorded in a shearography setup, a setup in which the direction of shear is decided by the user *a priori*. Accordingly, a smaller window dimension is selected along the direction of the shear.

3.2.2 Windowed Fourier transform based filtering

The ubiquity of WFT based fringe processing is applicable to denoising the noisy EPFs. The WFF and WFR based algorithms are both suitable for denoising noisy phase fringe patterns. It is of interest to note that the procedure of real fringe pattern denoising based on the WFT principle is applicable as it is without modification to phase fringe pattern denoising.

3.2.3 Noise subspace inflation

Since the local simple sinusoidal deterministic structure of the noiseless fringe signal sets itself apart from the random nature of the noise, most of the transform based fringe denoising algorithms either perform as adaptive low pass filters or apply certain approximations on the local phase to filter out the unnecessary noise contributions [14]. In this subsection, we will discuss a phase fringe pattern denoising algorithm based on the principle of noise subspace inflation (NSI), which can be looked upon as a local fringe signal dimensionality reduction technique. This technique performs the localized decomposition of an EPF into its deterministic

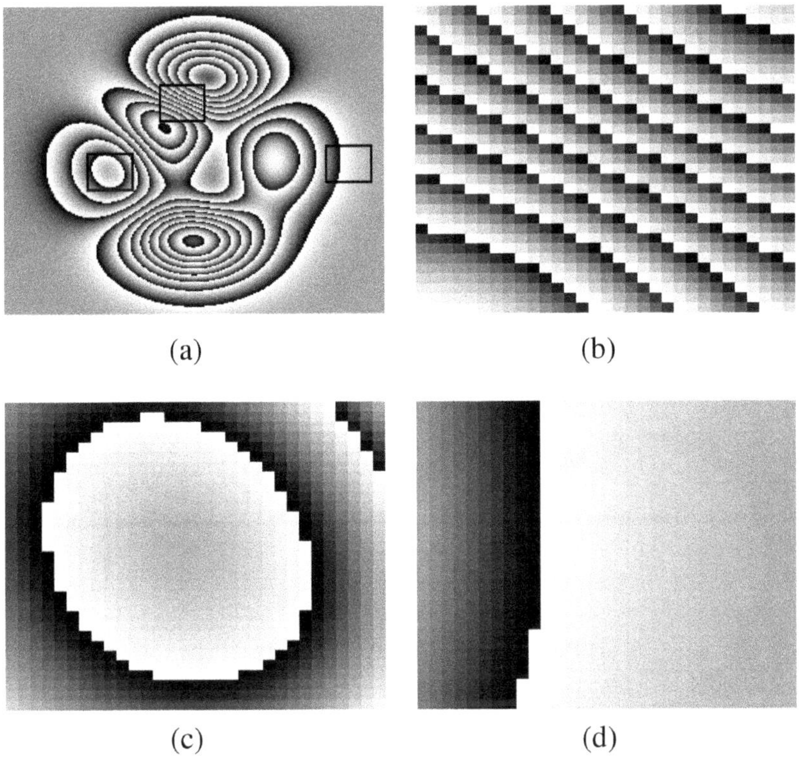

Figure 3.9. (a) An example of a phase fringe pattern. (b), (c) and (d) display the phases in the selected regions pointed out in figure (a).

signal and random noise component parts, and the EPF is reconstructed using only the deterministic signal component.

An example of phase fringe pattern is shown in figure 3.9(a). Three different regions of the phase fringe pattern highlighted in figure 3.9(a) are displayed separately in figures 3.9(b)–(d), respectively. While the overall structure of the phase in figure 3.9(a) looks complex, the local structures of the phases in figures 3.9(b)–(d) are simple in nature. Effectively employed in filtering the noise from a phase fringe pattern, this feature is fundamental to many fringe analysis techniques. Consider the phase fringe pattern of size 30 × 30 in figure 3.9(c). We would like to decompose the corresponding EPF into its deterministic signal and noise component parts. Similar to the dimensionality reduction based fringe denoising described in section 3.1.5, the singular value decomposition here too provides a unique decomposition of the data matrix. The SVD of the EPF is computed as

$$\mathbf{\Gamma}_w = \boldsymbol{U}_w \mathbf{\Lambda}_w \boldsymbol{V}_w^*,$$

where the subscript w indicates the windowed data of the fringe pattern. The singular values on the diagonal of the singular matrix $\mathbf{\Lambda}_w$, plotted in figure 3.10(a), show the presence of two dominant singular values. The fact that the remaining singular

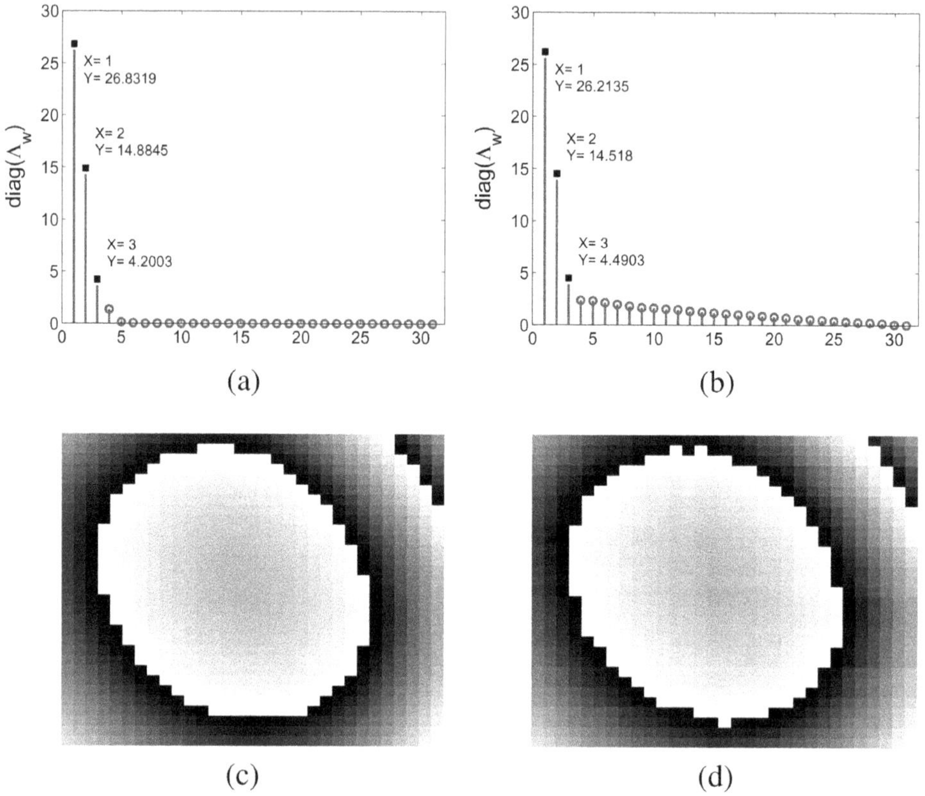

Figure 3.10. Diagonal singular values in (a) noiseless and (b) noisy conditions. Reconstructed phase fringe patterns in (c) noiseless and (d) noisy conditions.

values tend to be of low to very low magnitudes, indicates the lower dimensionality of the local EPF signal. The example shown in figure 3.10(a) is performed in the noiseless condition. By comparison, the corresponding singular values on the diagonal of the singular matrix for a similar exercise performed in a noisy condition with SNR=10 dB is shown in figure 3.10(b). The notion of two dominant singular values still holds, albeit the magnitudes of the remaining singular values are not insignificant anymore. By retaining only the dominant singular values along with their associated singular vectors, the noiseless EPF can be reconstructed. The phases of the reconstructed EPF in the noiseless and noisy conditions are shown in figures 3.10(c) and 3.10(d), respectively. This basic principle forms the basis of the NSI based fringe denoising. It is important to note that the windowed EPF data with window size 30×30 is considered only for the purpose of illustration. In practice, the norm is to select a small window size, say, 9×9. In this case, the notion of a single dominant singular value holds.

The described principle can be implemented by any of the two approaches outlined below.

3.2.3.1 Moving window method

In this method of fringe denoising, the local EPF is considered around each pixel within a window of pre-defined size $L \times L$:

$$\boldsymbol{\Gamma}_w = \begin{bmatrix} \Gamma_w(0,0) & \Gamma_w(0,1) & \cdots & \Gamma_w(0,L-1) \\ \Gamma_w(1,0) & \Gamma_w(1,1) & \cdots & \Gamma_w(1,L-1) \\ \vdots & \vdots & \ddots & \vdots \\ \Gamma_w(L-1,0) & \Gamma_w(L-1,1) & \cdots & \Gamma_w(L-1,L-1) \end{bmatrix}. \tag{3.8}$$

The window is designed to be symmetric with respect to the pixel location. Accordingly, the window size L is selected to have an odd value. Further, a simple phase structure is assumed within the windowed EPF and the noise component can be removed by computing its singular value decomposition. The noise-free local EPF is then reconstructed utilizing the first singular value and its associated singular vectors. The denoising capability of this procedure can be effectively enhanced by utilizing the noise subspace inflation of an enhanced data matrix. The enhanced data matrix is generated by the partition and stacking of the rows of the windowed EPF, $\boldsymbol{\Gamma}_w$, as follows [15]:

$$\boldsymbol{\Gamma}_e = \begin{bmatrix} \boldsymbol{\Gamma}_0 & \boldsymbol{\Gamma}_1 & \cdots & \boldsymbol{\Gamma}_{L-K} \\ \boldsymbol{\Gamma}_1 & \boldsymbol{\Gamma}_2 & \cdots & \boldsymbol{\Gamma}_{L-K+1} \\ \vdots & \vdots & \boldsymbol{\Gamma}_l & \vdots \\ \boldsymbol{\Gamma}_{K-1} & \boldsymbol{\Gamma}_K & \cdots & \boldsymbol{\Gamma}_{L-1} \end{bmatrix}, \tag{3.9}$$

where,

$$\boldsymbol{\Gamma}_l = \begin{bmatrix} \Gamma_w(l,0) & \Gamma_w(l,1) & \cdots & \Gamma_w(l,L-P) \\ \Gamma_w(l,1) & \Gamma_w(l,2) & \cdots & \Gamma_w(p,L-P+1) \\ \vdots & \vdots & \ddots & \vdots \\ \Gamma_w(l,P-1) & \Gamma_w(l,P) & \cdots & \Gamma_w(l,P-1) \end{bmatrix}. \tag{3.10}$$

Let us now look into the improvement this matrix enhancement operation provides to the noise filtering. The size of the enhanced matrix $\boldsymbol{\Gamma}_e$, $KP \times (L-K+1) \times (L-P+1)$, is decided by the tuning parameters K and P. The range of values of these parameters is $K = [1, L]$ and $P = [1, L]$. In the presence of noise, the rank of $\boldsymbol{\Gamma}_w$ is equal to L, and the rank of $\boldsymbol{\Gamma}_e$ is equal to $\min(KP, (L-K+1)(L-P+1))$. Under the assumption of a simple signal phase structure of $\boldsymbol{\Gamma}_w$, only a single dominant singular value is found to be present in the SVD of $\boldsymbol{\Gamma}_w$ along the diagonal of the singular matrix of size $L \times L$. The remaining singular values are associated with the noise component. Now consider the singular value decomposition of the enhanced matrix $\boldsymbol{\Gamma}_e$,

$$\boldsymbol{\Gamma}_e = \boldsymbol{U}_e \boldsymbol{\Lambda}_e \boldsymbol{V}_e^*. \tag{3.11}$$

It should be noted that in the SVD of $\boldsymbol{\Gamma}_e$ the size of the singular matrix $\boldsymbol{\Lambda}_e$ is $KP \times (L-K+1)(L-P+1)$. Since the enhanced matrix is generated using the windowed EPF, in principle, it carries the same fringe signal information, that is, a single fringe signal component with a simple phase structure in $\mathbf{\Gamma}_e$ despite it being of a size different than that of $\mathbf{\Gamma}_w$. It can thus be argued that while a single dominant singular value occurs along the diagonal of $\mathbf{\Lambda}_e$, the remaining singular values correspond to the noise. Thus, similar to $\mathbf{\Gamma}_w$, the signal subspace dimension in the decomposition of $\mathbf{\Gamma}_e$ is 1. However, the noise subspace dimension rather than being equal to $L - 1$ becomes equal to $\min(KP, (L - K + 1)(L - P + 1)) - 1$. This is the most important and distinctive aspect of the enhanced matrix and which as a result gives the possibility of tuning the noise subspace dimension of the enhanced matrix depending on the selection of K and P. For simplicity, let us represent the rank of the enhanced matrix as $\mathscr{R}_e = \min(KP, (L - K + 1)(L - P + 1))$. It thus becomes evident that we will not gain anything if we set the parameters K and L such that $\mathscr{R}_e \leqslant L$. On the other hand, if we set K and P such that $\mathscr{R}_e > L$ then we would achieve the inflated noise subspace with an increment of $\mathscr{R}_e - L$ in the dimension of the noise subspace. This effect is at the core of the working principle of the noise subspace inflation based fringe denoising algorithm. The inflated noise subspace allows for an effective decomposition of the noise component and which, in turn, results in a more effective windowed EPF denoising. To achieve the noise subspace inflation, the parameters K and P are selected such that

$$2 \leqslant K, P \leqslant \frac{1}{2}(L + 1). \tag{3.12}$$

Thus an improvement in fringe denoising using the enhanced matrix relative to the denoising performed using the windowed EPF could be achieved if appropriate values of K and P are selected in accordance with (3.12). The following steps outline the procedure of the moving window method of fringe denoising:

1. At a pixel location (x,y), a windowed EPF, $\mathbf{\Gamma}_w$, is created according to (3.8).
2. The enhanced matrix $\mathbf{\Gamma}_e$ is generated using $\mathbf{\Gamma}_w$ according to (3.9) and (3.10).
3. The SVD of the enhanced matrix is computed according to (3.11).
4. The denoised enhanced matrix $\Gamma_e^{(f)}$ is computed using the singular value and singular vectors associated with the signal subspace. Thus, we have

$$\mathbf{\Gamma}_e^{(f)} = \boldsymbol{u}_e^{(1)}\mathbf{\Lambda}_e^{(1)}\left(\boldsymbol{v}_e^{(1)}\right)^*,$$

   where $\mathbf{\Lambda}_e^{(1)}$, and $\boldsymbol{u}_e^{(1)}$ and $\boldsymbol{v}_e^{(1)}$ represent the first dominant singular value, and its associated singular vectors, respectively.
5. Finally, the denoised EPF at the pixel location (x,y) is computed as

$$\Gamma^{(f)}(x, y) = \mathbf{\Gamma}_e^{(f)}(1, (L - P + 1)(q - 1) + q),$$

   where, $q = \frac{L+1}{2}$.
6. The above steps are repeated at all the pixels to obtain the complete denoised phase fringe pattern.

3.2.3.2 Patchwise method

Leaving aside the manner in which the pixels of the phase fringe pattern are processed, the patchwise method of fringe denoising is, in principle, similar to the moving window method of fringe denoising. Instead of pixelwise denoising in the moving window method, the fringe pattern is denoised in a patchwise manner in this method. Thus, the EPF is divided into small sized patches and the NSI based fringe denoising is performed within each patch. The EPF data within each patch is restored upon denoising. An important advantage offered by patchwise denoising over the moving window approach lies in its highly improved computational efficiency. The compution time required in patchwise fringe denoising is at least ten times lower than that required for the moving window method.

There are two ways that can be adopted for dividing a phase fringe pattern into patches: non-overlapping and overlapping. These two approaches are displayed in figure 3.11. In the non-overlapping approach of patchwise fringe denoising, no EPF data is shared between any two patches. Thus, denoising of a particular patch is not affected by its neighboring patch EPF data. On the other hand, in the overlapping approach of patchwise fringe denoising, one half of the denoised EPF data in the current patch constitutes the EPF data in the subsequent neighboring patch. It is thus normal that the denoising in a given patch affects the denoising results in its neighboring patches. This latter approach offers better denoising capability compared to the non-overlapping patch based approach.

Various phase fringe denoising algorithms are described in the literature [2,13,16,17,18]. The relationship between these algorithms is established in [19]. Apart from these algorithms, other approaches of phase fringe pattern denoising include the 2-D continuous wavelet transform [20], local histogram [21] and localized Fourier trasform [22]. The comparison of different phase fringe pattern denoising methods can be found in [6,14,19].

An example of phase fringe pattern denoising obtained with the SCAF, WFF and NSI based methods is provided in figure 3.12. Figure 3.12(a) shows a noisy phase fringe pattern with wide spatial variations in the fringe density. The repeated application of the SCAF filter with filter window of size 7 × 3 provides the denoised fringe pattern shown in figure 3.12(b). Although the low density fringe areas are effectively denoised, the quality of high density fringe areas deteriorates due to the repeated application of the averaging operation. The WFF based fringe denoising algorithm is implemented with the frequency bands selected as [−2, 2] in both x and y directions. The WFF algorithm provides a high quality denoised phase fringe

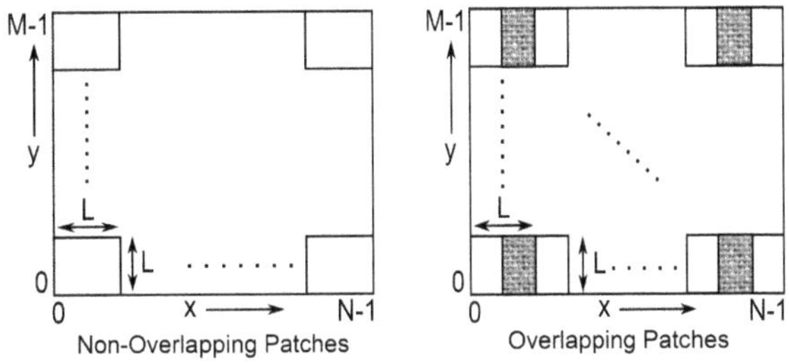

Figure 3.11. Patchwise denoising of phase fringe patterns with non-overlapping and overlapping patches.

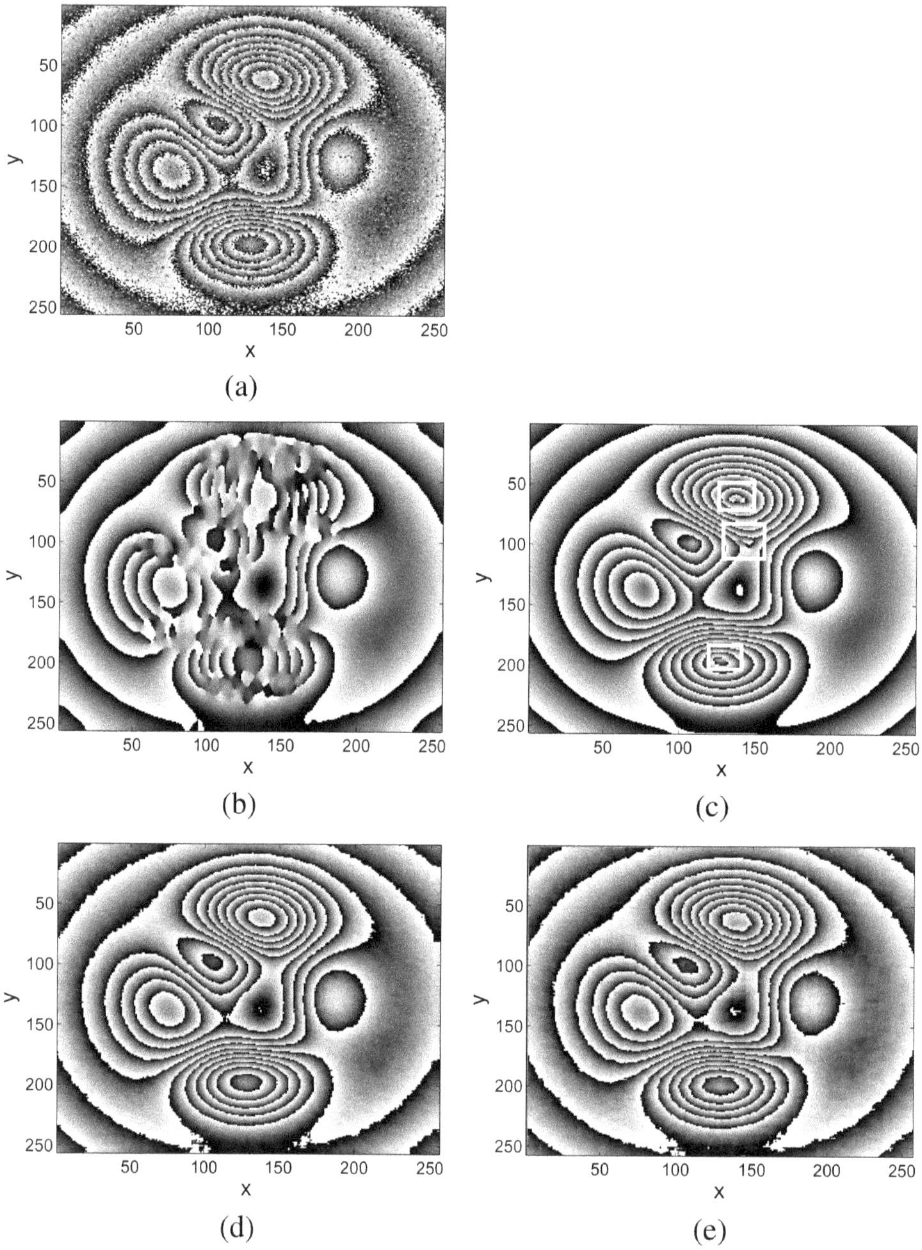

Figure 3.12. (a) Simulated phase fringe pattern at SNR = 0 dB. Denoised fringe patterns obtained using (b) sine–cosine average filter with a window size of 7 × 3 and five repeated applications of the filter, (c) windowed Fourier filter with frequency range of [−2, 2] in both the x and y directions and a frequency resolution of 0.05, (d) noise subspace inflation based moving window method with $L \times L = 9 \times 9$. (e) and a noise subspace inflation based patch-wise method and a patch size of 8 × 8.

pattern, as can be observed in figure 3.12(c). However, for the selected range of frequencies, certain regions of the fringe pattern indicated by rectangles have not been possible to reconstruct properly. This example once again illustrates the importance of appropriate selection of frequency range to obtain a reliable denoised fringe pattern.

The NSI based fringe denoising algorithm offers the unique advantage of obtaining fringe denoising independent of fringe densities. Since the algorithm is based on the assumption of a simple phase structure of the windowed EPF, the fringe frequencies do not play any role in deciding the performance of fringe denoising. This fact is illustrated in figure 3.12(d) in the form of a denoised phase fringe pattern obtained using the moving window method with $L = 9$. The complete phase fringe pattern is seen to be effectively filtered. A similar result observed in figure 3.12(e) is obtained using the patchwise method with $L = 8$.

The phase fringe pattern in figure 3.13(a) is recorded in an ESPI setup for the measurement of in-plane displacement of a fiber reinforced wood specimen under compression. The pattern is observed to be highly corrupted by the speckle noise. The fringe denoising results obtained using the SCAF, WFF and NSI based methods are provided in figures 3.13(b)–(d).

3.3 Phase unwrapping

In general, the two-dimensional phase that we wish to retrieve by phase unwrapping is a continuous function of spatial coordinates x and y. Under this assumption, which has guided many phase unwrapping algorithms [23], the phase unwrapping seemingly reduces to the trivial task of identifying and adding appropriate integer multiples of 2π at each pixel location. Thus

$$\phi(x, y) = \phi_w(x, y) + 2\pi I(x, y),$$

where the subscript w indicates the wrapped phase measurement; $I(x, y)$ indicates the integer values. The applicability of this simple procedure is limited by two important factors: noise and discontinuity. The phase noise affects the accurate identification of the integer multiples of 2π and the error in this identification gets propagated in the unwrapping of subsequent pixels. On the other hand, the discontinuity here is more related to the object as the phase data may not be available over the entire frame as a result of a loss in the fringe data arising, for example, due to the presence of holes or shadows on the object surface. Versatile phase unwrapping algorithms are those which have the ability to take care of such regions during the unwrapping process. Among a large collection of phase unwrapping algorithms reported in the literature, an important section includes those based on the polynomial phase approximation. For example, the phase unwrapping methods based on piecewise polynomial approximation of the phase [24,25,26] are based on the principle of Weierstrass approximation theorem [27], which states that *a continuous function over a fixed interval can be accurately represented as a polynomial of sufficiently higher order*. This theorem thus allows us to approximate a continuous phase along a row or column as a polynomial of order, say, P. Whether

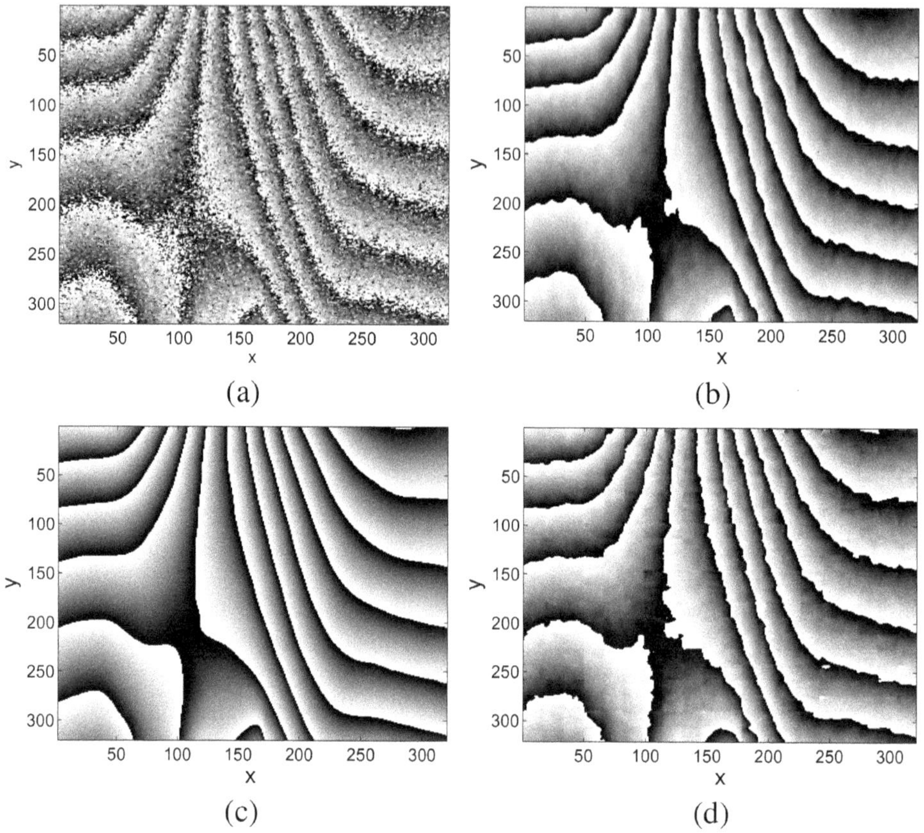

Figure 3.13. (a) Experimentally recorded phase fringe pattern in an ESPI setup. Denoised fringe patterns computed using (b) sine–cosine average filter method with a window size of 7 × 3 with five repeated applications of the filter, (c) windowed Fourier filter with a frequency range of [−1, 1] in both the x and y directions with a frequency resolution of 0.1, and (d) noise subspace inflation based patchwise method with a patch size of 16 × 16.

the order of the polynomial function is sufficient in fulfilling this task cannot be ascertained *a priori*. To overcome this limitation, the fixed interval of the EPF data can be divided into a number of small segments. Within each segment, the phase can be approximated with a polynomial function of a lower and pre-defined order. A similar argument also holds in the case where the EPF is divided into 2D blocks of small sizes instead of its 1D segmentation along rows and columns. In such a case, the phase within each block is represented as a 2D polynomial of x and y but again of a lower and pre-defined order. The basic idea to perform phase unwrapping based on local polynomial phase approximation at each pixel is straightforward, with the phase being typically approximated with a first order polynomial function within a small window around each pixel.

A successful implementation of this approach, in all its three versions cited above, passes through the ability of the phase unwrapping algorithm to accurately estimate the polynomial coefficients. The ability to estimate these coefficients imperatively requires the polynomial function to match as closely as possible to the phase within the defined intervals.

3.3.1 Segmentwise polynomial phase approximation

Basing ourselves on the assumption that the phase is approximated as a 1D polynomial function within a segment of row of the EPF, the phase unwrapping procedure is performed row-wise. The EPF along a row y can be represented as

$$\Gamma(x) = \exp[i\phi(x)].$$

Next, the EPF along the row is divided into N_w number of non-overlapping/overlapping segments such that each segment is of length $N_s = N/N_w$. The EPF within the sth segment can then be represented as

$$\Gamma_s(x) = \exp\left[i\phi_s(x)\right] \text{for } (s-1)N_s \leqslant x \leqslant sN_s - 1. \tag{3.13}$$

Considering the maximum-likelihood (ML) [26] method of polynomial phase coefficient estimation, we select the order of the polynomial as, say, $P = 2$. Accordingly, the EPF in (3.13) is written as

$$\Gamma_s(x) = \exp\left[i(c_0 + c_1x + c_2x^2)\right],$$

with,

$$\phi_s(x) = c_0 + c_1x + c_2x^2. \tag{3.14}$$

Equation (3.14) confirms that the accurate estimation of the coefficient set $\{c_0, c_1, c_2\}$ is crucial for obtaining an accurate estimation of the phase within the segment. The ML estimate of the coefficients [26] can be computed as

$$(\hat{c}_1, \hat{c}_2) = \arg\max_{c_1, c_2} \left| \sum_{x=1}^{N_s} \Gamma_s(x)\exp\left[-i(c_1x + c_2x^2)\right] \right| \tag{3.15}$$

$$\hat{c}_0 = \text{Arg}\left\{\frac{1}{N_s} \sum_{x=1}^{N_s} \Gamma_s(x)\exp\left[-i(\hat{c}_1x + \hat{c}_2x^2)\right]\right\}, \tag{3.16}$$

where $\hat{c}_0$, $\hat{c}_1$, $\hat{c}_2$ represent the estimates of c_0, c_1, c_2, respectively; $\text{Arg}\{\cdot\}$ represents the argument of its complex input variable. Equation (3.15) indicates that the estimation of coefficients c_1 and c_2 essentially involves a 2D maximization procedure. The accurate estimation of these coefficients requires an exhaustive search in two dimensions which could be computationally demanding. This search can be performed in two steps. First, the coarse estimate of these coefficients is obtained using (3.15). Second, these estimates are used as initial guess in the Nelder–Mead simplex search optimization algorithm [28]. This latter step provides the refined estimates, $\hat{c}_1$ and $\hat{c}_2$. In the coarse coefficient estimation using (3.15), the coefficients c_1 and c_2 are searched in the range $[0, 2\pi]$ and $[0, 2\pi/N_s]$, respectively.

Note that (3.16) involves the Arg function to compute the coefficient estimate $\hat{c}_0$ which wraps its value in the range $[-\pi, \pi)$. Consequently, discontinuities exist in the estimated phase at the boundaries of the segments over a given row. To overcome this issue, the c_0 coefficient is computed only in the first segment of each row. The c_0 coefficients in the subsequent segments can be computed as follows

$$\hat{c}_0^{(s+1)} = \hat{c}_0^{(s)} + \hat{c}_1^{(s)} N_s + \hat{c}_2^{(s)} N_s^2,$$

for $s = 0, \ldots, N_w - 1$. Another approach to deal with the c_0 coefficient ambiguity is to compute the difference between the unwrapped phase computed at the border pixel of the current and the previous segments. This difference is added as an offset to the unwrapped phase estimate of the current segment. This procedure of dealing with the phase discontinuity at the segment border is applied to all the phase estimation techniques based on segmentwise polynomial phase approximation. Note that at the end of the above procedure, the phase unwrapping is achieved only along the rows. A simple unwrapping procedure then gives a complete 2D unwrapped phase estimate by removing the 2π phase discontinuities along the columns.

A one-dimensional example of EPF of length $N = 256$ and at SNR=30 dB is shown in figure 3.14(a). The ML technique is implemented with $N_w = 8$. Figure 3.14(b) shows the estimated phase computed with the zeroth order coefficient c_0 obtained using (3.16). Figure 3.14(a) shows the estimated phase computed without and with 50% segment overlapping approach [29]. In the segment overlapping approach, the EPF in each row is divided into a number of overlapping segments. The errors in the estimation of phase without segment overlapping and with 50% segment overlapping approach are plotted in figure 3.14(c). The error values indicate that the overlapping approach provides a more accurate phase estimation compared to the non-overlapping approach. The plot of errors in the estimated phase in function of the number of segments (N_w) in figure 3.14(d) shows that the selection of a higher number of segments allows for better phase approximation with lower order polynomials.

Although the ML approach of polynomial coefficient estimation provides an optimal solution, the search procedure involved in it results in a huge computation cost. This problem worsens with the increase in the order of the polynomial phase. Moreover, if the estimated initial guess is not accurate, it is quite possible that the subsequent optimization algorithm may not in a given number of iterations converge towards the true solution. This is why a number of computationally efficient sub-optimal techniques of polynomial coefficient estimation have been proposed in the signal processing literature. A computationally efficient method of phase unwrapping has been proposed [30] based on cubic phase approximation ($P = 3$). In this method, the phase is approximated as a third order polynomial function in each segment of each row of the EPF. That is

$$\Gamma_s(x) = \exp\left[i(c_0 + c_1 x + c_2 x^2 + c_3 x^3)\right], \tag{3.17}$$

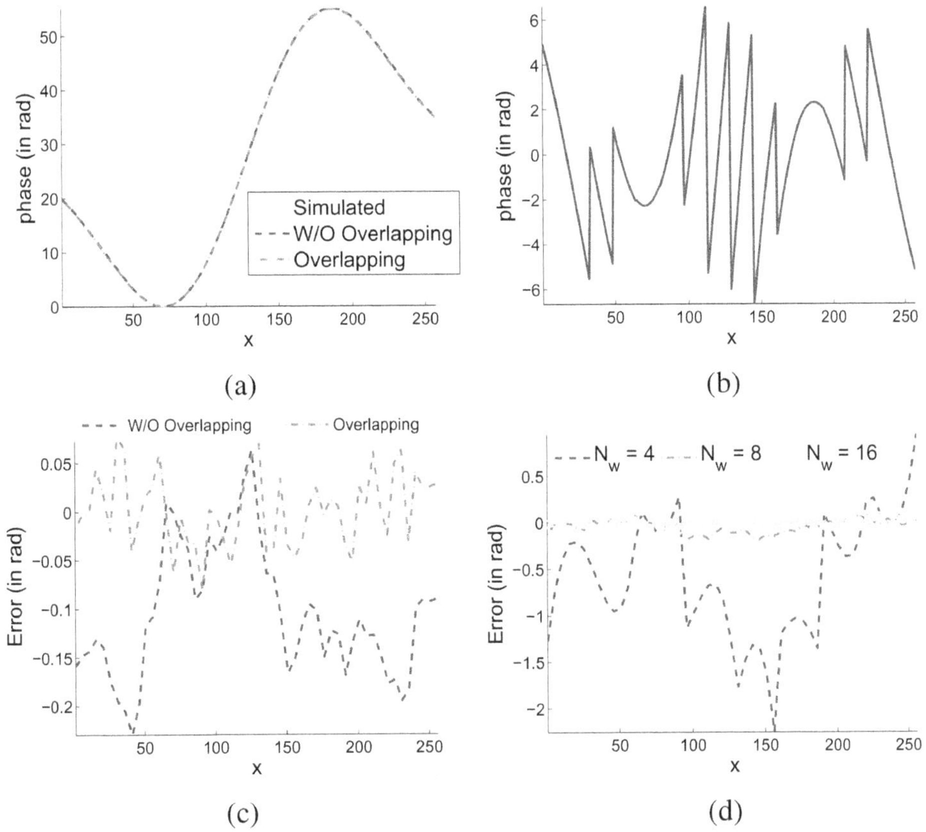

Figure 3.14. (a) Simulated and estimated phase with offset compensation; (b) estimated phase computed without offset compensation; (c) error in phase estimation with and without segment overlapping approach; (d) error in phase estimation in function of number of segments.

with,

$$\phi_s(x) = c_0 + c_1x + c_2x^2 + c_3x^3. \tag{3.18}$$

The cubic phase function (CPF) is defined as [31],

$$CPF_s(x, \omega) = \sum_{l=0}^{(N_s-1)/2} \Gamma_s(x + l)\Gamma_s(x - l)\exp(-i\omega l^2). \tag{3.19}$$

An important property that the CPF possesses is that the position of the peak observed in the plot of a CPF function with respect to ω corresponds to the second order derivative of phase. The value of ω at which the peak is observed is referred to as the instantaneous frequency rate in signal processing. According to (3.18)

$$\partial^2\phi_s/\partial x^2 = 2(c_2 + 3c_3x).$$

Substituting (3.17) into (3.19), we have

$$2(\hat{c}_2 + 3\hat{c}_3 x) = \underset{\omega}{\arg\max} \mid CPF_s(x, \omega) \mid . \tag{3.20}$$

Since in (3.20) there are two unknowns c_2 and c_3, we need to compute the CPF function at at least two distinct values of x. Let us consider that CPF is computed at $x = x_1$ and $x = x_2$,

$$2(\hat{c}_2 + 3\hat{c}_3 x_1) = \underset{\omega}{\arg\max} \mid CPF_s(x_1, \omega) \mid$$
$$2(\hat{c}_2 + 3\hat{c}_3 x_2) = \underset{\omega}{\arg\max} \mid CPF_s(x_2, \omega) \mid .$$

Solving the above equations, we obtain the coefficient estimates $\hat{c}_2$ and $\hat{c}_3$. The optimal values of x, $x_1 = 0$ and $x_2 = 0.11N_s$, have been shown to provide minimum coefficient estimation errors. Subsequently, the first order polynomial coefficient is estimated by computing the discrete Fourier transform (DFT) of the dechirped signal given below:

$$\begin{aligned} \Gamma_{1s}(x) &= \Gamma_s(x)\exp\left[-i(\hat{c}_2 x^2 + \hat{c}_3 x^3)\right] \\ \hat{c}_1 &= \underset{\omega}{\arg\max} \left| DFT\{\Gamma_{1s}(x)\} \right| \end{aligned}.$$

Finally, the zero order coefficient can be estimated using an operation similar to the one described in (3.16).

3.3.2 Blockwise polynomial phase approximation

We now assume that the EPF is divided into a number of 2D blocks each of size $M_y \times N_x$ as represented in figure 3.15. The phase within each block is represented as a 2D polynomial function of x and y [32]:

$$\phi_{mn}(x, y) = \sum_{(k,l)\in W} c(k, l) x^k y^l, \tag{3.21}$$

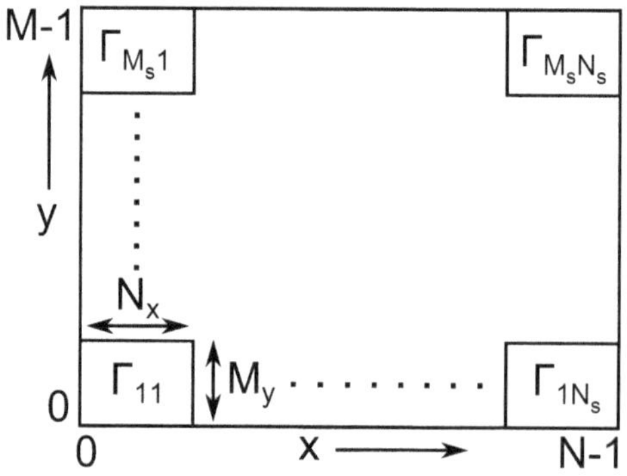

Figure 3.15. Scheme of two-dimensional blockwise phase unwrapping.

where, $W = \{0 \leqslant k, l \text{ and } 0 \leqslant k + l \leqslant P\}$, $m \in [1, M_s]$ and $n \in [1, N_s]$. To estimate the 2D polynomial coefficients, the EPF within each block Γ_{mn} is operated with a phase differencing operator (PD) as follows:

$$PD_{x^{(Q)},y^{(P-Q-1)}}[\Gamma_{mn}(x, y)] = \prod_{q=0}^{P-Q-1} \left\{ \prod_{p=0}^{Q} \left\{ \Gamma_{mn}^{(\dagger(p+q))}(x + p\tau_x, y + q\tau_y) \right\}^{\binom{Q}{p}} \right\}^{\binom{P-Q-1}{q}} \tag{3.22}$$

with

$$\Gamma_{mn}^{(\dagger(p+q))}(x, y) = \begin{cases} \Gamma_{mn}(x, y), & p + q \text{ even} \\ \Gamma_{mn}^{*}(x, y), & p + q \text{ odd} \end{cases}.$$

where, $Q \in [0, P - 1]$, $\tau_x = N_x/(Q + 1)$, $\tau_y = M_y/(P - Q)$ and $\binom{n}{r} = n!/(r!(n - r)!)$.

$PD_{x^{(Q)},y^{(P-Q-1)}}[\Gamma_{mn}(x, y)]$ is also termed the two-dimensional high order instantaneous moment (2D-HIM) of the EPF. In principle, the phase differencing operator transforms the EPF, with a 2D polynomial phase of degree P, into an EPF with a linear phase, that is, the 2D first order polynomial phase [33]. Accordingly, the simplified form of the expression given in (3.22) can be written as

$$PD_{x^{(Q)},y^{(P-Q-1)}}[\Gamma_{mn}(x, y)] = \exp\left[i(\omega_x x + \omega_y y)\right]. \tag{3.23}$$

The slopes of the 2D linear phase, that is, ω_x and ω_y carry the information on the 2D polynomial coefficients as

$$c(Q, P - Q) = \frac{\omega_y}{(-1)^{P-1} Q!(P - Q)! \tau_x^Q \tau_y^{P-Q-1}} \tag{3.24}$$

$$c(Q + 1, P - Q - 1) = \frac{\omega_x}{(-1)^{P-1}(Q + 1)!(P - Q - 1)! \tau_x^Q \tau_y^{P-Q-1}}. \tag{3.25}$$

It can be inferred from (3.23)–(3.25) that the frequencies at which the discrete FT of (3.23) attains maximum amplitude provide the estimate of the coefficient set $[c(0, P), c(1, P - 1), \ldots, c(P - 1, 1), c(0, P)]$ corresponding to the highest polynomial order P. The DFT of the 2D-HIM is termed the two-dimensional high order ambiguity function (2D-HAF). Subsequently, to obtain the coefficient estimates associated with the polynomial order $P - 1$, the part of EPF corresponding to the polynomial phase of order P is peeled off using the respective coefficients. Thus

$$\Gamma_{mn}^{(P-1)}(x, y) = \Gamma_{mn}(x, y)\exp\left[-i \sum_{Q=0}^{P} c(Q, P - Q) x^Q y^{P-Q}\right].$$

Effectively, $\Gamma_{mn}^{(P-1)}(x, y)$ is the EPF with a polynomial phase of highest order $P - 1$. This procedure of polynomial coefficient estimation based on the PD operator and the subsequent peeling off operation is performed iteratively to obtain the estimates of all the 2D polynomial coefficients. The estimated coefficients are plugged into

(3.21) to obtain the unwrapped phase within the block. A similar procedure is repeated in all the blocks.

The zeroth order coefficient within each block is computed as

$$c(0, 0) = \mathrm{Arg}\left\{\frac{1}{N_x M_y} \sum_{x=0}^{N_x-1} \sum_{y=0}^{M_y-1} \Gamma_{mn}^{(0)}(x, y)\right\}. \tag{3.26}$$

Due to the Arg operation involved in (3.26), the zeroth order coefficient is wrapped in the range $[-\pi, \pi)$. As a result, constant offsets exist among the unwrapped phases in the different blocks. The offset correction procedure similar to the one followed in the segmentwise phase unwrapping is adopted here to be rid of the discontinuities existing in the estimated phase [32].

3.3.3 Pixelwise polynomial phase approximation

Whereas the ML technique provides simultaneous estimation of polynomial coefficients in an optimal way at the cost of high computational overload, the suboptimal CPF and high order ambiguity (HAF) based techniques provide the polynomial coefficient estimates in a computationally efficient manner. However, all these methods suffer from error propagation due to the sequential procedure of polynomial coefficient estimation inherent to them. This problem can be circumvented by the simultaneous estimation of polynomial coefficients. Moreover, in the segmentwise and blockwise phase unwrapping, it is essential that the EPF data is available throughout each segment and block, respectively. Since, as mentioned in section 3.3, the object related discontinuities restrict us from making such an assumption, it would be more appropriate to perform phase unwrapping pixelwise while assuming the availability of valid EPF data within a small window around each pixel.

The problem of phase unwrapping from the perspective of state space analysis has received much attention in recent years [34–38]. A collection of state space models based on the dynamics of spatial phase evolution has been proposed. Of these, considering a broadly suitable model involving the pixelwise first order polynomial phase approximation [39], the phase within a small window around each pixel is approximated as the first order 2D polynomial function of x and y. Thus,

$$\phi_l(x_l, y_l) = c_0 + c_1 x_l + c_2 y_l, \tag{3.27}$$

where $|x_l| \leqslant L$ and $|y_l| \leqslant L$ represent the spatial coordinates local to the current pixel (x, y); L represents the window size parameter. Under the assumption of a symmetric window configuration, the phase at (x,y) is obtained as

$$\phi(x, y) = \phi_l(0, 0) = c_0.$$

The coefficients c_1 and c_2 represent the slopes of the linear phase, i.e. the local fringe frequencies, along x and y directions, respectively. Since our aim is to accurately estimate the polynomial coefficient vector $[c_0, c_1, c_2]^T$ at each pixel using the noisy EPF data, we define the following state space model:

$$\begin{aligned}\boldsymbol{\mu}_l &= \boldsymbol{F}\boldsymbol{\mu}_{l-1}\\ \boldsymbol{\Gamma}_l &= \boldsymbol{h}_l(\boldsymbol{\mu}_l) + \epsilon_l\end{aligned},$$

where, at the lth step, the state vector $\boldsymbol{\mu}$ is defined as $\boldsymbol{\mu} = [c_0, c_1, c_2]^T$; $\boldsymbol{F}$ is the state transition matrix; the measurement vector $\boldsymbol{\Gamma}_l$ is defined as

$$\boldsymbol{\Gamma}_l = \begin{bmatrix}\cos(\phi_w(x_l, y_l))\\ \sin(\phi_w(x_l, y_l))\end{bmatrix}.$$

The estimate of the measurement vector is defined to be a nonlinear function $\boldsymbol{h}_l$ of the state vector $\boldsymbol{\mu}_l$ as

$$\boldsymbol{h}_l(\boldsymbol{\mu}_l) = \begin{bmatrix}\cos(\boldsymbol{\mu}_l^T\boldsymbol{p})\\ \sin(\boldsymbol{\mu}_l^T\boldsymbol{p})\end{bmatrix}, \tag{3.28}$$

where $\boldsymbol{p} = [1, x_l, y_l]$. According to (3.27), all the three elements of $\boldsymbol{\mu}$ are constant within a window. As a result, the matrix $\boldsymbol{F}$ is defined to be an identity matrix $\boldsymbol{I}$ of size 3×3. Since, the nonlinear relationship between the state vector elements and the estimate of measurement vector (see (3.28)) imposes the use of a nonlinear state estimator, the extended Kalman filter (EKF) presents itself as a good option for a robust state estimation. At each pixel, the EKF based state estimation can be implemented as given below [40]:

1. Initialize the estimate of the state vector and the covariance of the state estimation error as

$$\begin{aligned}\hat{\boldsymbol{\mu}}_{0|0} &= \mathbb{E}(\boldsymbol{\mu}_0),\\ \boldsymbol{P}_{0|0} &= \mathbb{E}\left[(\boldsymbol{\mu}_0 - \hat{\boldsymbol{\mu}}_{0|0})(\boldsymbol{\mu}_0 - \hat{\boldsymbol{\mu}}_{0|0})^T\right],\end{aligned}$$

   where $\mathbb{E}$ represents the expectation operator; the covariance matrix $\boldsymbol{P}_{0|0}$ defines the uncertainty in the initial state estimate.
2. Once the initialization of the EKF is performed, the process and measurement update equations are implemented for $l = 1, 2, \ldots, (2L-1)^2$, as

$$\begin{aligned}\hat{\boldsymbol{\mu}}_{l|l-1} &= \boldsymbol{F}\hat{\boldsymbol{\mu}}_{l-1|l-1},\\ \boldsymbol{P}_{l|l-1} &= \boldsymbol{F}\boldsymbol{P}_{l-1|l-1}\boldsymbol{F}^T + \boldsymbol{Q},\end{aligned}$$

   where the process noise covariance matrix, $\boldsymbol{Q}$, takes into account the uncertainty in the state estimation due to the error in the approximation of phase as the first order polynomial. The value of $\boldsymbol{Q}$, usually set as a constant matrix, say $\boldsymbol{I}$, can, however, be selected in accordance to the window size parameter L. The accuracy of local polynomial phase approximation decreases with the higher value of L, i.e. the window size. It is important to note at this point that l takes values from 1 to $(2L+1)^2$ because there are $(2L+1) \times (2L+1)$ pixels in the EPF window around the current pixel. On the other hand, the values of x_l and y_l in (3.28) take the values in the range of $[-L, L]$. This distinction should be clear from the context.

3. Linearize the nonlinear measurement estimate $\boldsymbol{h}_l(\boldsymbol{\mu}_l)$ around the current state estimate $\hat{\boldsymbol{\mu}}_{l|l-1}$ using its first order Taylor series expansion as

$$\boldsymbol{h}_l(\boldsymbol{\mu}_l) = \boldsymbol{h}_l\left(\hat{\boldsymbol{\mu}}_{l|l-1}\right) + \left.\frac{\partial \boldsymbol{h}_l}{\partial \boldsymbol{\mu}}\right|_{\hat{\boldsymbol{\mu}}_{l|l-1}} (\boldsymbol{\mu}_l - \hat{\boldsymbol{\mu}}_{l|l-1}),$$

where the partial derivative matrix $\boldsymbol{H}_l = \left.\frac{\partial \boldsymbol{h}_l}{\partial \boldsymbol{\mu}}\right|_{\hat{\boldsymbol{\mu}}_{l|l-1}}$ is defined according to (3.28).

4. Measurement update equations:

$$\boldsymbol{K}_l = \boldsymbol{P}_{l|l-1}\boldsymbol{H}_l^T\left(\boldsymbol{H}_l\boldsymbol{P}_{l|l-1}\boldsymbol{H}_l^T + \boldsymbol{R}_\Gamma\right)^{-1},$$
$$\hat{\boldsymbol{\mu}}_{l|l} = \hat{\boldsymbol{\mu}}_{l|l-1} + \boldsymbol{K}_l\left[\boldsymbol{\Gamma}_l - \boldsymbol{h}_l(\hat{\boldsymbol{\mu}}_{l|l-1})\right],$$
$$\boldsymbol{P}_{l|l} = (\boldsymbol{I} - \boldsymbol{K}_l\boldsymbol{H}_l)\boldsymbol{P}_{l|l-1}(\boldsymbol{I} - \boldsymbol{K}_l\boldsymbol{H}_l)^T + \boldsymbol{K}_l\boldsymbol{R}_\Gamma\boldsymbol{K}_l^T,$$

where $\boldsymbol{R}_\Gamma$ represents the noise covariance matrix; the phase noise power in the EPF is not known *a priori*. In order to perform phase unwrapping over a wide range of noise power, we consider the worst case scenario which pleads in favor of setting a high value of noise covariance matrix indicative of a lower confidence on the EPF data; the Kalman gain is represented as $\boldsymbol{K}_l$. Considering a row-by-row scanning of the EPF data in the given window, $\hat{\boldsymbol{\mu}}_{l|l-1}$ and $\hat{\boldsymbol{\mu}}_{l|l}$ represent the state estimates at the lth step given the EPF data within the window up to and including the $(l-1)$th and lth step, respectively.

5. Once all the EPF data within the window is consumed, the final state vector $\hat{\boldsymbol{\mu}}_{final}$ carries the required estimates of the polynomial coefficients.

Note that, for the unwrapped phase estimation, the estimated phase information has to be carried over in the processing of the subsequent pixels. This is achieved by performing the initialization of the state vector, at a given pixel, as follows:

$$\hat{\boldsymbol{\mu}}_{0|0} = \hat{\boldsymbol{\mu}}_{\text{final}}^{(\text{neighbor})}$$

where $\hat{\boldsymbol{\mu}}_{\text{final}}^{(\text{neighbor})}$ represents the final state vector estimated at one of the neighboring pixels of the current pixel. This procedure of initialization of the state vector ensures the estimation of a continuous unwrapped phase. A priority criterion can be set to decide which neighboring pixel data should be selected in the initialization step of the current pixel. Such a criterion is represented in figure 3.16.

The fringe data is processed in a row-by-row manner. In the first forward pass starting from pixel [0, 0] to $[M-1, N-1]$, the neighboring pixels, for the purpose of initialization, are selected at the current pixel in accordance with figure 3.16(a). In

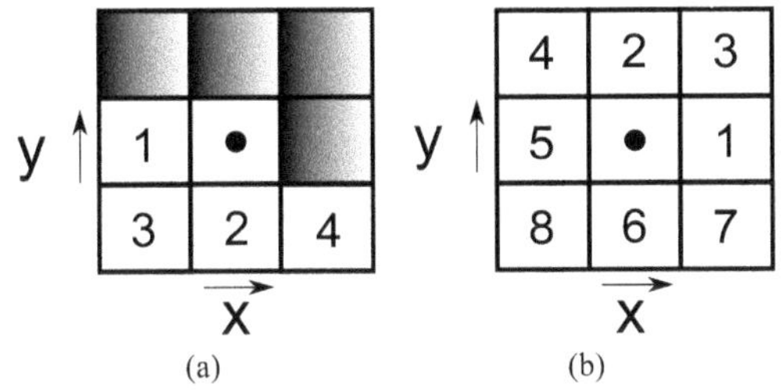

Figure 3.16. Pixel selection strategy in (a) forward and (b) backward directions. Pixel corresponding to number 1 is assigned the highest priority for the state vector initialization at the current pixel represented by a dot.

this figure, the current pixel is represented as a dot. Any of the four unshaded neighboring pixels can be selected for the purpose of initialization. The shaded pixels are those which have not yet been processed. The numbers assigned to the pixels shown in the figure indicate the priorty of their selection, with 1 indicating the highest priority.

The pixelwise phase unwrapping method provides the advantage of handling the object related discontinuities. In general, such discontinuities are identified with a binary mask separating the valid EPF data from the invalid data. Such a binary mask is used as an input to the pixelwise phase unwrapping algorithm. If the EPF data is masked, then it could be possible that no state vector information is available for the initialization of the current pixel in the forward pass. In this case, some pixels belonging to the valid EPF data region remain unprocessed. To obtain the phase estimates at these pixels, the fringe processing is performed in a row-by-row manner but in a backward direction starting from pixel $[M - 1, N - 1]$ to $[0, 0]$, with the pixel selection being performed according to figure 3.16(b). This procedure of forward and backward treatment of fringe data allows for obtaining the unwrapped phase estimation even in the presence of object related discontinuities.

A simulation example of phase unwrapping is provided in figure 3.17. Figure 3.17(a) shows the phase fringe pattern simulated at SNR=10 dB. The phase unwrapping is performed using segment-wise, block-wise and pixel-wise phase unwrapping algorithms. The errors in phase unwrapping are shown in figures 3.17(b)–(d), respectively. The effects of segment-wise, block-wise and pixel-wise methods of phase unwrapping are clearly visible in the error maps. Figure 3.18 shows an example of the capability of the pixel-wise method of phase unwrapping in handling object related discontinuities. The phase fringe pattern shown in figure 3.18(a) is simulated such that the fringe data is available only on the surface of the circular disk. A binary mask is generated to distinguish between the valid and invalid portions of the fringe data. Pixel-wise phase unwrapping is applied using a binary mask to obtain the unwrapped phase estimate shown in figure 3.18(b).

Figure 3.19(a) shows a phase fringe pattern corresponding to the out-of-plane displacement of an aluminum plate recorded in a digital holographic interferometry

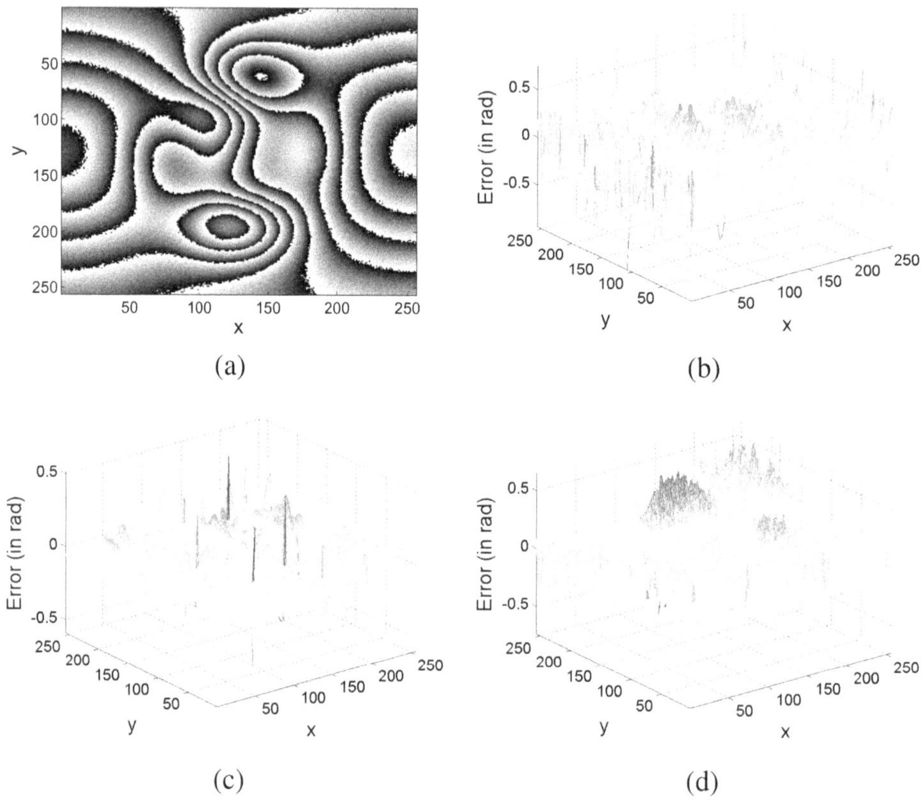

Figure 3.17. (a) Simulated phase fringe pattern at SNR = 10 dB. Errors in the unwrapped phase estimation obtained using (b) segment-wise, (c) block-wise, and (d) pixel-wise phase unwrapping methods. All the error values are in radians.

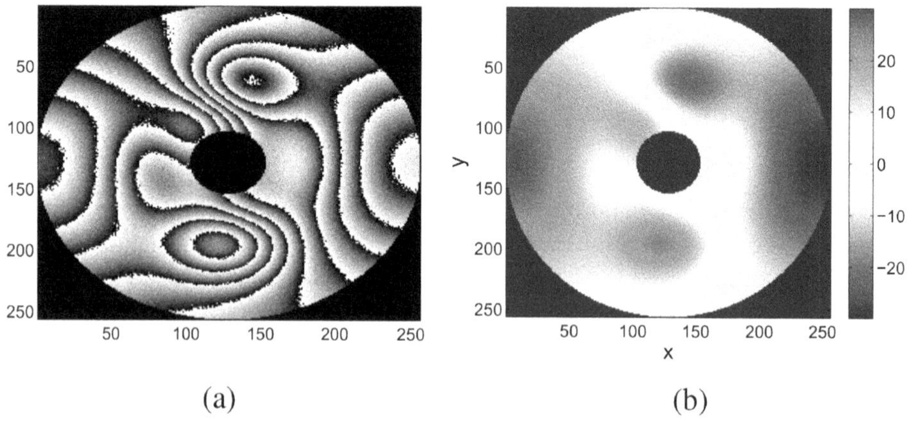

Figure 3.18. (a) Simulated phase fringe pattern carrying both valid and invalid fringe regions (b) unwrapped phase estimate obtained using the pixel-wise phase unwrapping method. The phase values are in radians.

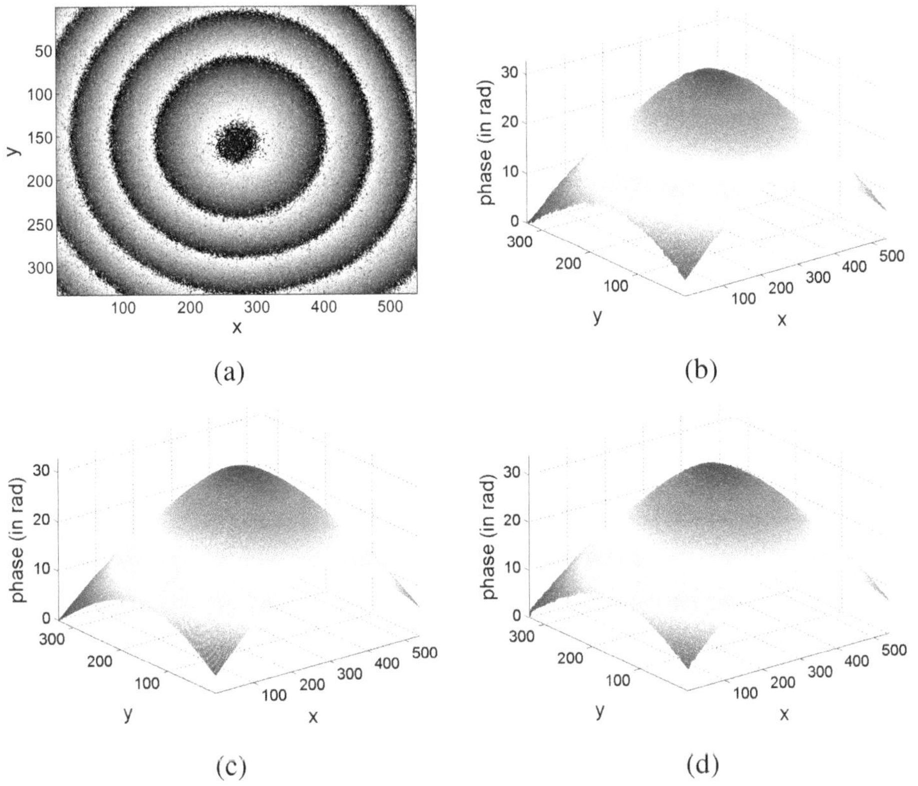

Figure 3.19. (a) Example of a phase fringe pattern corresponding to the out-of-plane displacement of an aluminum plate. Phase estimates computed using (b) segment-wise, (c) block-wise, and (d) pixel-wise phase unwrapping algorithms.

setup. The phase fringe pattern is corrupted by the speckle noise. Figures 3.19(b)–(d) show the unwrapped phase estimates obtained using segment-wise, block-wise and pixel-wise methods, respectively. Note that no denoising algorithm is applied prior to the phase unwrapping operation.

3.3.4 Quality guided phase unwrapping

Apart from the object related discontinuities which are usually masked prior to the application of phase unwrapping, inherent discontinuities may still be seen to occur on a phase map. In such situations, the phase unwrapping strategies described in sections 3.3.1–3.3.3 are not applicable directly, as they fall under the category of path independent phase unwrapping algorithms wherein the phase unwrapping result is independent of the path of fringe scanning during the unwrapping process. These methods are limited to providing a continuous unwrapped phase estimate from a noisy phase fringe pattern. In order to handle phase fringe patterns carrying phase discontinuous regions, extra information in the form of a quality map is essential to guide the unwrapping algorithm. A quality map provides a path following strategy to the unwrapping algorithm which enables the phase fringe pattern to be unwrapped along a path unique to it. Accordingly, such phase unwrapping methods are categorized as quality guided phase unwrapping algorithms. The quality map is usually generated based on the *a priori* knowledge of the application at hand. In cases where only the wrapped phase information is available, the phase derivative variance provides the required quality map. Since the phase discontinuous regions exhibit abrupt phase variations, the phase derivatives at these locations are of high magnitudes. The phase derivative variance map is thus capable of identifying such locations. The phase derivative variance based quality map is computed as

$$QualityMap(x, y) = \frac{\sqrt{\sum_{|x_l, y_l| \leqslant L} \left(\Delta^x_{x_l, y_l} - \overline{\Delta^x}_{x,y}\right)^2} + \sqrt{\sum_{|x_l, y_l| \leqslant L} \left(\Delta^y_{x_l, y_l} - \overline{\Delta^y}_{x,y}\right)^2}}{(2L + 1)^2},$$

where,

$$\Delta^x_{x_l, y_l} = \mathscr{W}\left\{\phi_w(x_l + 1, y_l) - \phi_w(x_l, y_l)\right\},$$
$$\Delta^y_{x_l, y_l} = \mathscr{W}\left\{\phi_w(x_l, y_l + 1) - \phi_w(x_l, y_l)\right\},$$

represent the wrapped partial derivatives of phase with respect to x and y, respectively; $|x_l| \leqslant L$ and $|y_l| \leqslant L$ represent the pixels within a window around the pixel (x, y); L represents the odd valued window size parameter; $\overline{\Delta^x}_{x,y}$ and $\overline{\Delta^y}_{x,y}$ represent the averages of the wrapped partial phase derivatives with respect to x and y, respectively, within the window around the pixel under consideration; $\mathscr{W}\{\cdot\}$ represents the wrap operator which wraps its argument into the range $[-\pi, \pi)$.

A simulated phase fringe pattern containing a phase discontinuous region is shown in figure 3.20(a). Figure 3.20(b) shows the unwrapped phase estimate obtained by the pixel-wise polynomial phase approximation based method. Since in this method, the unwrapped phase estimate is computed by row-by-row or column-by-column pixel scanning, the unwrapped phase is estimated accurately in the continuous region of the phase map. The polynomial phase approximation becomes invalid as soon as a pixel in the discontinuous region is encountered. From this step on, the error in the unwrapped phase estimate propagates to the subsequent pixels, giving rise to an unreliable unwrapped phase estimate. This aspect illustrated in figure 3.20(b) shows the importance of the quality guided phase unwrapping to obtain reliable phase estimates. Figure 3.20(c) shows the quality map of the phase fringe pattern in the form of a normalized phase derivative variance computed with the window size $L = 1$. The actual phase derivative variance values are negated such that high value in the quality map indicates good quality of the pixel and vice versa. It can be observed in figure 3.20(c) that low quality pixels lie at the boundary of the phase discontinuity. The quality guided phase unwrapping algorithms utilize this quality map as input to guide the phase unwrapping algorithm so that higher quality pixels are unwrapped prior to the lower quality pixels. The unwrapped phase estimate obtained using

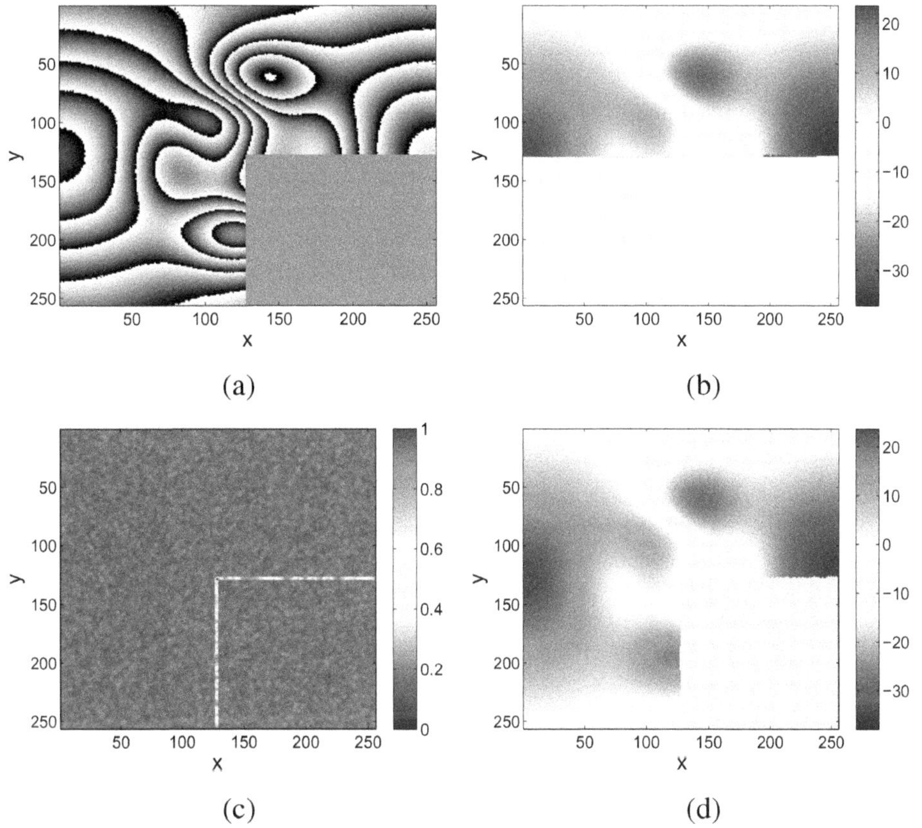

Figure 3.20. (a) Simulated phase fringe pattern with inherent phase discontinuity; (b) unwrapped phase estimate based on the pixel-wise polynomial phase approximation; (c) phase derivative variance based quality map; (d) unwrapped phase estimate using path-following based quality guided phase unwrapping algorithm.

flood-fill algorithm based quality guided phase unwrapping is shown in figure 3.20(d).

References

[1] Kaufmann G H and Galizzi G E 1996 Speckle noise reduction in television holography fringes using wavelet thresholding *Opt. Eng.* **35** 9–14

[2] Kemao Q 2007 Two-dimensional windowed Fourier transform for fringe pattern analysis: principles, applications and implementations *Opt. Lasers Eng.* **45** 304–17

[3] Tang C, Han L, Ren H, Zhou D, Chang Y, Wang X and Cui X 2008 Second-order oriented partial-differential equations for denoising in electronic-speckle-pattern interferometry fringes *Opt. Lett.* **33** 2179–81

[4] Bernini M B, Federico A and Kaufmann G H 2008 Noise reduction in digital speckle pattern interferometry using bidimensional empirical mode decomposition *Appl. Opt.* **47** 2592–8

[5] Federico A and Kaufmann G H 2001 Comparative study of wavelet thresholding methods for denoising electronic speckle pattern interferometry fringes *Opt. Eng.* **40** 2598–604

[6] Tang C, Wang L, Yan H and Li C 2012 Comparison on performance of some representative and recent filtering methods in electronic speckle pattern interferometry *Opt. Lasers Eng.* **50** 1036–51
[7] Guerrero J A, Marroquin J L, Rivera M and Quiroga J A 2005 Adaptive monogenic filtering and normalization of ESPI fringe patterns *Opt. Lett.* **30** 3018–20
[8] Fodor I K 2002 A survey of dimension reduction techniques *Technical report* (Lawrence Livermore National Laboratory) UCRL-ID-148494
[9] Vargas J, Sorzano C O S, Antonio Quiroga J, Estrada J C and Carazo J M 2013 Fringe pattern denoising by image dimensionality reduction *Opt. Lasers Eng.* **51** 921–8
[10] Wang H, Kemao Q, Gao W, Lin F and Seah H S 2009 Fringe pattern denoising using coherence-enhancing diffusion *Opt. Lett.* **34** 1141–3
[11] Wang H and Kemao Q 2011 Comparative analysis on some spatial-domain filters for fringe pattern denoising *Appl. Opt.* **50** 1687–96
[12] Zhou X, Yang T, Zou H and Zhao H 2012 Multivariate empirical mode decomposition approach for adaptive denoising of fringe patterns *Opt. Lett.* **37** 1904–6
[13] Aebischer H A and Waldner S 1999 Simple and effective method for filtering speckle-interferometric phase fringe patterns *Opt. Commun.* **162** 205–10
[14] Chen X, Chen T, Wenjun X, Su Y and Su K 2016 General construction of transform-domain filters, filtering methods for electronic speckle pattern interferometry, and comparative analyses *Appl. Opt.* **55** 2214–22
[15] Hua Y 1992 Estimating two-dimensional frequencies by matrix enhancement and matrix pencil *IEEE Trans. Sig. Proc.* **40** 2267–80
[16] Servin M, Cuevas F J, Malacara D, Marroquin J L and Rodriguez-Vera R 1999 Phase unwrapping through demodulation by use of the regularized phase-tracking technique *Appl. Opt.* **38** 1934–41
[17] Yun H Y, Hong C K and Chang S W 2003 Least-squares phase estimation with multiple parameters in phase-shifting electronic speckle pattern interferometry *J. Opt. Soc. Am.* **20** 240–7
[18] Trouve E, Caramma M and Maitre H 1996 Fringe detection in noisy complex interferograms *Appl. Opt.* **35** 3799–806
[19] Kemao Q, Nam L T H, Feng L and Soon S H 2007 Comparative analysis on some filters for wrapped phase maps *Appl. Opt.* **46** 7412–8
[20] Escalante N, Villa J, de la Rosa I, de la Rosa E, González- Ramírez E, Gutiérrez O, Olvera C and Araiza M 2013 2-D continuous wavelet transform for espi phase-maps denoising *Opt. Lasers Eng.* **51** 1060–5
[21] Huang M J and Sheu W 2005 Histogram-data-orientated filter for inconsistency removal of interferometric phase maps *Opt. Eng.* **44** 1–11
[22] Li C, Tang C, Yan C, Wang L and Zhang H 2011 Localized Fourier transform filter for noise removal in electronic speckle pattern interferometrywrapped phase patterns *Appl. Opt.* **50** 4903–11
[23] Estrada J C, Servin M and Vargas J 2012 2D simultaneous phase unwrapping and filtering: A review and comparison *Opt. Lasers Eng.* **50** 1026–9
[24] Friedlander B and Francos J M 1996 Model based phase unwrapping of 2-D signals *IEEE Trans. Sig. Proc.* **44** 2999–3007
[25] Liang Z 1996 A model-based method for phase unwrapping *IEEE Trans. Med. Imag.* **15** 893–7

[26] Gorthi S S and Rastogi P 2009 Piecewise polynomial phase approximation approach for the analysis of reconstructed interference fields in digital holographic interferometry *J. Opt. A: Pure Appl. Opt.* **11** 065405
[27] Rudin W 1976 *Principles of Mathematical Analysis* 3rd edn (New York: McGraw-Hill)
[28] Lagarias J C, Reeds J A, Wright M H and Wright P E 1998 Convergence properties of the nelder-mead simplex method in low dimensions *SIAM J. Optim.* **9** 112–47
[29] Gorthi S and Sand Rastogi P 2009 Discrete chirp Fourier transform based analysis of reconstructed interference fields in digital holographic interferometry *J. Mod. Opt.* **56** 1317–22
[30] Gorthi S and Sand Rastogi P 2010 Phase estimation in digital holographic interferometry using cubic-phase-function based method *J. Mod. Opt.* **57** 595–600
[31] O'Shea P 2004 A fast algorithm for estimating the parameters of a quadratic FM signal *IEEE Trans. Sig. Proc.* **52** 385–93
[32] Rajshekhar G and Rastogi P 2012 Fringe demodulation using the two-dimensional phase differencing operator *Opt. Lett.* **37** 4278–80
[33] Francos J M and Friedlander B 1998 Two-dimensional polynomial phase signals: parameter estimation and bounds *Multidimens. Sys. Sig. Proc* **9** 173
[34] Loffeld O, Nies H, Knedlik S and Yu W 2008 Phase unwrapping for sar interferometry–a data fusion approach by Kalman filtering *IEEE Trans. Geosci. Remote Sensing* **46** 47–58
[35] Gurov I and Volynsky M 2012 Interference fringe analysis based on recurrence computational algorithms *Opt. Lasers Eng.* **50** 514–21
[36] Rajshekhar G and Rastogi P 2013 Phase estimation using a state-space approach based method *Opt. Lasers Eng.* **51** 1004–7
[37] Xie Y M and Li Y H 2014 Enhanced phase unwrapping algorithm based on unscented Kalman filter, enhanced phase gradient estimator, and path-following strategy *Appl. Opt.* **53** 4049–60
[38] Cheng Z, Liu D, Yang Y, Ling T, Chen Y, Zhang L, Bai J, Shen Y, Miao L and Huang W 2015 Practical phase unwrapping of interferometric fringes based on unscented Kalman filter technique *Opt. Exp.* **23** 32337–49
[39] Kulkarni R and Rastogi P 2017 Local-polynomial-approximation-based phase unwrapping using state space analysis *Appl. Opt.* **56** 29–34
[40] Simon D 2006 *Optimal State Estimation: Kalman, H Infinity, and Nonlinear Approaches* (Hoboken, NJ: Wiley) pp 395–433

Chapter 4

Phase estimation: multicomponent spatial fringe analysis

Chapters 2 and 3 discuss one of the major achievements of digital signal processing—its use in reliably estimating the phase from a fringe pattern. The phase estimation methods mentioned therein are in general effective for the measurement of single components of displacement, that is, either in-plane or out-of-plane, at a time, undergone by an object submitted to load. However, for a complete characterization of the deformation process, it is desirable to obtain the measurement of all three, that is, two orthogonal in-plane and one out-of-plane displacement components. It is possible to obtain such measurements in electronic speckle pattern interferometry or digital holographic interferometry setups by illuminating the object surface *sequentially* from different directions and computing the phases corresponding to each illumination beam. These phases together allow for computing *a posteriori* the measurement of the displacement vector. Due to the sequential recording of the fringe patterns, this procedure of displacement measurement is suitable only in the case of static loading.

However, for the nondestructive optical methods to prevail and prosper, it is important for them to seize upon the opportunity that the digitization era provides, of developing in their folds the capacity to obtain these measurements in a *simultaneous* manner from a single recording of a fringe signal or exponential phase field. We will restrict our attention to the EPFs that are produced in digital holographic interferometry.

This chapter discusses spatial fringe analysis techniques based on *amplitude discrimination* for the simultaneous estimation of phases from a single frame of an EPF containing multiple signal components recorded in a multi-beam digital holographic interferometry setup. These methods can be particularly useful for enabling the simultaneous measurement of three displacement components of an object submitted to an arbitrary load.

doi:10.1088/978-0-7503-1469-5ch4

4.1 Background

We will first describe in brief how the interference field carrying information on the multidimensional deformation is recorded. Here for the sake of simplicity we will restrict ourselves to the recording of two phase informations. A two-beam DHI setup is shown in figure 4.1. The object surface is illuminated *simultaneously* by two light beams placed in the x–z plane symmetrically with respect to the object surface normal. The setup consists of a single reference beam directly illuminating the CCD surface (not shown in the figure). The object and the CCD camera are placed at $(x, y, z = 0)$ and $(x', y', z = D)$ planes, respectively. Let us represent the scattered object and reference beams at the CCD plane as $O_1(x', y')$, $O_2(x', y')$ and $R(x', y')$, respectively. The intensity of the hologram recorded by the CCD is

$$\begin{aligned} I &= |R + O_1 + O_2|^2, \\ &= (R + O_1 + O_2)(R + O_1 + O_2)^*, \\ &= I_0 + R(O_1^* + O_2^*) + R^*(O_1 + O_2), \end{aligned} \tag{4.1}$$

where $I_0 = |R|^2 + |O_1|^2 + |O_2|^2 + O_1O_2^* + O_1^*O_2$. The spatial dependence of the quantities in (4.1) is omitted for the sake of brevity. Considering the Fresnel approximation of wave propagation, the wavefront scattered by the object at $z = 0$ plane is given by [1]

$$\begin{aligned} \Gamma(x, y) = \frac{i}{\lambda D} \exp\left(-i\frac{2\pi}{\lambda}D\right) \int_{-\infty}^{\infty} \int_{-\infty}^{\infty} R(x', y')I(x', y') \\ \times \exp\left[-i\frac{\pi}{\lambda D}\left((x - x')^2 + (y - y')^2\right)\right] dx'dy'. \end{aligned} \tag{4.2}$$

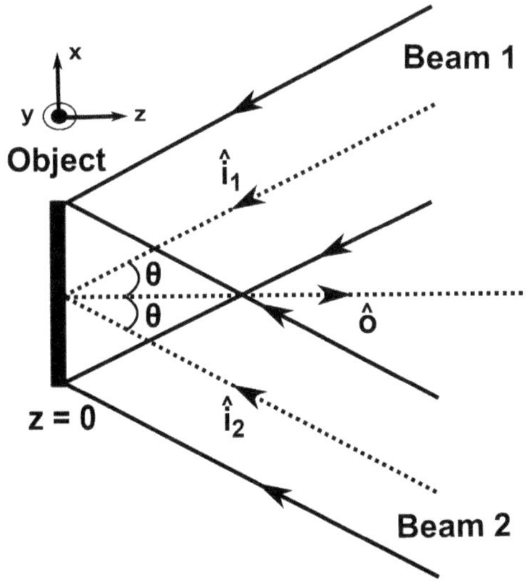

Figure 4.1. Schematic of a two-beam digital holographic interferometry experimental setup for the measurement of in-plane and out-of-plane displacements along x and z directions of an object subjected to loading (for example, plate bending).

Substituting (4.1) into (4.2), we have

$$\Gamma(x, y) = \frac{i}{\lambda D} \exp\left(-i\frac{2\pi}{\lambda}D\right) \int_{-\infty}^{\infty} \int_{-\infty}^{\infty} \left[RI_0 + R^2\left(O_1^* + O_2^*\right) + |R|^2(O_1 + O_2)\right]$$
$$\times \exp\left[-i\frac{\pi}{\lambda D}\left((x - x')^2 + (y - y')^2\right)\right] dx' dy'.$$

In the *in-line* digital holographic configuration, the contributions due to RI_0, $R^2(O_1^* + O_2^*)$ and $|R|^2(O_1 + O_2)$ in the reconstructed wavefront overlap spatially. On the other hand, in the *off-axis* digital holographic configuration, these terms are spatially separated due to the introduction of a carrier by a reference beam falling obliquely on the CCD plane. This allows us during the reconstruction to select the contribution only due to $|R|^2(O_1 + O_2)$. The complex amplitude of the scattered object wave at $z = 0$ is

$$\Gamma_b(x, y) = A_{1b}(x, y)\exp[i\psi_1(x, y)] + A_{2b}(x, y)\exp[i\psi_2(x, y)].$$

Note that the phase terms ψ_1 and ψ_2 associated with the two beams are random in nature on account of the object surface roughness. Upon object deformation, the scattered wave can be represented as

$$\Gamma_a(x, y) = A_{1a}(x, y)\exp[i(\psi_1(x, y) + \phi_1(x, y))] + A_{2a}(x, y)\exp[i(\psi_2(x, y) + \phi_2(x, y))]$$

where ϕ_1 and ϕ_2 are the deterministic phase terms introduced by the object deformation into the scattered beams, 1 and 2, respectively. We are interested in estimating these two terms simultaneously.

The EPF containing the two signal components with phases ϕ_1 and ϕ_2 can be written as

$$\Gamma(x, y) = A_1(x, y)\exp\left(i\phi_1(x, y)\right) + A_2(x, y)\exp\left(i\phi_2(x, y)\right) + \varepsilon(x, y). \tag{4.3}$$

The digital holographic interferometry setup shown in figure 4.1 is sensitive to the object surface displacement along the x and z directions. Let us represent the displacement vector as $\vec{\mathbf{d}} = d_x\hat{x} + d_y\hat{y} + d_z\hat{z}$ with $\hat{x}$, $\hat{y}$ and $\hat{z}$ representing the unit vectors along x, y and z directions, respectively. The relationship between the phases, ϕ_1 and ϕ_2, and displacement components, d_x and d_z, is obtained using the sensitivity vectors $\hat{s}_1$ and $\hat{s}_2$, defined as

$$\hat{s}_1 = \hat{o} - \hat{i}_1$$
$$\hat{s}_2 = \hat{o} - \hat{i}_2,$$

where $\hat{i}_1$ and $\hat{i}_2$ are the unit vectors along the directions of object beams 1 and 2, respectively; and $\hat{o}$ is the unit vector along the observation direction. From figure 4.1 we have,

$$\hat{i}_1 = -\sin(\theta)\hat{x} - \cos(\theta)\hat{z},$$
$$\hat{i}_2 = \sin(\theta)\hat{x} - \cos(\theta)\hat{z},$$
$$\hat{o} = \hat{z}.$$

Based on these definitions and figure 4.1 we obtain,

$$\phi_1 = \frac{2\pi}{\lambda}\vec{\mathbf{d}} \cdot \hat{s}_1 = \frac{2\pi}{\lambda}[d_z(1 + \cos(\theta)) + d_x \sin(\theta)]$$

and

$$\phi_2 = \frac{2\pi}{\lambda}\vec{\mathbf{d}} \cdot \hat{s}_2 = \frac{2\pi}{\lambda}[d_z(1 + \cos(\theta)) - d_x \sin(\theta)],$$

where λ is the wavelength of the laser source. The components of the displacement vector, d_x and d_z, follow as

$$d_x = \frac{\lambda}{4\pi}\frac{(\phi_1(x, y) - \phi_2(x, y))}{\sin(\theta)}, \tag{4.4}$$

$$d_z = \frac{\lambda}{4\pi}\frac{(\phi_1(x, y) + \phi_2(x, y))}{(1 + \cos(\theta))}. \tag{4.5}$$

Several techniques based on the use of multiple object-reference beam pairs in digital holographic interferometry have been investigated to accomplish this goal. A multiplexing–demultiplexing technique in digital holography involving an incoherent mixing of two waves scattered from an object surface has been developed [2]. A separate reference beam is assigned to each of the scattered object beams. The incoherence between the two pairs of object-reference beams is established by setting their polarizations orthogonal with respect to each other. Such incoherent pairs of object-reference beams can also be achieved by the use of delay lines [3] or the use of different wavelength laser sources [4, 5]. Another approach based on the use of phase shifting and involving the recording of multiple phase shifted holograms in digital holography is reported in [6]. In the two-beam DHI setup shown in figure 4.1, a single EPF containing two signal components is obtained. In order to accurately evaluate the phases $\phi_1(x, y)$ and $\phi_2(x, y)$ from $\Gamma(x, y)$ given in (4.3), we need to make certain assumptions and establish a criterion whereby we will be able to discriminate the two signal components from each other. In addition, the phases are assumed to be spatially continuous. It is important to note that since there are two signal components present in the EPF, the *arctan* operation does not provide the wrapped phase estimates as it does in the case of an EPF containing a single signal component.

4.2 Segment-wise multiple phase estimation

In section 3.3.3, we have discussed the *Weierstrass approximation theorem* based on which the spatially continuous phases can be approximated as one-dimensional or two-dimensional polynomial functions of x and y of pre-defined polynomial order. Continuing on the same lines, this section focuses on the segment-wise multiple phase estimation approach, wherein the fringe analysis is performed in a row-wise or column-wise manner. Considering that the fringe analysis is performed in a row-wise manner, the EPF in a given row y can be written as

$$\Gamma(x) = A_1(x)\exp\left[i\phi_1(x)\right] + A_2(x)\exp\left[i\phi_2(x)\right] + \varepsilon(x). \tag{4.6}$$

Along this row, the phases ϕ_1 and ϕ_2 can be approximated with the polynomial functions of x. Since the appropriate order of the polynomial is not known *a priori*, $\Gamma(x)$ is first divided into N_w number of non-overlapping/overlapping segments. Within each segment of the EPF of length $N_s(= N/N_w)$, the phases are approximated as the polynomial functions of x of pre-defined polynomial order P. Thus we have,

$$\Gamma_s(x) = A_1 \exp\left[i\phi_{1_s}(x)\right] + A_2 \exp\left[i\phi_{2_s}(x)\right] + \varepsilon_s(x),$$

where,

$$\phi_{1_s}(x) = a_{s0} + a_{s1}x + \cdots + a_{sP}x^P, \tag{4.7}$$

$$\phi_{2_s}(x) = b_{s0} + b_{s1}x + \cdots + b_{sP}x^P. \tag{4.8}$$

Note that within the segment, the amplitudes of the signal components are assumed to be constant. It can be observed from (4.7) and (4.8) that the accurate estimation of the polynomial coefficients $(a_{s0}, a_{s1}, \ldots, a_{sP})$ and $(b_{s0}, b_{s1}, \ldots, b_{sP})$ is essential for the accurate estimation of phases ϕ_1 and ϕ_2, respectively. Various signal processing techniques have been developed to analyze multicomponent polynomial phase signals arising in applications, such as radar and sonar surveillance, the goal being to obtain the accurate estimation of the polynomial coefficients. These techniques can be effectively applied for phase estimation in a multicomponent signal.

Note that the sequential procedure of a row-wise or column-wise segmentation of the EPF, phase estimation, offset computation and phase stitching outlined in section 3.3.3 is also applicable to the case of polynomial phase approximation based multiple phase estimation techniques described in this section.

4.2.1 Product cubic phase function

Only a few approaches have been developed for multicomponent spatial fringe analysis. The first method to be described here considers the phase approximation with a second order polynomial, that is, $P = 2$. From (4.7) and (4.8), we have

$$\phi_{1_s}(x) = a_{s0} + a_{s1}x + a_{s2}x^2, \tag{4.9}$$

$$\phi_{2_s}(x) = b_{s0} + b_{s1}x + b_{s2}x^2. \tag{4.10}$$

The approach, termed the product cubic phase function, has been proposed in [7] for the estimation of the second order coefficients from a multicomponent polynomial phase signal. It works by computing the cubic phase function (*CPF*) of $\Gamma_s(x)$ for different values of x. The expression of *CPF* is

$$CPF(x, \Omega) = \sum_{l=0}^{\frac{N_s-1}{2}} \Gamma_s(x+l)\Gamma_s(x-l)\exp(-i\Omega l^2),$$

where $-\frac{N_s-1}{2} \leqslant x \leqslant \frac{N_s-1}{2}$; and the length of segment N_s is considered to be an odd number. In the case of a single component second order polynomial phase EPF, with amplitude A_1, the location of the peak in the plot of *CPF* with respect to Ω provides the estimate of a_{s2}. However, unwanted cross-terms are seen to appear along with the auto-terms associated with the signal components in the case of an EPF containing two second order polynomial phase components. The presence of cross-terms makes it difficult to reliably estimate the second order polynomial coefficients from the *CPF*. This difficulty can be overcome by the use of the product cubic phase function (*PCPF*) which is defined as a product of *CPF*s computed for a given set of values of x as [7]

$$PCPF(\Omega) = \prod_{k=1}^{K} CPF(x_k, \Omega),$$

where K is the number of x values considered in the computation of *PCPF*. The plot of PCPF with respect to Ω exhibits two peaks, each corresponding to the second order polynomial coefficients, that is, a_{s2} and b_{s2}. Due to this product form of *CPF*s, the auto-terms are magnified as they are independent of x, the cross-terms are diminished and the noise rejection capability is improved. However, since there is no *a priori* information available on the signal components, the second order polynomial coefficients cannot be computed unambiguously from the values of Ω associated with the peaks. The said *a priori* information is associated with the knowledge on the assignment of peak locations to their corresponding signal components. This difficulty is always present in the process of estimating polynomial coefficients from a multicomponent polynomial phase signal. To resolve this difficulty, an *amplitude discrimination criterion* which sets the amplitudes of the signal components A_1 and A_2 as unequal is developed. This is achieved by setting different intensities to the object illumination beams. Thus for $A_1 > A_2$, that is, with the intensity of beam 1 being greater than that of beam 2, the *PCPF* will have a dominant peak corresponding to a_{s2}. Consequently, $\hat{a}_{s2}$ can be estimated by locating the position of the dominant peak, $\Omega_{\max}$, using the relationship $\Omega_{\max} = 2\hat{a}_{s2}$. The lower order polynomial coefficients can be computed by peeling off the second order polynomial phase term from the EPF. To estimate a_{s1}, the EPF is peeled off as follows

$$\Gamma_{s1}(x) = \Gamma_s(x)\exp(-i\hat{a}_{s2}x^2).$$

Now, the highest order polynomial phase coefficient of the signal component with amplitude A_1 is a_{s1}. Consequently, the estimate of a_{s1} can be obtained by computing the discrete Fourier transform (DFT) of $\Gamma_{s1}(x)$ as

$$\mathscr{F}_{s1}(\omega) = \sum_{x=-\frac{N_s-1}{2}}^{\frac{N_s-1}{2}} \Gamma_{s1}(x)\exp(-i\omega x)$$

$$\hat{a}_{s1} = \arg\max_{\omega} |\mathscr{F}_{s1}(\omega)|.$$

The DFT of $\Gamma_{s1}(x)$ is computed using the fast Fourier transform. The dominant peak of the DFT spectrum gives a coarse estimate of a_{s1}. A fine estimate $\hat{a}_{s1}$ is obtained thereafter by applying IFEIF (*iterative frequency estimation by interpolation on Fourier coefficients*) [8]. The signal $\Gamma_{s1}(x)$ is next peeled off to remove the phase contribution due to a_{s1} to obtain $\Gamma_{s0}(x)$ as

$$\Gamma_{s0}(x) = \Gamma_{s1}(x)\exp(-i\hat{a}_{s1}x).$$

$\hat{a}_{s0}$ and $\hat{A}_1$, the respective zero order coefficient and amplitude estimates are computed as

$$\hat{a}_{s0} = \operatorname{Arg}\left\{\frac{1}{N_s}\sum_{x=-\frac{N_s-1}{2}}^{\frac{N_s-1}{2}} \Gamma_{s0}(x)\right\}, \tag{4.11}$$

$$\hat{A}_1 = \left|\frac{1}{N_s}\sum_{x=-\frac{N_s-1}{2}}^{\frac{N_s-1}{2}} \Gamma_{s0}(x)\right|. \tag{4.12}$$

Once the coefficient $[\hat{a}_{s2}, \hat{a}_{s1}, \hat{a}_{s0}]$ and amplitude, $\hat{A}_1$, estimates are obtained, the corresponding EPF signal component is reconstructed as

$$\Gamma_{A_{1s}}(x) = \hat{A}_1 \exp\left[i(\hat{a}_{s0} + \hat{a}_{s1}x + \hat{a}_{s2}x^2)\right].$$

The signal component $\Gamma_{A_{1s}}(x)$ is subtracted from the EPF signal $\Gamma_s(x)$ so that the signal component with amplitude A_1 is removed from $\Gamma_s(x)$. Thus $\Gamma_{A_{2s}}(x)$ can be written as

$$\Gamma_{A_{2s}}(x) = \Gamma_s(x) - \Gamma_{A_{1s}}(x).$$

The new signal $\Gamma_{A_{2s}}(x)$ primarily contains the second signal component, that is, the signal component with amplitude A_2. This procedure of coefficient and amplitude estimation is repeated to obtain the polynomial coefficient $[\hat{b}_{s2}, \hat{b}_{s1}, \hat{b}_{s0}]$ and amplitude, $\hat{A}_2$, estimates corresponding to the second signal component. The estimates of $\phi_{1s}(x)$ and $\phi_{2s}(x)$ can be obtained by substituting the estimated coefficients into (4.9) and (4.10), respectively. It is interesting to note that the phase estimates computed within each segment are in an unwrapped form. The phases $\phi_{1s}(x)$ and $\phi_{2s}(x)$ are estimated in all the segments and in all the rows.

A one-dimensional EPF simulated at SNR of 20 dB containing two signal components with embedded second order polynomial phases is shown in figure 4.2(a). The two phases are simulated with polynomial coefficients $a_0 = 1.0472$, $a_1 = 0.0733$, $a_2 = 0.002$ and $b_0 = 0.7854$, $b_1 = -0.1222$, $b_2 = 0.005$, respectively. At first, the amplitudes of the two signal components are set as $A_1 = A_2 = 1$. The *PCPF* of the EPF is computed and the plot of the absolute value of *PCPF* with respect to Ω is provided in figure 4.2(b). Two peaks corresponding to a_2 and b_2 with equal amplitudes are visible in the figure. This clearly signifies that without any prior knowledge it would be impossible to assign the estimated second order polynomial coefficients to their respective signal components. The simulation is also performed with $A_1 = 2$ and $A_2 = 1$. The plot of the absolute value of *PCPF* with respect to Ω provided in figure 4.2(c) shows the presence of a single dominant peak corresponding to a_2 due to the fact, $A_1 > A_2$. On the other hand, if we set $A_2 > A_1$, a dominant peak corresponding to b_2 will appear. Thus, the amplitude discrimination criterion allows us to unambiguously estimate the polynomial coefficients and assign them to their respective signal components. The polynomial coefficient estimation procedure

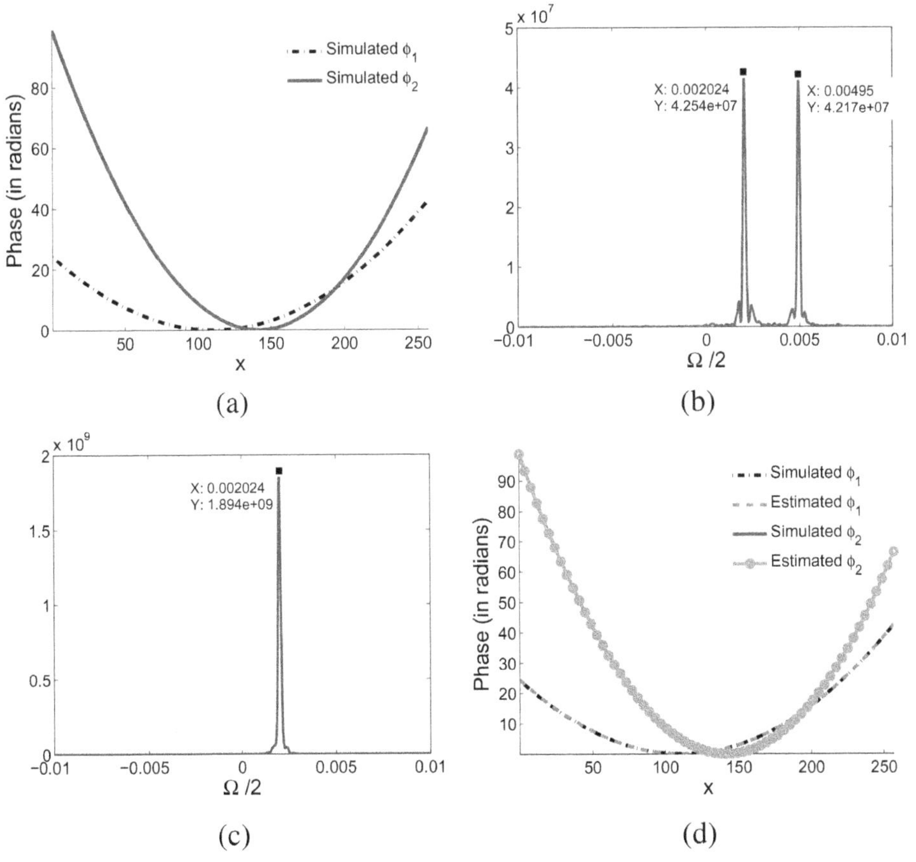

Figure 4.2. (a) Simulated ϕ_1 and ϕ_2 (in radians); (b) *PCPF* magnitude with respect to $\Omega/2$ for $A_1 = A_2 = 1$; (c) *PCPF* magnitude with respect to $\Omega/2$ for $A_1 = 2$ and $A_2 = 1$; (d) simulated and estimated ϕ_1 and ϕ_2 (in radians).

is implemented for $A_1 = 2$ and $A_2 = 1$, and the plots of the estimated and simulated unwrapped phases are shown in figure 4.2(d).

To achieve better accuracy in the estimation of second order polynomial coefficients, the *PCPF* based method can be implemented in an iterative manner to update the coefficient estimate around the previously estimated coefficient value. Iterations can also be performed to alternately reconstruct each signal component, removing its contribution from the multicomponent EPF at each iteration. These iterative operations are helpful in improving the multiple phase estimation accuracy.

Let us consider an example of a two-dimensional EPF of size, 257×257, simulated at SNR = 20 dB. The EPF is assumed to contain two signal components. The real part of the EPF, generally termed as a *moiré fringe pattern*, is shown in figure 4.3(a). Fourier spectrum of the EPF shown in figure 4.3(b) indicates the presence of two spectrally overlapping signal components. The multiple phase estimation is performed in a row-wise manner. In each row, the EPF is divided into four segments, each of lengths $N_s = 65$. The amplitude discrimination criterion is set with $A_2 = 2$ and $A_1 = 1$. The unwrapped estimates of ϕ_1 and ϕ_2 are shown in figures 4.3(c) and (d), respectively. For the sake of illustration, the wrapped forms of estimated ϕ_1 and ϕ_2 are shown in figures 4.3(e) and (f), respectively.

4.2.2 Pseudo–Wigner–Hough transform

A second order polynomial coefficient estimation technique based on the combination of Wigner–Ville-Distribution (WVD) and Hough transform, termed the Wigner–Hough transform (WHT), has also been developed to resolve such problems [9]. The WVD of a second order polynomial phase signal produces a distribution of energy that is concentrated along a straight line in the space–frequency (s–f) plane. The polynomial phase coefficients can thus be estimated by computing the parameters associated with the straight line. The Hough transform, a popular technique used in image processing for the detection of lines in images is applied for the detection of a linear pattern in the s–f plane. The Hough transform of the WVD of a signal essentially transforms it from the s–f domain to the phase parameter domain. The location of peaks observed in the phase parameter domain provide the estimates of the polynomial phase coefficients. The polynomial coefficient estimation using the WHT is equally applicable in the multicomponent second order polynomial phase signal analysis. The WVD of a multicomponent second order polynomial phase signal produces a distribution of energy corresponding to the multiple signal components in the s–f plane along with the unwanted cross terms. The presence of cross terms arising on account of the bilinear property of the WVD can seriously hamper our ability to analyze the multicomponent signal. Whereas the energy distribution corresponding to the multiple signal components is positive, the energy distribution of cross terms is observed to oscillate between positive and negative values. This attribute is effectively exploited by the WHT. In the WHT of a multicomponent second order polynomial phase signal, the effect of cross terms

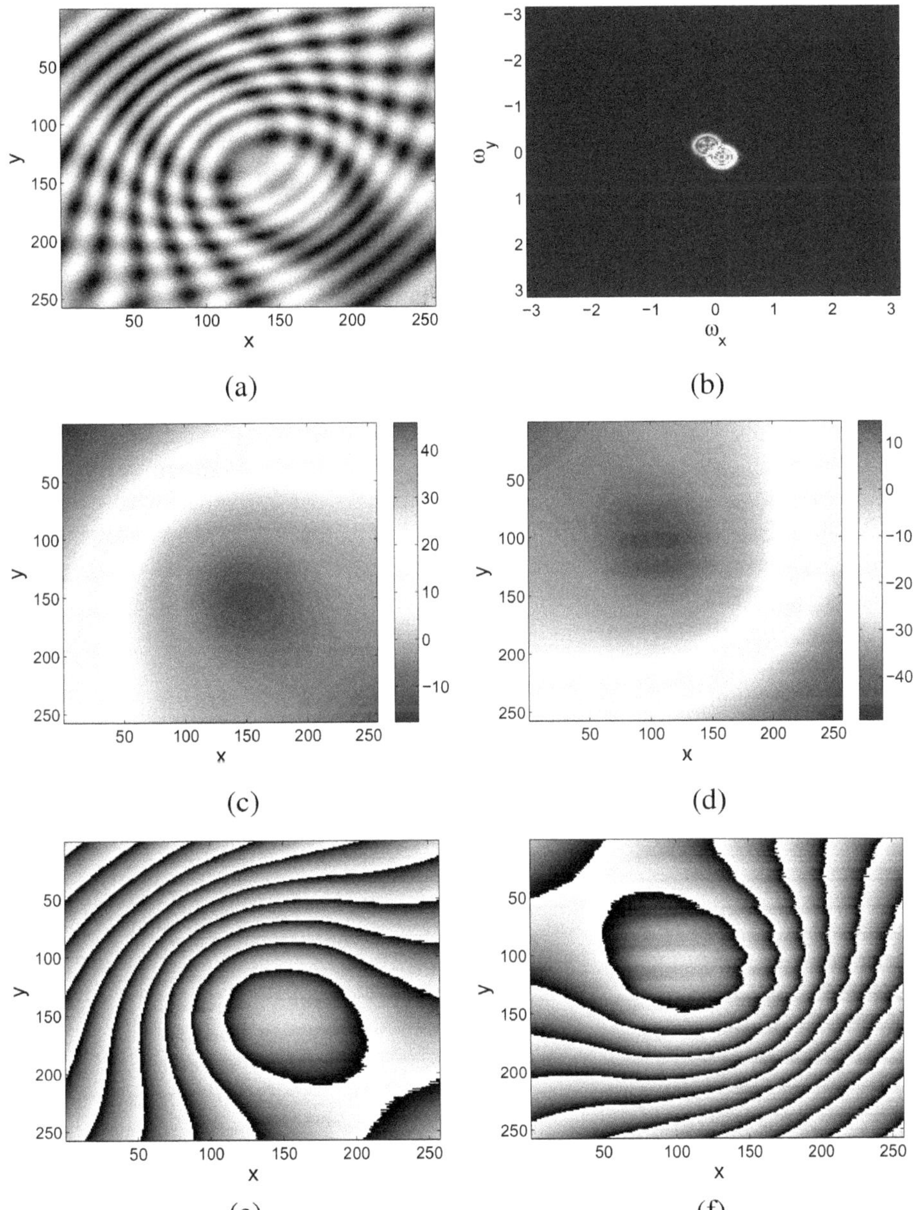

Figure 4.3. (a) Simulated moiré fringe pattern; (b) Fourier spectrum of the EPF. Unwrapped phase estimates of (c) ϕ_1 and (d) ϕ_2 (in radians). Rewrapped phase estimates of (e) ϕ_1 and (f) ϕ_2.

produced by WVD is removed by the Hough transform due to the inherent integration process involved in it. This allows us to use the WHT to accurately estimate the phase parameters of all the signal components in a *simultaneous* manner.

The fact that the advantages of the WHT come at a price of high computational cost has led researches to combine pseudo-Wigner–Ville-distribution (pWVD) and Hough transform for developing an approach, appropriately termed pseudo-Wigner–Hough-transform (PWHT) [10], to address the problem of multicomponent spatial fringe analysis [11]. The PWHT of the segmented EPF, $\Gamma_s(x)$, can be represented as

$$PWHT(\boldsymbol{a}) = \sum_{x=L}^{N_s-L-1} \sum_{l=-L}^{L} \Gamma_s(x+l)\Gamma_s^*(x-l)\exp(i2\omega(x;\boldsymbol{a})),$$

where L is a parameter defining the window length, $W = 2L + 1$; $\boldsymbol{a} = (a_1, a_2)$ represents the domain of phase parameters (a_{s1}, b_{s1}) and (a_{s2}, b_{s2}); $\omega = a_1 + a_2 x$. The value of $L = 0.1N_s$ has been shown to provide the optimal estimation of the phase parameters [10]. Whereas the number of peaks observed in the plot of PWHT is an indicator of the presence of the number of signal components, their locations provide the estimates of phase parameters. In comparison to the WHT, the PWHT is computationally more efficient due to the windowing involved in the pWVD. Additionally, the fact that the width of the peak observed in the PWHT parameter space is much larger than that observed in the WHT parameter space suggests that the PWHT parameter space offers improved numerical properties over that offered by the WHT for parameter initialization during optimization [10]. It is important to ensure that the initial estimates of the phase parameters calculated using the PWHT are sufficiently accurate for the optimization algorithm to converge to true phase parameter values. In order to achieve this, peak detection in the PWHT is performed on the grid of phase parameters a_1 and a_2 with appropriate grid spacings of Δ_{a_1} and Δ_{a_2}, respectively. The values of Δ_{a_1} and Δ_{a_2} can be calculated as [10]

$$\Delta_{a_1} = \frac{\pi}{1.4W}.$$

$$\Delta_{a_2} = \frac{\pi}{(-1.19L^2 + 1.2LN_s - 3L + 0.4N_s + 17.5)}.$$

Taking advantage of the fact that better accuracy in the coefficient estimation is achieved by utilizing smaller grid spacings, the use of these estimated coefficients as initial values for Nelder–Mead simplex optimization algorithm [12] contributes to a further refinement of the phase parameters. Similar to the PCPF, the PWHT based method too employs the amplitude discrimination criterion to assign the estimated phase parameters to their respective signal components. However, it should be noted that the phase parameters of all the signal components are estimated *simultaneously* as opposed to the sequential estimation of the polynomial phase coefficients in the *PCPF* based method. Consequently, the inherent error propagation effect observed in the sequential phase parameter estimation is avoided in the PWHT based phase estimation method.

Although the error propagation effect is eliminated in the PWHT, an inherent bias caused by the presence of multiple peaks might still exist in the parameter space in the case of a multicomponent EPF. These peaks disturb one another due to their mutual spreading. As a result, the peaks' locations get shifted from their true phase parameter values. To avoid this problem, signal compensation operation can be performed as described below.

1. Set up the amplitude discrimination criterion with $A_1 > A_2$.
2. Select only the estimates of $\{a_{s1}, a_{s2}\}$ associated with the first signal component by computing the PWHT of $\Gamma_s(x)$.
3. The estimates of a_{s0} and A_1 are computed using (4.11) and (4.12), and the corresponding signal component is reconstructed.
4. A new signal is computed by removing from the multicomponent EPF the signal reconstructed in step 3. This new signal primarily contains the signal component with amplitude A_2.
5. The procedure of signal reconstruction and signal removal is performed in an iterative manner to obtain improved estimates of the polynomial coefficients.
6. These coefficients are used to compute the unwrapped phase estimates using (4.9) and (4.10).

Experimental validation of this approach is shown in figure 4.4. The object is an aluminum plate deformed using a point load applied to its center and which is also rotated in its plane to generate both the out-of-plane and in-plane displacements on the object surface. The corresponding moiré fringe pattern is shown in figure 4.4(a). The Fourier spectrum of the EPF is shown in figure 4.4(b). The PWHT based phase estimations of ϕ_1 and ϕ_2 in their unwrapped forms are shown in figures 4.4(c) and (d), respectively. Based on the relationship between the estimated phases and the displacement components, the sum and the difference of the estimated phases ϕ_1 and ϕ_2 are shown in figures 4.4(e) and (f), respectively.

4.3 Block-wise multiple phase estimation

Since the two techniques just presented operate either in the row-by-row or column-by-column manner, only up to two phase estimates can be obtained based on the segment-wise polynomial phase approximation. Thus an important restriction of the techniques described in section 4.2 is that the EPF cannot contain more than two signal components. The second restriction of these techniques is that their performance is limited by the second order polynomial phase approximation which may be by far not applicable universally.

A two-dimensional EPF analysis based on the two-dimensional product high order ambiguity function (2D-PHAF) has been proposed [13] to obtain multiple phase estimates from a single frame of an EPF containing two or more signal components. Such a multicomponent phase estimation technique has an important relevance for those optical methods which have inherent potential though not the means of data processing for generating at least three phase maps from their encoded signal components in the EPF, for example, in deformation

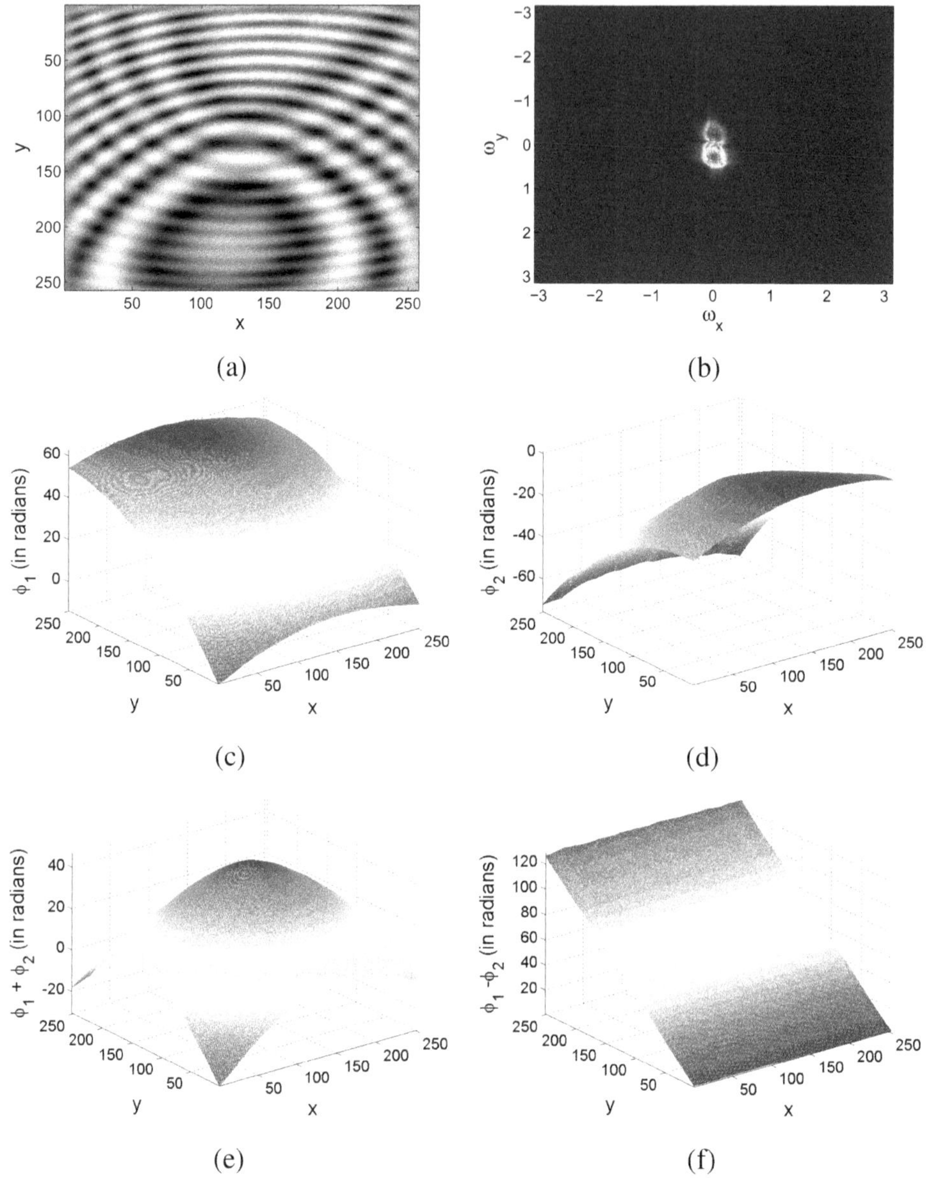

Figure 4.4. (a) Experimentally recorded moiré fringe pattern; (b) Fourier spectrum of the EPF. Unwrapped phase estimates of (c) ϕ_1 and (d) ϕ_2 (in radians). Estimates of (e) $\phi_1 + \phi_2$ and (f) $\phi_1 - \phi_2$ (in radians).

measurement, for the estimation from a single data frame of all three components of displacement of a deformed object. The schematic of such a three-beam digital holographic interferometry setup, with three light beams simultaneously illuminating the object surface from different directions and a single reference beam

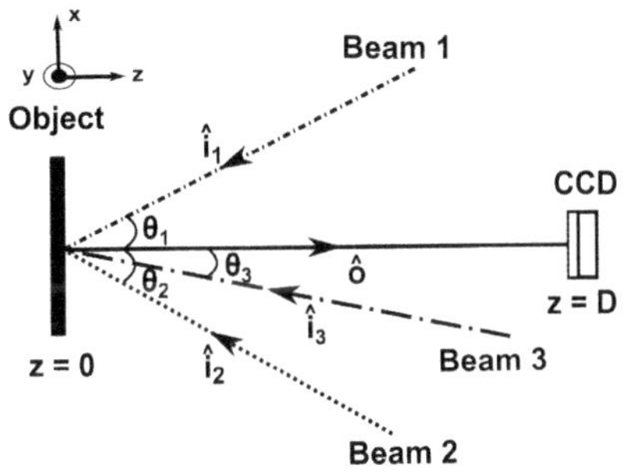

Figure 4.5. Schematic of a digital holographic interferometry experimental setup with three object illumination beams for the simultaneous measurement of d_x, d_y and d_z.

(not shown) directly illuminating the CCD, is shown in figure 4.5. The three beams termed beams 1, 2 and 3 fall on the object surface at respective angles of θ_1, θ_2, and θ_3 with respect to the z axis. The three illumination directions are non-coplanar. A typical EPF recorded in such a setup contains three signal components, each one associated with a single light beam. The EPF can be represented as

$$\Gamma(x, y) = \sum_{k=1}^{3} A_k(x, y)\exp[i\phi_k(x, y)] + \varepsilon(x, y).$$

Let $\hat{i}_1$, $\hat{i}_2$ and $\hat{i}_3$ represent the unit vectors along the three illumination directions. The sensitivity vectors $\hat{s}_1$, $\hat{s}_2$ and $\hat{s}_3$ are defined as,

$$\hat{s}_j = \hat{i}_j - \hat{o}, \text{ for } j = 1, 2, 3.$$

The phases, displacement vector $\vec{\mathbf{d}}$ and the sensitivity vectors are related as

$$\phi_j = \frac{2\pi}{\lambda}\hat{s}_j \cdot \vec{\mathbf{d}}, \text{ for } j = 1, 2, 3. \tag{4.13}$$

From (4.13), the orthogonal components of displacement **d**, that is, d_x, d_y, d_z can be expressed as

$$\begin{bmatrix} \phi_1 \\ \phi_2 \\ \phi_3 \end{bmatrix} = \frac{2\pi}{\lambda}\begin{bmatrix} s_{1x} & s_{1y} & s_{1z} \\ s_{2x} & s_{2y} & s_{2z} \\ s_{3x} & s_{3y} & s_{3z} \end{bmatrix}\begin{bmatrix} d_x \\ d_y \\ d_z \end{bmatrix}, \tag{4.14}$$

where (s_{jx}, s_{jy}, s_{jz}) represent the orthogonal components of $\hat{s}_j$ for $j = 1, 2, 3$. It is evident from (4.14) that for an accurate measurement of the displacement vector, all three phases have to be estimated accurately. To do so, a parametric phase representation approach is developed which is similar to the blockwise phase unwrapping method discussed in section 3.3.2 that divides the EPF into a number of two-dimensional blocks. Under the assumption of the spatial continuity of the phases, the phase of each signal component of the EPF within each block is approximated with a 2D polynomial function of x and y of a pre-defined polynomial order. Referring to figure 3.15, let us represent the EPF within a block of size $M_y \times N_x$ as,

$$\Gamma_{mn}(x, y) = \sum_{k=1}^{3} A_k(x, y)\exp[i\phi_k(x, y)] + \varepsilon(x, y),$$

where, $x \in [0, N_x - 1]$ and $y \in [0, M_y - 1]$. The phases can be represented as,

$$\phi_k(x, y) = \sum_{p=0}^{P} \sum_{q=0}^{p} c_{p-q,q}^{(k)} x^{p-q} y^q, \ k = 1, 2, 3,$$

where P is the order of the 2D-polynomial. A multicomponent polynomial phase signal analysis technique based on 2D-PHAF [14] is used to accurately estimate the polynomial coefficients. The procedure of coefficient estimation and subsequent phase estimation within each block is outlined below.

1. First, we define a 2D multilag high order instantaneous moment (2D-ml-HIM) of order P using the following iterative steps,

$$\begin{aligned} \Gamma_1[x, y] &= \Gamma_{mn}(x, y), \\ \Gamma_2[x, y; \tau_1, \vartheta_1] &= \Gamma_1^*[x, y]\Gamma_1[x + \tau_1, y + \vartheta_1], \\ &\cdots \\ \Gamma_P[x, y; \boldsymbol{\tau}_{P-1}, \boldsymbol{\vartheta}_{P-1}] &= \Gamma_{P-1}^*[x, y; \boldsymbol{\tau}_{P-2}, \boldsymbol{\vartheta}_{P-2}] \\ &\quad \Gamma_{P-1}[x + \tau_{P-1}, y + \vartheta_{P-1}; \boldsymbol{\tau}_{P-2}, \boldsymbol{\vartheta}_{P-2}], \end{aligned} \tag{4.15}$$

where $\boldsymbol{\tau}_j := [\tau_1, \ldots, \tau_j]$ and $\boldsymbol{\vartheta}_j := [\vartheta_1, \ldots, \vartheta_j]$ are the sets of the delays used up to step j. As a result of the iterative procedure, the Pth order polynomial phase EPF gets transformed into a first order polynomial phase EPF, $\Gamma_P[\cdot]$. Subsequently, the 2D multilag high order ambiguity function (2D-ml-HAF) is computed as the FT of 2D-ml-HIM obtained in (4.15)

$$\boldsymbol{\Gamma}_P[\varsigma, \nu; \boldsymbol{\tau}_{P-1}, \boldsymbol{\vartheta}_{P-1}] = \mathscr{F}\{\Gamma_P[x, y; \boldsymbol{\tau}_{P-1}, \boldsymbol{\vartheta}_{P-1}]\}(\varsigma, \nu).$$

It can be shown that in the case of a single component 2D-polynomial phase EPF, the peak location of $\boldsymbol{\Gamma}_P[\varsigma, \nu; \boldsymbol{\tau}_{P-1}, \boldsymbol{\vartheta}_{P-1}]$ provides the estimate of the coefficients of the order P [14]. On the other hand, in the case of a multicomponent signal, the peaks associated with each signal component are seen to be present along with the unwanted peaks associated with the cross-terms. For a reliable estimation of the polynomial coefficients in the presence of the cross-terms, the 2D-product high order ambiguity function (2D-PHAF) is introduced in the case of a multicomponent EPF. In 2D-PHAF, whereas the peaks associated with the individual signal components are enhanced, the peaks in the spectrum associated with the cross-terms are suppressed. Considering that the EPF signal is divided into a number of small sized blocks, the phase within each block is approximated with a two-dimensional third order polynomial function ($P = 3$). The third order 2D-PHAF is

$$P_3[\varsigma, \nu; \boldsymbol{\tau}^{(2)}, 0] = \prod_{t=1}^{T} \Gamma_3\left(\frac{\tau_t^{(1)}\tau_t^{(2)}}{\tau_1^{(1)}\tau_1^{(2)}}\varsigma, \frac{\tau_t^{(1)}\tau_t^{(2)}}{\tau_1^{(1)}\tau_1^{(2)}}\nu; \boldsymbol{\tau}^{(2)}, 0\right) \tag{4.16}$$

$$P_3[\varsigma, \nu; 0, \boldsymbol{\vartheta}^{(2)}] = \prod_{t=1}^{T} \Gamma_3\left(\frac{\vartheta_t^{(1)}\vartheta_1^{(2)}}{\vartheta_1^{(1)}\vartheta_1^{(2)}}\varsigma, \frac{\vartheta_t^{(1)}\vartheta_t^{(2)}}{\vartheta_1^{(1)}\vartheta_1^{(2)}}\nu; 0, \boldsymbol{\vartheta}^{(2)}\right). \tag{4.17}$$

Here, $\{\tau_t^{(1)}, \vartheta_t^{(1)}\}$ and $\{\tau_t^{(2)}, \vartheta_t^{(2)}\}$ are the two different sets of lags used in the computation of 2D-PHAF. The frequency scaling operation involved in (4.16) and (4.17) is the standard frequency scaling property of the FT. This frequency scaling operation aligns the peaks observed for different sets of lags at the same location in the spectrum. In order to compute the third order polynomial coefficients, the set of lags should be selected such that the matrix Φ_3 defined below has a full-column rank.

$$\Phi_3 = \begin{pmatrix} 6\tau_1^{(1)}\tau_1^{(2)} & 2\eta_{12}^{(1)} & 2\vartheta_1^{(1)}\vartheta_1^{(2)} & 0 \\ 0 & 2\tau_1^{(1)}\tau_1^{(2)} & 2\eta_{12}^{(1)} & 6\vartheta_1^{(1)}\vartheta_1^{(2)} \\ \vdots & \vdots & \vdots & \vdots \\ 6\tau_T^{(1)}\tau_T^{(2)} & 2\eta_{12}^{(T)} & 2\vartheta_T^{(1)}\vartheta_T^{(2)} & 0 \\ 0 & 2\tau_T^{(1)}\tau_T^{(2)} & 2\eta_{12}^{(T)} & 6\vartheta_T^{(1)}\vartheta_T^{(2)} \end{pmatrix},$$

with $\eta_{12}^{(t)} := \tau_t^{(2)}\vartheta_t^{(1)} + \tau_t^{(1)}\vartheta_t^{(2)}$. In $P_3[\varsigma, \nu; \boldsymbol{\tau}^{(2)}, 0]$ and $P_3[\varsigma, \nu; 0, \boldsymbol{\vartheta}^{(2)}]$, three peaks are observed corresponding to the highest order polynomial coefficients of the three signal components. These coefficients can be estimated based on the peak locations as,

$$\hat{c}_{3,0}^{(k)} = \frac{\varsigma_1^{(k)}}{6\tau_1^{(1)}\tau_1^{(2)}}, \quad \hat{c}_{2,1}^{(k)} = \frac{\nu_1^{(k)}}{2\tau_1^{(1)}\tau_1^{(2)}}$$

$$\hat{c}_{1,2}^{(k)} = \frac{\varsigma_2^{(k)}}{2\vartheta_1^{(1)}\vartheta_1^{(2)}}, \quad \hat{c}_{0,3}^{(k)} = \frac{\nu_2^{(k)}}{6\vartheta_1^{(1)}\vartheta_1^{(2)}}$$

for $k = 1, 2, 3$. The subscripts 1 and 2 of ς and ν indicate the peaks observed in the computed 2D-PHAFs, $P_3[\varsigma, \nu; \boldsymbol{\tau}^{(2)}, 0]$ and $P_3[\varsigma, \nu; 0, \boldsymbol{\vartheta}^{(2)}]$, corresponding to the highest order coefficients of polynomials of x and y, respectively.

2. Once the coefficients are estimated, it is essential to assign them to their respective signal components. To do so, an amplitude discrimination criterion is utilized which sets different values to the amplitudes of the EPF signal components. This is achieved by setting different intensities to the three object illumination beams. As a result, peaks in the 2D-PHAF do not have the same magnitudes. Let us consider that the amplitudes of the signal components are set as $A_1 > A_2 > A_3$. Thus, the highest amplitude peak

observed in the 2D-PHAF corresponds to the third order polynomial coefficient associated with the signal component with amplitude A_1.

3. Using the estimated polynomial coefficients associated with the signal component with amplitude A_1, the respective polynomial phase EPF is peeled off from the EPF signal $\Gamma_{mn}(x, y)$ as

$$\Gamma_{mn}^{(1)}(x, y) = \Gamma_{mn}(x, y)\exp\left[-i\left(\hat{c}_{3,0}^{(1)}x^3 + \hat{c}_{2,1}^{(1)}x^2y + \hat{c}_{1,2}^{(1)}xy^2 + \hat{c}_{0,3}^{(1)}y^3\right)\right].$$

Noting that $\Gamma_{mn}^{(1)}(x, y)$ corresponds to the second order polynomial phase signal component with amplitude A_1, one can compute the second order 2D-PHAF to obtain the estimates of second order polynomial coefficients as

$$P_2[\varsigma, \nu; \boldsymbol{\tau}^{(1)}, 0] = \prod_{t=1}^{T} \boldsymbol{\Gamma}_2^{(1)}\left(\frac{\tau_t^{(1)}}{\tau_1^{(1)}}\varsigma, \frac{\tau_t^{(1)}}{\tau_1^{(1)}}\nu; \tau_t^{(1)}, 0\right),$$

$$P_2[\varsigma, \nu; 0, \boldsymbol{\vartheta}^{(1)}] = \prod_{t=1}^{T} \boldsymbol{\Gamma}_2^{(1)}\left(\frac{\vartheta_t^{(1)}}{\vartheta_1^{(1)}}\varsigma, \frac{\vartheta_t^{(1)}}{\vartheta_1^{(1)}}\nu; 0, \vartheta_t^{(1)}\right).$$

In order to compute the second order polynomial coefficients, a set of lags is selected such that the matrix Φ_2 defined below has a full-column rank

$$\Phi_2 = \begin{pmatrix} 2\tau_1^{(1)} & \vartheta_1^{(1)} & 0 \\ 0 & \tau_1^{(1)} & 2\vartheta_1^{(1)} \\ \vdots & \vdots & \vdots \\ 2\tau_T^{(1)} & \vartheta_T^{(1)} & 0 \\ 0 & \tau_T^{(1)} & 2\vartheta_T^{(1)} \end{pmatrix}.$$

4. The highest amplitude peaks observed in $P_2[\varsigma, \nu; \boldsymbol{\tau}^{(1)}, 0]$ and $P_2[\varsigma, \nu; 0, \boldsymbol{\vartheta}^{(1)}]$ correspond to the second order polynomial coefficients of the signal component with amplitude A_1. These coefficients can be estimated based on the peak locations as

$$\hat{c}_{2,0}^{(1)} = \frac{\varsigma_1^{(1)}}{2\tau_1^{(1)}}, \quad \hat{c}_{0,2}^{(1)} = \frac{\nu_1^{(1)}}{2\vartheta_1^{(1)}}$$

$$\hat{c}_{1,1}^{(1)} = \frac{\varsigma_2^{(1)}}{\vartheta_1^{(1)}}, \text{ or } \hat{c}_{1,1}^{(1)} = \frac{\nu_2^{(1)}}{\tau_1^{(1)}}.$$

5. Using the estimated coefficients, the second order polynomial phase term associated with the signal component with amplitude A_1 is peeled from $\Gamma_{mn}^{(1)}(x, y)$ as

$$\Gamma_{mn}^{(2)}(x, y) = \Gamma_{mn}^{(1)}(x, y)\exp\left[-i\left(\hat{c}_{2,0}^{(1)}x^2 + \hat{c}_{1,1}^{(1)}xy + \hat{c}_{0,2}^{(1)}y^2\right)\right].$$

Now, as the signal component with amplitude A_1 in $\Gamma_{mn}^{(2)}(x, y)$ has a first order polynomial phase, the first order coefficient can be computed using the FT,

$$\left(\hat{c}_{1,0}^{(1)}, \hat{c}_{0,1}^{(1)}\right) = \max_{\varsigma, \nu} \left|\mathcal{F}\left(\Gamma_{mn}^{(2)}(x, y)\right)(\varsigma, \nu)\right|.$$

6. The estimates of zero order coefficient and amplitude A_1 are obtained as

$$\hat{c}_{0,0}^{(1)} = \text{Arg}\left\{\left\langle \Gamma_{mn}^{(2)}(x, y)\exp\left[-i\left(\hat{c}_{1,0}^{(1)}x + \hat{c}_{0,1}^{(1)}y\right)\right]\right\rangle\right\}$$

$$\hat{A}_1 = \left|\left\langle \left\{\Gamma_{mn}^{(2)}(x, y)\exp\left[-i\left(\hat{c}_{1,0}^{(1)}x + \hat{c}_{0,1}^{(1)}y\right)\right]\right\} \right\rangle\right|,$$

   where $\langle\cdot\rangle$ represents the averaging operation.
7. Once the estimates of all the polynomial coefficients are obtained, the estimate of the signal component with amplitude A_1 is computed and then subtracted from the EPF signal as

$$\bar{\Gamma}_{mn}(x, y) = \Gamma_{mn}(x, y) - \hat{A}_1 \exp\left[i\hat{\phi}_1(x, y)\right],$$

   where $\hat{\phi}_1(x, y)$ is the estimate of $\phi_1(x, y)$ given as

$$\hat{\phi}_1(x, y) = \sum_{p=0}^{3} \sum_{q=0}^{p} \hat{c}_{p-q,q}^{(1)} x^{p-q} y^q.$$

8. The steps from 1 to 6 are repeated twice to obtain the phase estimates $\hat{\phi}_2(x, y)$ and $\hat{\phi}_3(x, y)$ using the EPF, $\bar{\Gamma}_{mn}(x, y)$, computed in step 7.
9. Note that two-dimensional phase estimates computed within each block are in unwrapped forms. Since the multiple phase estimation performed in any given block is independent of that performed in its adjacent blocks, there exists an inherent ambiguity in the estimation of the zero order coefficient due to the use of Arg operation. As a result, a constant phase offset is always present between the unwrapped phases computed in the adjacent blocks. Phase stitching operation described in [15] is performed to obtain the continuous unwrapped phase estimates.

That optical metrology and digital signal processing have proved a powerful combination in the extraction of multiple embedded phase components from a composite fringe pattern is illustrated by the two subsequent examples, one through simulation and the other through experiment. First, let us consider a two-dimensional EPF of size 512 × 512 containing three signal components simulated at SNR of 20 dB. The resulting composite fringe system is shown in figure 4.6(a). The Fourier spectrum of the EPF shown in figure 4.6(b) indicates the presence of three spectrally overlapping signal components. The EPF is divided

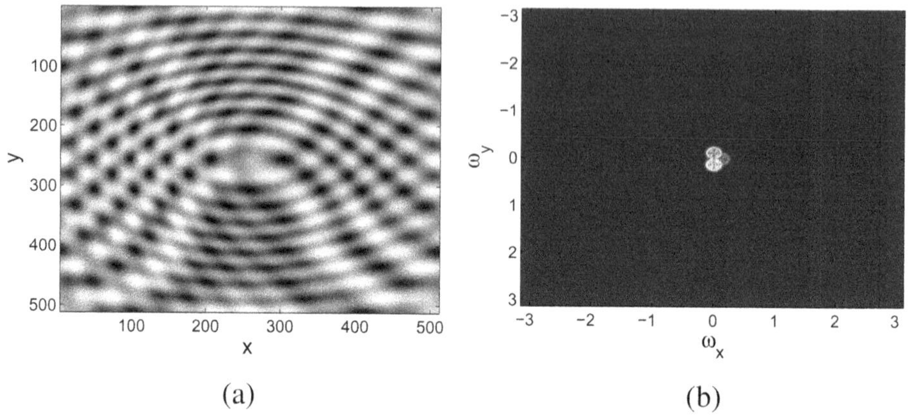

Figure 4.6. (a) Simulated composite fringe system and (b) Fourier spectrum of the EPF.

into 16 blocks, each of size 129 × 129, and a block-wise multiple phase estimation procedure is implemented to obtain the unwrapped estimates of ϕ_1, ϕ_2 and ϕ_3, as shown in figures 4.7(a), (c) and (e), respectively. For the sake of illustration, the wrapped forms of the estimated phases ϕ_1, ϕ_2 and ϕ_3 are shown in figures 4.7(b), (d) and (f), respectively,

Next, an equivalent experiment is performed using a three-beam digital holographic interferometry setup. A clamped plate is deformed using a point load applied to its center and is also rotated in its plane. The composite fringe system corresponding to the superposition of three fringe patterns is shown in figure 4.8(a). The multiple phase estimation method based on 2D-PHAF with a block size of 129 × 129 and $P = 3$ is used to demodulate the composite fringe system to obtain the unwrapped estimates of ϕ_1, ϕ_2 and ϕ_3. The estimated phases and the sensitivity vectors are plugged into (4.14). The maps of displacement components d_x, d_y and d_z (in meters) are shown in figures 4.8(b), (c) and (d), respectively.

4.4 Pixel-wise multiple phase estimation

In section 3.3.3, we have discussed the advantages of pixel-wise phase unwrapping over the segment-wise and block-wise phase unwrapping. Whereas the pixel-wise phase estimation is capable of handling rapid spatial phase variations, the segment-wise and block-wise approaches are limited to estimating the phases with low spatial variations. Likewise, this limitation is also a part of multiple phase estimation based on segment-wise and block-wise approaches discussed in sections 4.2 and 4.3. This section discusses a pixel-wise method of multiple phase estimation from an EPF containing two or more signal components, such as in the three-beam digital holographic interferometry setup shown in figure 4.5. However, for simplicity, we will restrict our analysis to an EPF containing two signal components.

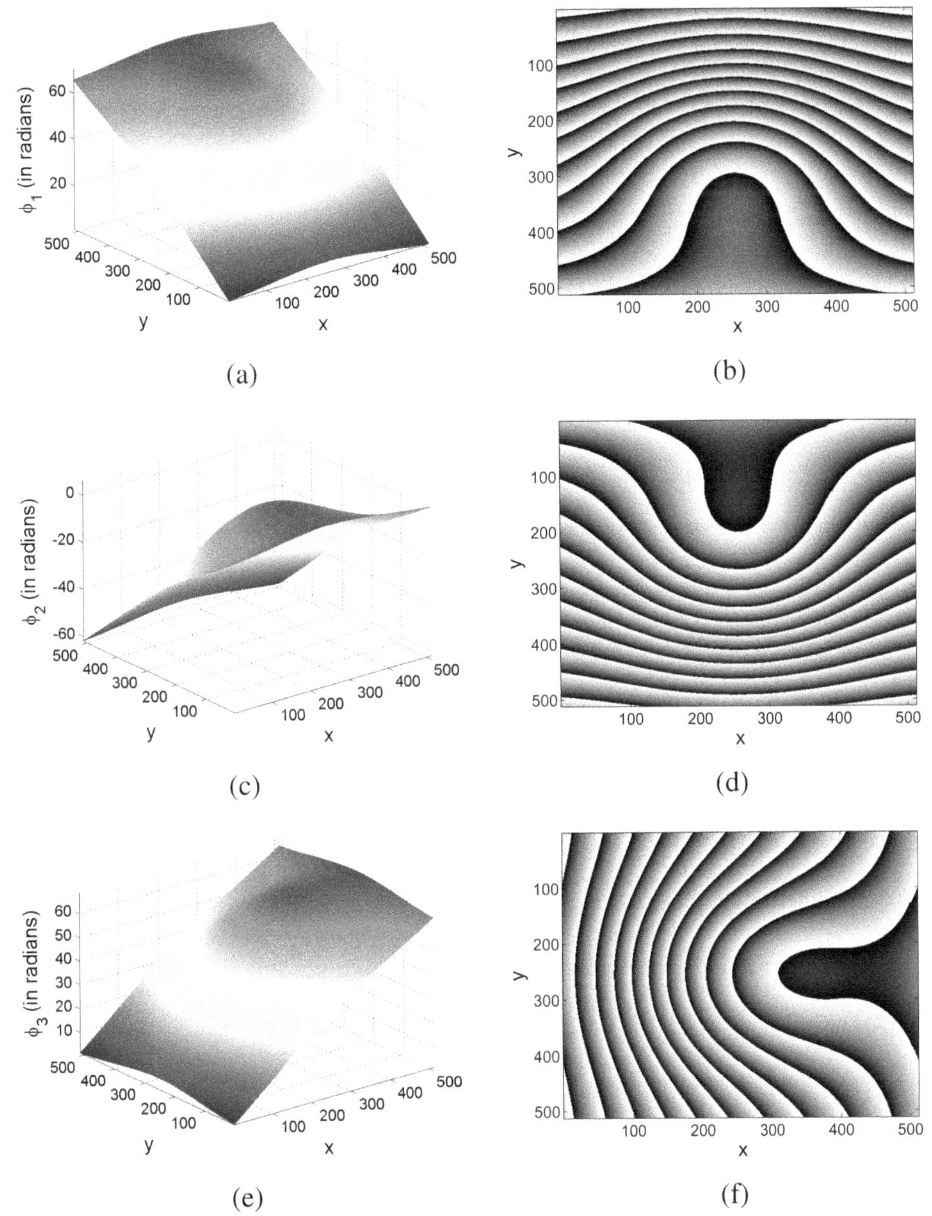

Figure 4.7. Unwrapped estimates of (a) ϕ_1, (c) ϕ_2, and (e) ϕ_3. Rewrapped estimated of (b) ϕ_1, (d) ϕ_2, and (f) ϕ_3.

In the pixel-wise approach of phase unwrapping, the absolute phase around each pixel is approximated as a first order polynomial function of x and y within a finite window. The phase around pixel (x, y) can thus be written as

$$\phi_l(x_l, y_l) = c_0 + c_1 x_l + c_2 y_l,$$

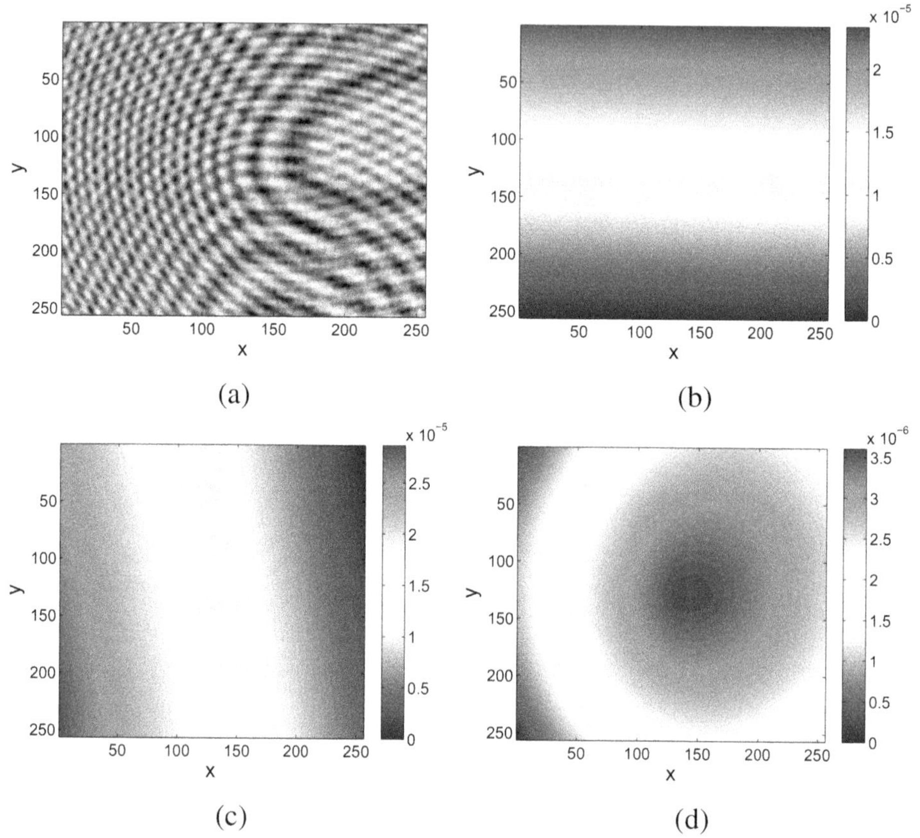

Figure 4.8. (a) Experimentally recorded composite fringe pattern in a three-beam digital holographic interferemetry setup. Display of (b) d_x, (c) d_y, and (d) d_z in meters.

where $|x_l| \leqslant L$ and $|y_l| \leqslant L$ represent the local spatial coordinates with respect to the current pixel (x, y); L represents the window size parameter. The localized first order polynomial phase representations of the two phases ϕ_1 and ϕ_2 are

$$\phi_{1L}(x_l, y_l) = a_0 + a_1 x_l + a_2 y_l, \tag{4.18}$$

$$\phi_{2L}(x_l, y_l) = b_0 + b_1 x_l + b_2 y_l. \tag{4.19}$$

Since a symmetric window is considered around each pixel, the phase estimates at the pixel (x, y) are obtained as

$$\begin{aligned} \phi_1(x, y) &= \phi_{1L}(0, 0) = a_0 \\ \phi_2(x, y) &= \phi_{2L}(0, 0) = b_0 \end{aligned}.$$

Based on phase approximations in (4.18) and (4.19), the EPF within the window can be represented as

$$\Gamma_L(x_l, y_l) = A_1 \exp(i\phi_{1L}(x_l, y_l)) + A_2 \exp(i\phi_{2L}(x_l, y_l)) + \varepsilon_L(x_l, y_l).$$

We assume the amplitudes of the two EPF signal components to be constant within the window. Following the development in section 3.3.3, the state space formulation for the multiple phase estimation is given as:

$$\boldsymbol{\mu}_l = \boldsymbol{F}\boldsymbol{\mu}_{l-1}, \tag{4.20}$$

$$\boldsymbol{\Gamma}_l = \boldsymbol{h}_l(\boldsymbol{\mu}_l) + \boldsymbol{\varepsilon}_l, \tag{4.21}$$

where the state vector is defined as

$$\boldsymbol{\mu}_l = [A_1, a_0, a_1, a_2, A_2, b_0, b_1, b_2]^T. \tag{4.22}$$

It is important to note the distinction in the definition of the state vector in the multiple phase estimation and the phase unwrapping (section 3.3.3) methods. In the case of multiple phase estimation, the state vector consists of the amplitudes of the signal components and the polynomial phase coefficients as given in (4.22) as its elements. The inclusion of the amplitudes of signal components as the elements of the state vector is essential for a reliable estimation and assignment of the polynomial coefficients to their respective signal components. According to the definition of the state vector given in (4.22), the state transition matrix $\boldsymbol{F}$ is an identity matrix of size 8 × 8. The measurement vector $\boldsymbol{\Gamma}_l$ is defined as

$$\boldsymbol{\Gamma}_l = \begin{bmatrix} \Re\{\Gamma_L(x_l, y_l)\} \\ \Im\{\Gamma_L(x_l, y_l)\} \end{bmatrix}. \tag{4.23}$$

The nonlinear function $\boldsymbol{h}_l$ of the state vector provides the estimate of the measurement vector as

$$\boldsymbol{h}_l(\boldsymbol{\mu}_l) = \begin{bmatrix} \boldsymbol{\mu}_l(1)\cos(\boldsymbol{\mu}_l^T \boldsymbol{p}_1) + \boldsymbol{\mu}_l(5)\cos(\boldsymbol{\mu}_l^T \boldsymbol{p}_2) \\ \boldsymbol{\mu}_l(1)\sin(\boldsymbol{\mu}_l^T \boldsymbol{p}_1) + \boldsymbol{\mu}_l(5)\sin(\boldsymbol{\mu}_l^T \boldsymbol{p}_2) \end{bmatrix}, \tag{4.24}$$

where $\boldsymbol{p}_1 = [0, 1, x_l, y_l, 0, 0, 0, 0]$ and $\boldsymbol{p}_2 = [0, 0, 0, 0, 0, 1, x_l, y_l]$. Since the estimate of the measurement vector is a nonlinear function of state vector elements, a nonlinear state estimation method, such as the extended Kalman filter outlined in section 3.3.3, is utilized for multiple phase estimation based on the state space formulation given in (4.20)–(4.24). Since the state estimate computed at a pixel is used as an initial estimate in the state estimation of its neighbouring pixels, the multiple phase estimates are obtained in an unwrapped form. Given that these unwrapped phase estimates are obtained in a pixel-wise manner, the post-processing operation of phase stitching required to remove the phase offsets in segment-wise and block-wise methods is not necessary in the pixel-wise method of multiple phase estimation.

Similar to the pixel-wise method of phase unwrapping, the pixel-wise method of multiple phase estimation is capable of operating with fringe patterns which may contain invalid fringe regions. This feature is not available explicitly either with the segment-wise or the block-wise methods of phase estimation. The pixel selection

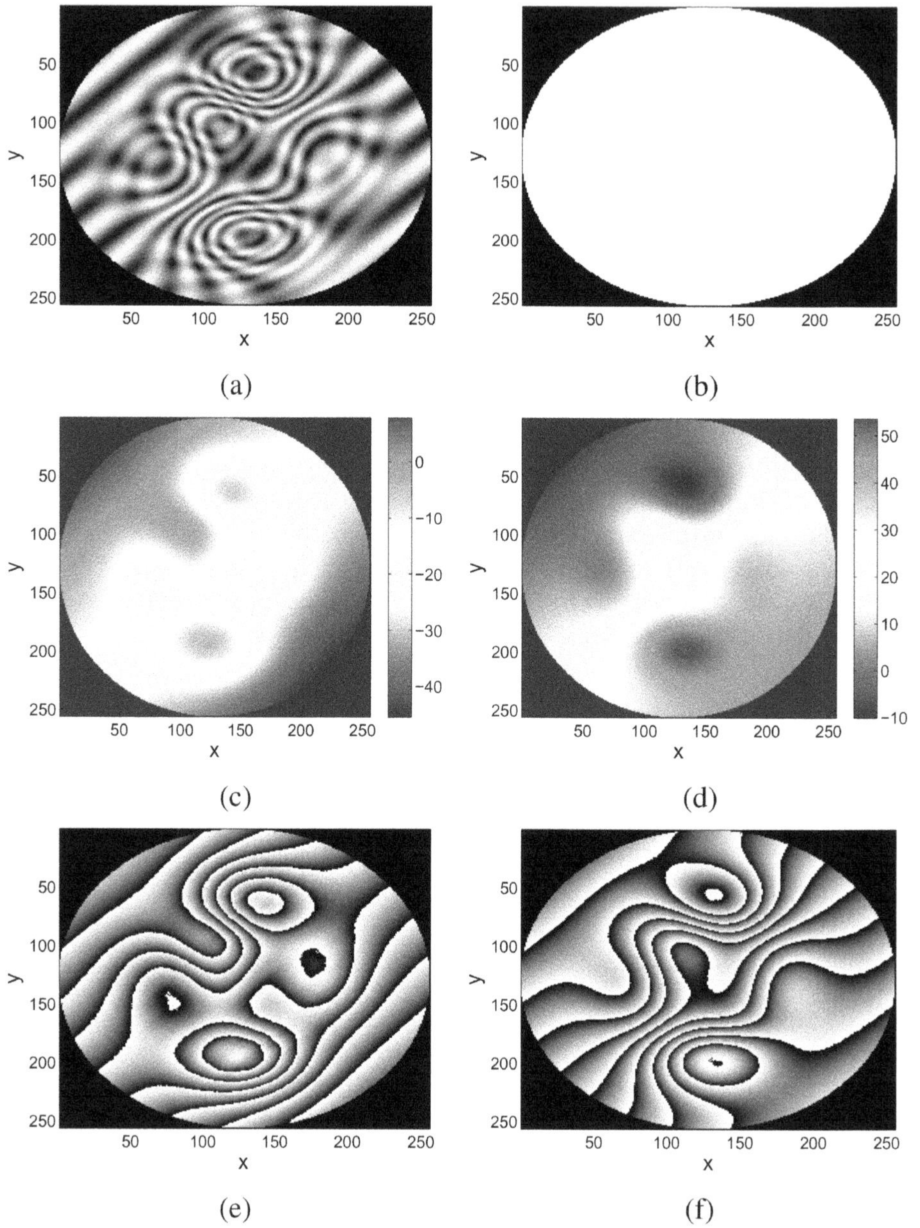

Figure 4.9. (a) Simulated moiré fringe pattern; (b) binary mask. Unwrapped estimates of (c) ϕ_1 and (d) ϕ_2 (in radians). Rewrapped estimates of (e) ϕ_1 and (f) ϕ_2.

strategy given in figure 3.16 is used to perform the appropriate initialization of the state vector in the state estimation procedure.

An example of multiple phase estimation with an EPF containing two signal components simulated at 20 dB is shown in figure 4.9. The corresponding moiré fringe pattern is shown in figure 4.9(a). The fringe pattern is simulated to include

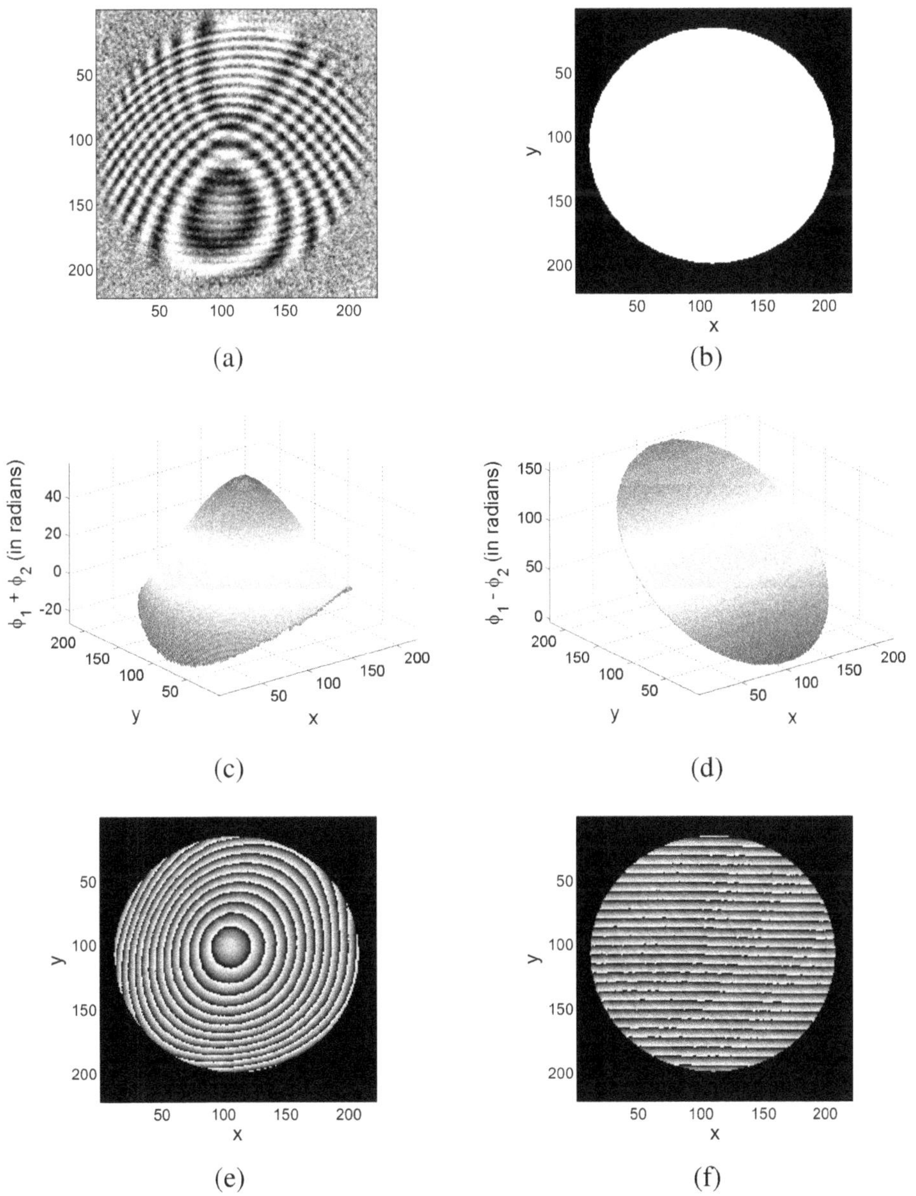

Figure 4.10. (a) Experimentally recorded moiré fringe pattern in a two-beam DHI setup; (b) binary mask separating the valid and invalid fringe regions; unwrapped estimates of (c) $\phi_1 + \phi_2$ and (d) $\phi_1 - \phi_2$ (in radians). Rewrapped estimates of (e) $\phi_1 + \phi_2$ and (f) $\phi_1 - \phi_2$.

certain fringeless areas where EPF data is not available. The corresponding binary mask is shown in figure 4.9(b) with white and black pixels indicating valid and invalid fringe areas, respectively. The phases ϕ_1 and ϕ_2, are simulated to show rapid spatial variations. Due to the rapid variations, it is difficult to approximate these phases with second or third order polynomials using the segment-wise or block-wise approaches. However, the pixel-wise method used in conjunction with the input binary mask is found to provide reliable estimation of ϕ_1 and ϕ_2. This is shown in figures 4.9(c) and (d), respectively, computed with $L = 4$. The corresponding wrapped forms of the phase estimates are shown in figures 4.9(e) and (f) for the sake of intuitive reasoning.

An interesting experimental validation of this method is shown in figure 4.10. The valid EPF data shown in figure 4.10(a) obtained in a two-beam digital holographic interferometry setup on a circular plate is observed to be available over a circular region in the rectangular frame of the EPF. A binary mask created manually to select the portion of the EPF data on the plate is shown in figure 4.10(b). The pixel-wise multiple phase estimation method is implemented with $L = 6$ to obtain the unwrapped estimates of ϕ_1 and ϕ_2. Since the out-of-plane, d_z, and the in-plane, d_x, displacement components are directly proportional to the sum and the difference of ϕ_1 and ϕ_2, respectively, the estimates of $\phi_1 + \phi_2$ and $\phi_1 - \phi_2$ are shown in figures 4.10(c) and (d), respectively. The wrapped forms of the estimates of $\phi_1 + \phi_2$ and $\phi_1 - \phi_2$ are shown in figures 4.10(e) and (f), respectively.

References

[1] Schnars U and Jüptner W P O 2002 Digital recording and numerical reconstruction of holograms *Meas. Sci. Tech.* **13** R85–101

[2] Picart P, Moisson E and Mounier D 2003 Twin-sensitivity measurement by spatial multiplexing of digitally recorded holograms *Appl. Opt.* **42** 1947–57

[3] Pedrini G and Tiziani H J 1997 Quantitative evaluation of two-dimensional dynamic deformations using digital holography *Opt. Laser Tech.* **29** 249–56

[4] Saucedo A T, de la Torre-Ibarra M H, Mendoza Santoyo F and Moreno I 2010 Digital holographic interferometer using simultaneously three lasers and a single monochrome sensor for 3D displacement measurements *Opt. Exp.* **18** 19867–75

[5] Picart P, Mounier D and Desse J M 2008 High-resolution digital two-color holographic metrology *Opt. Lett.* **33** 276–8

[6] Okazawa S, Fujigaki M, Morimoto Y and Matui T 2005 Simultaneous measurement of out-of-plane and in-plane displacements by phase-shifting digital holographic interferometry *Appl. Mech. Mat.* **3-4** 223–8

[7] Wang P and Yang J 2006 Multicomponent chirp signals analysis using product cubic phase function *Dig. Sig. Proc.* **16** 654–69

[8] Aboutanios E and Mulgrew B 2005 Iterative frequency estimation by interpolation on Fourier coefficients *IEEE Trans. Sig. Proc.* **53** 1237–42

[9] Barbarossa S 1995 Analysis of multicomponent LFM signals by a combined Wigner–Hough transform *IEEE Trans. Sig. Proc.* **43** 1511–5

[10] Cirillo L, Zoubir A and Amin M 2008 Parameter estimation for locally linear FM signals using a time-frequency Hough transform *IEEE Trans. Sig. Proc.* **56** 4162–75

[11] Kulkarni R and Rastogi P 2014 Simultaneous measurement of in-plane and out-of-plane displacements using pseudo-Wigner–Hough transform *Opt. Exp* **22** 8703–11

[12] Pham D S and Zoubir A M 2007 Analysis of multicomponent polynomial phase signals *IEEE Trans. Sig. Proc.* **55** 56–65

[13] Kulkarni R and Rastogi P 2015 Three-dimensional displacement measurement from phase signals embedded in a frame in digital holographic interferometry. *Appl. Opt.* **54** 3393–7

[14] Barbarossa S, di Lorenzo P, Vecchiarelli P, Silvi A and Bruner A 2013 Parameter estimation of 2D polynomial phase signals: an application to moving target imaging with SAR *Proc. IEEE International Conference on Acoustics, Speech and Signal Processing (Vancouver)* pp 4554–8

[15] Rajshekhar G and Rastogi P 2012 Fringe demodulation using the two-dimensional phase differencing operator *Opt. Lett.* **37** 4278–80

IOP Publishing

Single and Multicomponent Digital Optical Signal Analysis
Estimation of phase and its derivatives
Rishikesh Kulkarni and Pramod Rastogi

Chapter 5

Phase derivative estimation: spatial fringe analysis

Estimation of single or multiple phase derivatives is of utmost importance in many areas of science and engineering. We have organized this chapter to review approaches that provide the basis for obtaining phase derivatives from a fringe signal or EPF. The phase derivative estimation techniques are broadly classified into non-parametric and parametric methods. The non-parametric methods of phase derivative estimation are based on the computation of energy distribution of the EPF in function of both the time (or space) and frequency. While the non-parametric techniques are mainly designed for the estimation of first order phase derivatives in the form of instantaneous frequencies, the generalized non-parametric method discussed in this chapter is capable of providing the estimation of arbitrary order phase derivatives. On the other hand, of the two approaches of the parametric method of phase derivative estimation that we will consider, the first approach based on the spatially varying autoregressive modeling of the fringe signal or EPF allows us to estimate single or multiple phase derivatives with respect to rows or columns, while the second one based on the locally stationary EPF model along with the matrix enhancement and matrix pencil method for the high resolution frequency estimation provides us with the phase derivative estimates with respect to both the rows and columns, respectively.

5.1 Non-parametric phase derivative estimation: space frequency analysis

Nonparametric phase derivative estimation methods are unique in that they can analyze non-stationary sinusoidal signals in the absence of any specific assumption on the signal model. This section outlines the Wigner–Ville distribution and its improved variants based methods for the estimation of first order phase derivatives, and to complete the offer, a generalized phase derivative estimation method to obtain the estimation of arbitrary order phase derivatives.

doi:10.1088/978-0-7503-1469-5ch5

5.1.1 Wigner–Ville distribution

Wigner–Ville-distribution (WVD) has been applied to analyze non-stationary signals in a wide range of engineering applications such as speech processing, sonar, and engine diagnosis. Defined as a quadratic time (space)-frequency distribution of a one-dimensional signal, it is given by [1]

$$WVD_{\Gamma}(x,\,\omega) = \int_{-\infty}^{\infty} \Gamma\left(x + \frac{\tau}{2}\right)\Gamma^{*}\left(x - \frac{\tau}{2}\right)\exp(-i\omega\tau)d\tau. \tag{5.1}$$

It can be observed from (5.1) that the WVD utilizes auto-correlation of the signal to compute the power spectrum. In the context of fringe analysis, it is computed as the Fourier transform (FT) of the instantaneous auto-correlation function of the fringe signal. By applying the change of variable $\theta = \frac{\tau}{2}$ in (5.1), we obtain

$$WVD_{\Gamma}(x,\,\omega) = 2\int_{-\infty}^{\infty} \Gamma(x + \theta)\Gamma^{*}(x - \theta)\exp(-i2\omega\theta)d\theta.$$

Implemented digitally, the WVD of a one-dimensional fringe signal of length N gives,

$$WVD_{\Gamma}(x,\,k) = 2\sum_{|u|<N/2} \Gamma(x + u)\Gamma^{*}(x - u)\exp(-i2\omega_k u), \tag{5.2}$$

where ω_k represents the discrete frequency $\frac{\pi k}{N}$ with k being an integer. Fast Fourier transform can be used in the digital implementation of the WVD. Let us consider the one-dimensional EPF in a given row y, $\Gamma(x) = \exp(i\phi(x))$. The local auto-correlation of the EPF is given by

$$\begin{aligned}\Gamma(x + u)\Gamma^{*}(x - u) &= \exp(i\phi(x + u))\exp(-i\phi(x - u)) \\ &= \exp(i(\phi(x + u) - \phi(x - u))).\end{aligned} \tag{5.3}$$

For the simplicity of analysis, we assume the EPF to be noiseless. Considering the Taylor series expansion of the phase up to second order, we obtain

$$\phi(x + u) \approx \phi(x) + u\frac{\partial\phi(x)}{\partial x} + \frac{u^2}{2}\frac{\partial^2\phi(x)}{\partial x^2} \tag{5.4}$$

$$\phi(x - u) \approx \phi(x) - u\frac{\partial\phi(x)}{\partial x} + \frac{u^2}{2}\frac{\partial^2\phi(x)}{\partial x^2}. \tag{5.5}$$

Substituting (5.3)–(5.5) into (5.2), we get

$$WVD_{\Gamma}(x,\,k) = \sum_{|u|<N/2} \exp\left(i\dot{\phi}_x(x)u\right)\exp(-i\omega_k u), \tag{5.6}$$

where $\dot{\phi}_x(x) = \frac{\partial\phi(x)}{\partial x}$; and the multiplicative constant 2 has been neglected without any loss of generality. The WVD distribution of the EPF given in (5.6) encodes the information on the phase derivative at each pixel location. Thus, a single component EPF, has a single fringe frequency ω_k at each pixel x. This frequency which corresponds to the peak observed in the WVD spectrum of the EPF provides the estimate of the phase derivative as

$$\dot{\phi}_x(x) = \arg\max_{\omega_k} WVD_\Gamma(x, k). \tag{5.7}$$

The WVD based phase derivative estimation is demonstrated using a simulation example of an EPF signal with a quadratic phase $\phi(x) = a_1x + a_2x^2$, where a_1 and a_2 are constants. Since, as per (5.7), the peaks in the WVD spectrum are observed at $\omega_k = a_1 + 2a_2x$, we can state that the peak in the spectrum traces a linear pattern with respect to x. Given that in the case of the quadratic phase, the approximations in (5.4) and (5.5) are exact, the WVD of such an EPF provides the most optimal estimates of its phase derivative. On the other hand, in the case of a highly nonlinear phase, the approximations in (5.4) and (5.5) are loosely met. As a result, interference terms associated with higher order phase derivatives are observed in the WVD of the EPF. The presence of the interference terms adversely affects the accuracy of the phase derivative estimation.

Figure 5.1 shows the plots of the WVD of a non-stationary signal with a quadratic phase (figure 5.1(a)) and a highly nonlinear phase (figure 5.1(b)). Whereas the WVD contains a sharp linear trace in the case of quadratic phase, it contains a highly nonlinear trace together with unnecessary interference terms in the case of nonlinear phase. In the case of nonlinear phase, the overlapping of the interference terms with the noiseless EPF contribution in the WVD makes it difficult to obtain the estimate of the phase derivative. Various quadratic time-frequency distribution techniques have been proposed over the years to minimize the effects of the interference terms. One of these techniques involving the application of windowing in the lag (τ) direction in the computation of the instantaneous autocorrelation of the EPF is described next.

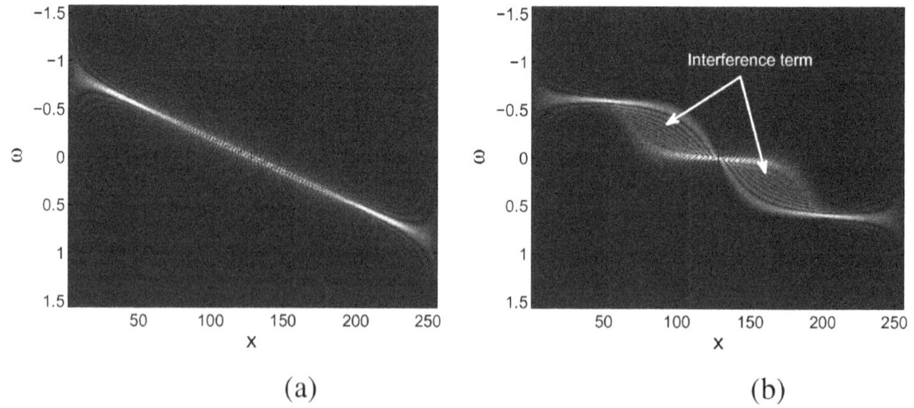

Figure 5.1. Wigner–Ville distribution computed for (a) quadratic phase, and (b) highly nonlinear phase.

Pseudo Wigner–Ville-distribution

Pseudo-Wigner–Ville-distribution (pWVD) is one of the important smoothing WVD distributions that have been developed to reduce the interference effect observed in the WVD. In the pWVD, the instantaneous autocorrelation of the signal is multiplied with a window function $w(l)$ in the lag direction. Thus, the pWVD of the fringe signal can be represented as

$$pWVD_{\Gamma}(x, k) = \sum_{|l|<L/2} w(l)\Gamma(x + l)\Gamma^{*}(x - l)\exp(-i\omega_{k}l), \tag{5.8}$$

where L is the window length. Substituting the phase approximations given in (5.4) and (5.5), we obtain

$$pWVD_{\Gamma}(x, k) = \sum_{|l|<L/2} w(l)\exp(i\dot{\phi}_{x}(x)l)\exp(-i\omega_{k}l), \tag{5.9}$$

Equation (5.9) is the FT of the window function $w(l)$ modulated by $\exp(i\dot{\phi}_{x}(x)l)$. Using the frequency shifting property of the FT, we obtain

$$pWVD_{\Gamma}(x, k) = \hat{w}(\omega_{k} - \dot{\phi}_{x}(x)),$$

where, $\hat{w}(\omega)$ indicates the FT of the window function. In the implementation of the pWVD, a window function possessing low pass frequency filtering characteristics is used, with which the maximum value of $\hat{w}(\omega)$ appears at $\omega = 0$. The phase derivative estimate is then given by

$$\dot{\phi}_{x}(x) = \arg\max_{\omega_{k}} pWVD_{\Gamma}(x, k).$$

For the sake of illustration, we compute the pWVD of the EPF with quadratic and highly nonlinear phases considered in the example given in figure 5.1. Whereas the plot of the pWVD shown in figure 5.2(a) is similar to the one shown in figure 5.1(b), figure 5.2(b) clearly shows the superiority of the pWVD over the WVD in the tracking of the nonlinear phase derivatives. The interference terms are effectively suppressed in the pWVD plot albeit at the cost of loss in the frequency resolution. However, the phase derivative estimation accuracy is not affected significantly due to this loss in frequency resolution. This is testified in figures 5.2(c) and (d).

Going one step further, it might be interesting to study the effect of windowing on the phase derivative estimation. Whereas a small window size is important for capturing the rapid variations in the phase derivatives, a large size window provides improved noise robustness in the phase derivative estimation though at the cost of lower sensitivity to the rapid variations in the phase derivatives. That is, the window size is inversely proportional to the rate of change of the phase derivative. To elucidate this point, phase derivative estimates computed using the Gaussian window function with different window lengths at SNR = 30 dB are shown in figure 5.3. Figure 5.3(a) shows that the small window size, $L = 4$, helps in effectively tracking rapidly varying phase derivatives. However welcome this result is, the use

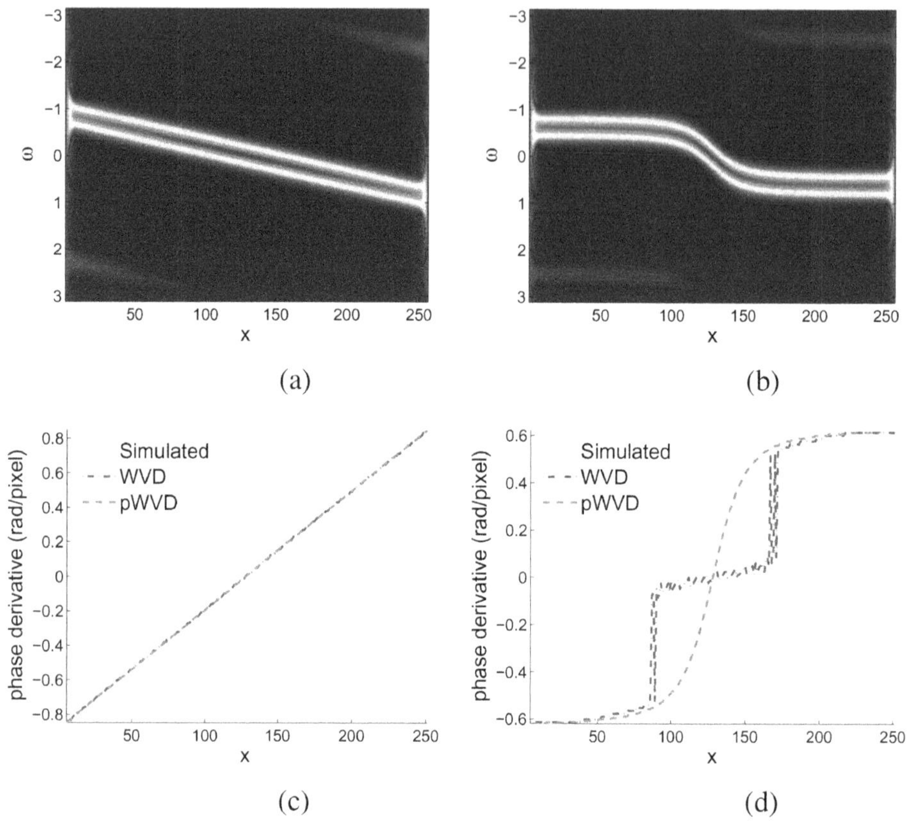

Figure 5.2. Space–frequency representation of the EPF computed with pWVD for (a) quadratic phase and (b) highly nonlinear phase. Estimated phase derivatives with WVD and pWVD for (c) quadratic phase and (d) highly nonlinear phase.

of a small window size is at the cost of ushering in higher sensitivity to noise in the estimated phase derivatives. The phase derivative estimate computed with a large size window of size $L = 128$ is shown in figure 5.3(b). It can be observed from this plot that the selection of a large size window makes it difficult to track rapid variations in the phase derivative. Finally, figure 5.3(c) shows the phase derivative estimate computed with $L = 32$. For this particular example, the selection of window size $L = 32$ provides more accurate phase derivative estimates than that obtained for $L = 4$ or $L = 128$. Depending on the rate of change of the phase derivative, it is clear that for any arbitrarily nonlinear phase derivative the appropriate window size cannot be selected *a priori*.

Next, the phase derivative estimates are computed with $L = 32$ with different window types such as Gaussian, rectangular, Hamming and Hanning. The estimated phase derivatives in figure 5.3(d) are a pointer to the fact that the use of these window types provides equivalent results in the phase derivative estimation. We have selected Gaussian window function for further analysis. An example of phase derivative estimation using a simulated phase fringe pattern is shown in figure 5.4(a).

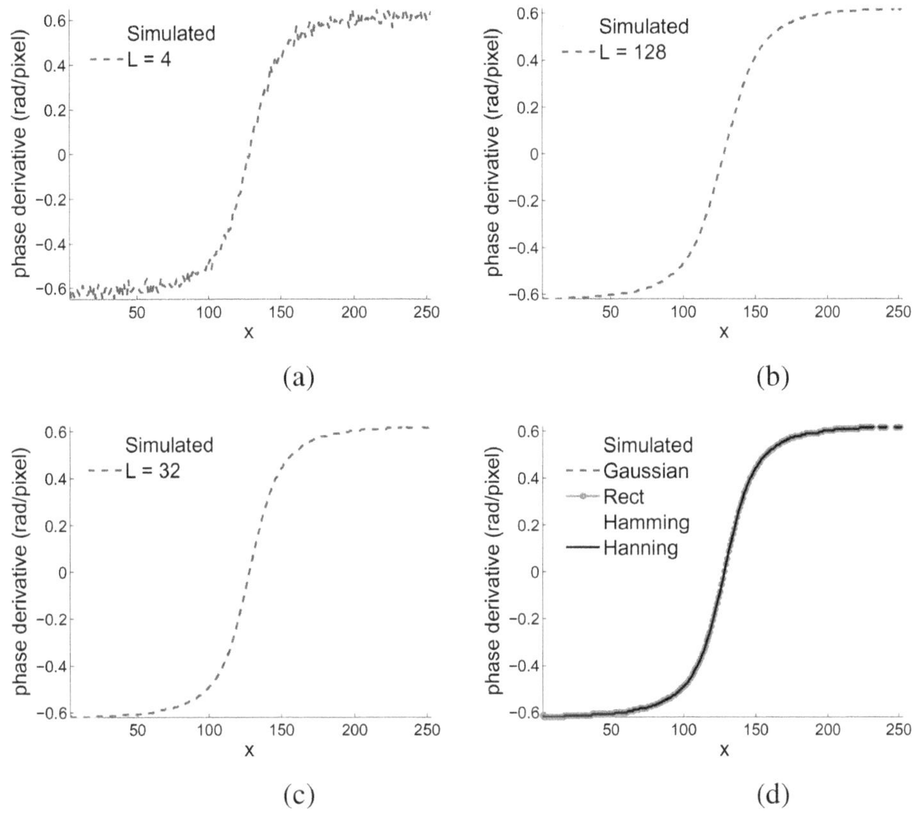

Figure 5.3. Phase derivative estimates computed with window lengths of (a) $L = 4$, (b) $L = 128$ and (c) $L = 32$. (d) Phase derivative estimates computed with L = 32 using different window types.

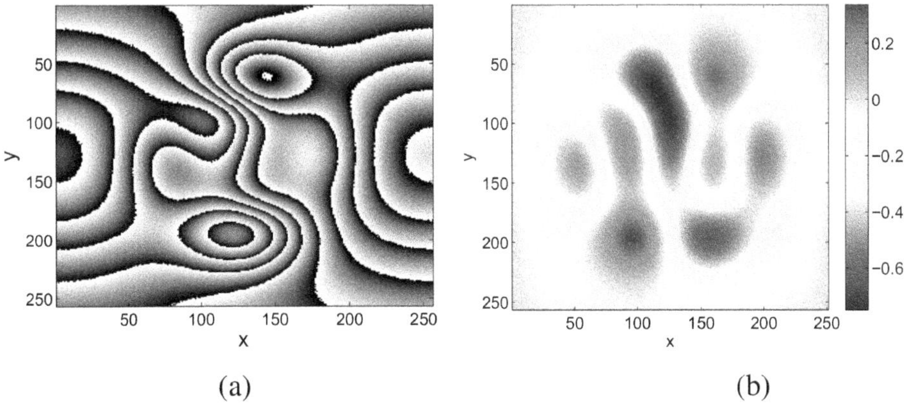

Figure 5.4. (a) Simulated phase fringe pattern; (b) phase derivative estimate, with respect to x, computed using pWVD with a Gaussian window of size $L = 32$.

The pWVD is implemented with $L = 32$ using the Gaussian window function. The estimated phase derivative with respect to x is shown in figure 5.4(b).

The result shown in figure 5.3 suggests that the accuracy of phase derivative estimation can be significantly improved by opting for an adaptive selection of the window size at each pixel location. Such an adaptive window pWVD technique has been proposed [2] wherein the optimal window size is computed at each pixel location from a pre-defined set of window sizes. It is important to note that the window size has the same effect in the case of both the pWVD and spectrogram (which is nothing but the one-dimensional windowed FT, that is, short-time FT) based phase derivative estimation methods [2, 3]. Consequently, the adaptive window size selection method is equally applicable to both the techniques.

Adaptive window Wigner–Ville distribution

While applying the pWVD method to phase derivative estimation, let us consider that the phase derivative estimate at x computed with a window of size L is represented as $\omega_k^{(L)}(x)$. The superscript L represents the dependence of the phase derivative estimate on the window size. Here we ask what is the optimal size of the window so that the error in the estimation, i.e., $\Delta\omega_k^{(L)}(x) = \dot{\phi}_x(x) - \omega_k^{(L)}(x)$ is minimized, knowing well that the error magnitude depends on the window size and the noise present in the EPF. Since the noise is assumed to be additive white Gaussian, we can treat the estimate $\omega_k^{(L)}(x)$ as a random variable and compute quantities, such as its bias (B), variance (σ_ω^2) and the mean square estimation error (MSE) associated with it. This gives

$$B = \mathbb{E}\left\{\dot{\phi}_x(x) - \omega_k^{(L)}(x)\right\},$$

$$\sigma_\omega^2 = \mathbb{E}\left\{\left[\dot{\phi}_x(x) - \mathbb{E}\left(\omega_k^{(L)}(x)\right)\right]^2\right\},$$

$$MSE = \mathbb{E}\left\{\left(\dot{\phi}_x(x) - \omega_k^{(L)}(x)\right)^2\right\}.$$

The expression for the bias and variance of the estimate obtained using the spectrogram and pWVD are given in [2, 3]. The bias is computed in function of the window size L as

$$B(L) = \sum_{s=1}^{\infty} L^{2s}(\omega_k(x))^{2s} b_s. \tag{5.10}$$

The variance in the phase derivative estimate obtained using spectrogram is given as

$$\sigma_\omega^2(L) = \frac{\sigma_\varepsilon^2 T E_1}{2\,|A|^2\, L^3 F_1^2}. \tag{5.11}$$

The variance in the phase derivative estimate obtained using the pWVD is given as

$$\sigma_\omega^2(L) = \frac{\sigma_\varepsilon^2}{2\,|A|^2}\left(1 + \frac{\sigma_\varepsilon^2}{2\,|A|^2}\right)\frac{TE_1}{L^3 F_1^2}. \tag{5.12}$$

The sampling interval T is equal to 1. The quantities b_s, E_1 and F_1 are defined as

$$E_1 = \int_{-1/2}^{1/2} w^2(x)x^2 dx,$$

$$F_1 = \int_{-1/2}^{1/2} w(x)x^2 dx,$$

$$b_s = \frac{-1}{(2s+1)!F_1} \int_{-1/2}^{1/2} w(x)x^{2s+2} dx.$$

The estimates of EPF amplitude A and noise variance σ_ε^2 can be computed as [2],

$$\hat{A}^2 = \left\{ \left| 2\left(\frac{1}{N} \sum_{x=0}^{N-1} |\Gamma(x)|^2 \right)^2 - \frac{1}{N} \sum_{x=1}^{N-1} |\Gamma(x)|^4 \right| \right\}^{1/2}$$

$$\hat{\sigma}_\varepsilon^2 = \left| \frac{1}{N} \sum_{x=1}^{N-1} |\Gamma(x)|^2 - \hat{A}^2 \right|.$$

Equations (5.10)–(5.12) suggest that increasing the window length gives rise to an increase in the bias and a decrease in the variance. On the other hand, decreasing the window length results in a decrease in the bias and an increase in the variance. This can be intuitively understood as follows: since selecting a smaller window length puts restrictions on the number of data samples with which to work, the phase derivative estimate has the tendancy to exhibit high variance due to its increased susceptiblity to noise. However, the use of small size window ensures that the phase within the window is approximated as a quadratic with a higher level of confidence. As a result, the phase derivative is estimated with a lower value of bias. On the other hand, a large size window provides better noise robustness in the phase derivative estimation because of the availability of a higher number of EPF samples. Thus, the phase derivative is estimated with a lower variance. However, since in this case the phase within the window is approximated with lower confidence, the use of a large window results in an increased bias in the phase derivative estimate. The optimal solution thus passes through a trade-off between the bias and variance. A scheme based on the intersection of confidence interval (ICI) has been proposed to this end [4, 5].

This method, treating the phase derivative estimate as a random variable, $\omega_k^{(L)}$, assigns for a given window length L an arbitrary probability density function $P_\kappa(\omega)$. A typical probability density function is given in figure 5.5(a). The area under the probability density function in the range $\omega_k^{(L)}(x) \in [\mathbb{E}(\omega_k^{(L)}(x)) - \kappa\sigma_\omega(L), \mathbb{E}(\omega_k^{(L)}(x))] + \kappa\sigma_\omega(L)]$ indicates the probability $P_\kappa = P(|\hat{\omega}_k^L(x) - \mathbb{E}(\omega_k^{(L)}(x))| \leqslant \kappa\sigma_\omega(L))$. The probability P_κ increases, that is, $P_\kappa \to 1$ with the increase in κ. The conditions associated with $\omega_k^{(L)}(x)$ are

$$\left| \omega_k^{(L)}(x) - E\left(\omega_k^{(L)}(x)\right) \right| \geqslant \left| \omega_k^{(L)}(x) - \dot{\phi}_x(x) \right| - |B(L)| \,.$$

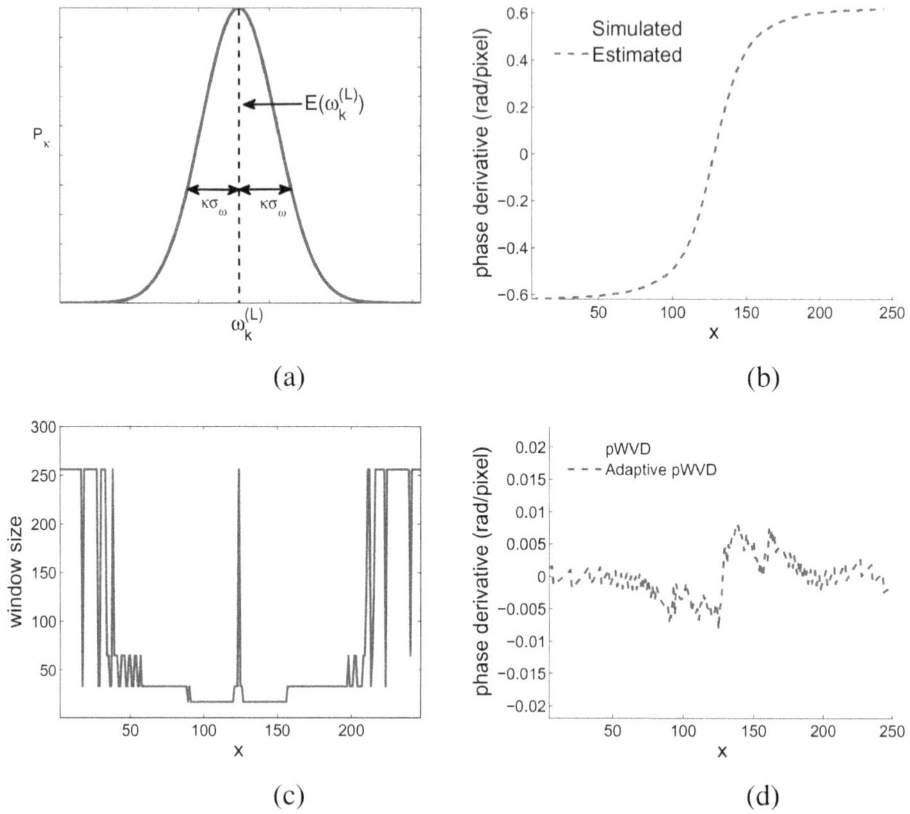

Figure 5.5. (a) Probability density function associated with the phase derivative estimate considered to be a random variable. (b) simulated and estimated phase derivative; (c) optimal window size computed as a function of x; (d) errors in the estimation of phase derivative computed using fixed window pWVD and adaptive window pWVD.

With the above definition of P_κ at a given parameter κ, we obtain,

$$\left|\omega_k^{(L)}(x) - \dot{\phi}_x(x)\right| - |B(L)| \leqslant \kappa\sigma_\omega(L). \tag{5.13}$$

In accordance with (5.10)–(5.12), the bias is small and the variance is dominant for a small window size L. That is,

$$|B(L)| \leqslant \kappa\sigma_\omega(L).$$

In consequence, from (5.13), we obtain

$$\left|\omega_k^{(L)}(x) - \dot{\phi}_x(x)\right| \leqslant 2\kappa\sigma_\omega(L). \tag{5.14}$$

Finally, we can express (5.14) as

$$\omega_k^{(L)}(x) - 2\kappa\sigma_\omega(L) \leqslant \dot{\phi}_x(x) \leqslant \omega_k^{(L)}(x) + 2\kappa\sigma_\omega(L). \tag{5.15}$$

Equation (5.15) indicates the lower and the upper bounds of the estimated phase derivative around the true phase derivative. Thus, for a given value of L_i, it is known that the true phase derivative lies in the confidence interval $D_i = [U_i, V_i]$ where, $U_i = \omega_k^{(L_i)}(x) - 2\kappa\sigma_\omega(L_i)$ and $V_i = \omega_k^{(L_i)}(x) + 2\kappa\sigma_\omega(L_i)$. The adaptive window size is selected from a set of pre-defined window sizes. The confidence intervals are computed for each selection of the window size. If the condition in (5.15) is satisfied for two consecutive window sizes, then the true phase derivative lies in both the corresponding confidence intervals. In this case, we have $D_{i-1} \cap D_i \neq \phi$ where ϕ represents an empty set. Based on this, the confidence intervals are computed with increasing window sizes and the condition in (5.15) is evaluated for the consecutive window sizes. The optimal window size is the maximum window size L_{opt} for which the condition $D_{i-1} \cap D_i \neq \phi$ is satisfied. This means that for window sizes greater than L_{opt}, two consecutive confidence intervals do not have the true phase derivative in common. In addition, for any window size greater than L_{opt}, the bias in the phase derivative estimate is always greater than that associated with L_{opt}. Thus, by selecting $L = L_{opt}$ at a given pixel location, we achieve the bias and variance trade-off in the phase derivative estimate. The ICI algorithm is briefly summarized as follows:

1. A set of windows is selected such that $L_1 < L_2 < \cdots < L_{\max}$.
2. For each window size, the phase derivative estimate is computed using either spectrogram or pWVD. The variance of the phase derivative is computed as in (5.12). The confidence interval is computed by selecting a fixed value of $\kappa = 1.75$.
3. Based on the rule of intersection of the confidence interval, the optimal window size L_{opt} is obtained. The phase derivative estimate computed with L_{opt} is selected as the optimal phase derivative estimate.
4. The above procedure is repeated at each pixel location to obtain the complete two-dimensional phase derivative estimate.

A one-dimensional example of phase derivative estimation obtained with the adaptive window pWVD is given in figure 5.5(b). Phase derivative estimation is performed with the adaptive window pWVD based method with a set of windows of sizes [2, 4, ..., 256]. The optimal window sizes L_{opt} computed as a function of x are shown in figure 5.5(c). It is readily observed that the ICI algorithm selects large window sizes at the pixels where the phase derivative is slowly varying and small window sizes where the phase derivative is rapidly varying. The errors in the phase derivative estimate are shown in figure 5.5(d) which clearly indicates the superiority of the adaptive window pWVD based phase derivative estimation.

A two-dimensional example of phase derivative estimation with respect to y, performed with the adaptive window pWVD based method on the phase fringe pattern given in figure 5.4(a), is shown in figure 5.6(a). The optimal window size computed in a column $x = 128$ in function of y is shown in figure 5.6(b).

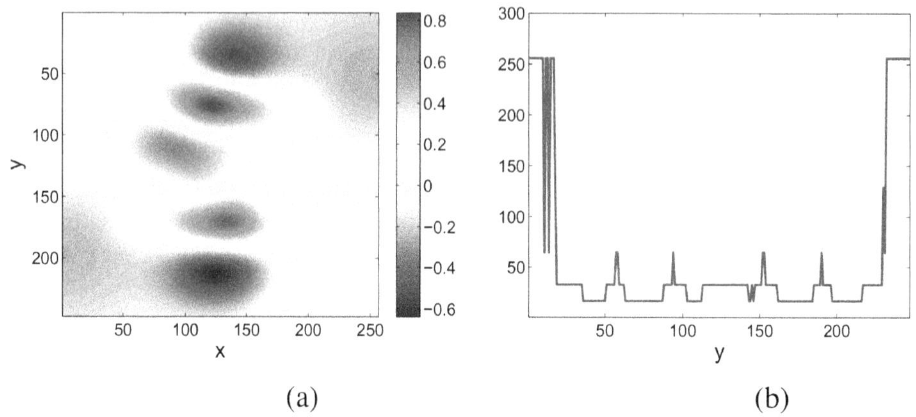

Figure 5.6. (a) Two-dimensional estimated phase derivative (of size 256 × 256) with respect to y; (b) optimal window size computed as a function of y at x = 128.

5.1.2 Generalized phase derivative estimation using complex lag argument

A limitation of the WVD based methods discussed in the previous subsection is that their use is restricted to the estimation of first order phase derivatives only. We believe that the estimation of high order derivatives, including the second, is more important than is generally supposed. Particularly in deformation analysis, for example, the second order phase derivatives provide the information on the curvature and twist developed on the object surface subjected to loading. The WVD based methods are not capable of explicitly estimating the second and higher order phase derivatives. Although, one may apply the WVD based phase derivative estimation in a repeated manner to estimate the higher order derivatives, the accuracy of the derivative estimates is adversely affected by the propagation of error due to the sequential operation. These complications have led to the desire to obtain second or higher order phase derivative estimates directly from the EPF. An approach based on the introduction of complex-lag argument in space-frequency distribution has been proposed for generalized phase derivative estimation. In this approach, the parameters in the newly defined quadratic energy distribution formulation of the EPF are tuned to obtain the estimate of phase derivative of a specified order. The generalized phase derivative of order p with respect to x can be represented as

$$\phi^{(p)}(x) = \frac{\partial^p \phi(x)}{\partial x^p}.$$

In this approach, the generalized complex-lag moment (GCM) of the EPF is defined as [6]

$$GCM_q^p(\Gamma)(x, l) = \prod_{k=0}^{q-1} \Gamma^{\omega_{q,k}^{q-p}}\left(x + \omega_{q,k}\sqrt[p]{\frac{p!}{q}}l\right),$$

where $\omega_{q,k} = \exp(i2\pi k/q)$. The local phase derivative variations are accurately captured by applying the windowing along the lag direction as follows

$$GCM_q^p(\Gamma)(x, l) = w(l)\prod_{k=0}^{q-1} \Gamma^{\omega_{q,k}^{q-p}}\left(x + \omega_{q,k}\sqrt[p]{\frac{p!}{q}l}\right). \tag{5.16}$$

Here, q is a parameter such that $q \geqslant p$ and the window w is a real symmetric window. Substituting $\Gamma(x) = \exp(i\phi(x))$ into (5.16), we obtain

$$GCM_q^p(\Gamma)(x, l) = w(l)\exp\left(i\sum_{k=0}^{q-1} \phi\left[x + \omega_{q,k}\sqrt[p]{\frac{p!}{q}l}\right]\omega_{q,k}^{q-p}\right). \tag{5.17}$$

Using the Taylor series expansion of the phase, we have

$$\phi\left[x + \omega_{q,k}\sqrt[p]{\frac{p!}{q}l}\right] = \sum_{r=0}^{\infty} \phi^{(r)}(x)\frac{\omega_{q,k}^{r}}{r!}\left(\sqrt[p]{\frac{p!}{q}l}\right)^{r}. \tag{5.18}$$

Substituting (5.18) into (5.17), we obtain

$$GCM_q^p(\Gamma)(x, l) = w(l)\exp\left(i\sum_{k=0}^{q-1}\sum_{r=0}^{\infty} \phi^{(r)}(x)\frac{\omega_{q,k}^{r+q-p}}{r!}\left(\sqrt[p]{\frac{p!}{q}l}\right)^{r}\right). \tag{5.19}$$

Noting the fact that $\omega_{q,k}^{r+q-p} = \exp(i2\pi k[r + q - p]/q)$, we have

$$\sum_{k=0}^{q-1} \omega_{q,k}^{r+q-p} = \begin{cases} q, \text{ if r = 0 and q = p} \\ q, \text{ if r = qs + p} \\ 0, \text{ otherwise} \end{cases},$$

where $s \in [0, 1, 2, \ldots]$. The above expression indicates that the argument of GCM in (5.19) becomes zero unless $r = qs + p$ or $r = 0$ and $q = p$. Considering the case when $q \neq p$, we have

$$\begin{aligned} GCM_q^p(\Gamma)(x, l) &= w(l)\exp\left(iq\sum_{r=0}^{\infty} \frac{\phi^{(qs+p)}(x)}{(qs + p)!}\left(\sqrt[p]{\frac{p!}{q}l}\right)^{qs+p}\right) \\ &= w(l)\exp\left(i\phi^{(p)}(x)l\right)\exp\left(iU(x, l)\right), \end{aligned} \tag{5.20}$$

where

$$U(x, l) = q\sum_{r=1}^{\infty} \frac{\phi^{(qs+p)}(x)}{(qs + p)!}\left(\sqrt[p]{\frac{p!}{q}l}\right)^{qs+p}.$$

Since the analysis is performed over the windowed segment of the EPF, it is reasonable to assume a slow spatial variation of phase. Consequently, the $(qs + p)$th order derivative of the phase, that is, $\phi^{(qs+p)}(x)$ can be considered to be negligible for $s \geqslant 1$. Therefore, $U(x, l)$ is set to 0 and (5.20) reduces to

$$GCM_q^p(\Gamma)(x, l) = w(l)\exp\left(i\phi^{(p)}(x)l\right). \tag{5.21}$$

Equation (5.21) suggests that the FT of GCM provides us with the estimate of $\phi^{(p)}(x)$. The FT of GCM is termed the generalized complex-lag distribution (GCD) of the EPF

$$GCD_q^p(\Gamma)(x, \Omega) = \sum_{|l|<L/2} w(l)\exp(i\phi^{(p)}(x))\exp(-i\Omega l). \tag{5.22}$$

The above expression is quite similar to the one obtained in the case of pWVD based phase derivative estimation given in (5.9). From (5.22), we obtain

$$GCD_q^p(\Gamma)(x, \Omega) = \hat{w}(\Omega - \phi^{(p)}(x)), \tag{5.23}$$

$$\phi^{(p)}(x) = \arg\max_{\Omega} \left| GCD_q^p(\Gamma)(x, \Omega) \right|. \tag{5.24}$$

It should be noted that (5.24) is obtained in the case of $q \neq p$. For $q = p$, an extra phase term associated with $r = 0$ appears in (5.19). Accordingly, (5.21) and (5.23) are modified as

$$GCM_q^p(\Gamma)(x, l) = w(l)\exp(i\phi^{(p)}(x)l)\exp(iq\phi(x))$$

$$GCD_q^p(\Gamma)(x, \Omega) = \exp(iq\phi(x))\hat{w}(\Omega - \phi^{(p)}(x)). \tag{5.25}$$

There is an additional scaling term $\exp(iq\phi(x))$ in the GCD of the EPF signal in the case, $q = p$. However, this term has no effect on the peak-detection scheme shown in (5.25) because it is independent of the lag variable l and $|\exp(iq\phi(x))| = 1$. The complex-lag argument based generalized phase derivative estimation method can be outlined as follows:

1. The GCM of the EPF signal for the specified phase derivative order p is computed at the pixel location x as in (5.16).
2. Compute GCD as the FT of the GCM of the EPF.
3. Compute the pth order phase derivative using the peak location observed in the GCD spectrum (5.24).
4. Repeat the above steps at each pixel and in each row to obtain a complete two-dimensional pth order phase derivative estimate with respect to x.

The parameter $p = 1$ is set to compute the first order phase derivative estimate. If we set $q = 2$, the GCD of the EPF signal is computed as,

$$GCD_2^1(\Gamma)(x, l) = w(l)\Gamma(x + l/2)\Gamma^{-1}(x - l/2)\exp(-i\Omega l). \tag{5.26}$$

Comparing (5.26) with (5.8), we can deduce that for $p = 1$ and $q = 2$, the complex-lag argument based generalized phase derivative estimation method is equivalent to the pWVD based method. The second order phase derivative estimate can be obtained by setting $p = 2$ and $q = 2$. In this case, the GCD of the EPF signal is computed as

$$GCD_2^2(\Gamma)(x, l) = w(l)\Gamma(x + \sqrt{l})\Gamma(x - \sqrt{l})\exp(-i\Omega l).$$

The complex-lag argument based method is implemented to compute the first order and second order phase derivative estimates with respect to y from the phase fringe

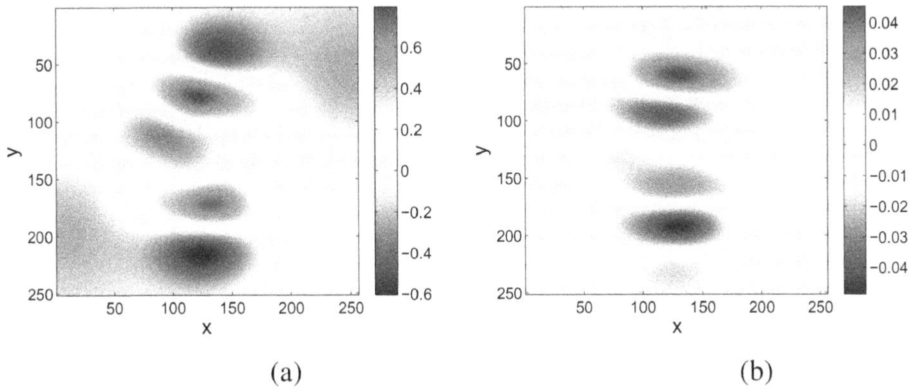

Figure 5.7. (a) First order and (b) second order phase derivatives, with respect to y, estimated using the complex-lag argument based generalized phase derivative estimation method.

pattern shown in 5.4(a). Figures 5.7(a) and (b) show the estimated first order and second order phase derivatives with respect to y.

5.2 Parametric phase derivative estimation: autoregressive modeling of fringe signal

Autoregressive (AR) modeling of the sinusoidal signal has found use in many applications, such as speech analysis, seismology, radar and sonar systems, and vibration monitoring. The realization that a fringe signal can be modeled as a non-stationary sinusoid has given rise to the development of a generalized method of phase derivative estimation based on the autoregressive modeling of the fringe signal [7–10]. This approach is generalized in the sense that it is applicable to the analysis of both the real valued carrier modulated fringe signal and the complex valued EPF containing single/multiple signal components.

Let us start with a simple example of AR modeling of a one-dimensional constant frequency sinusoidal signal, $\Gamma(x) = \cos(\omega x)$. This signal can be represented as a constant coefficient AR process

$$\Gamma(x) = a_1\Gamma(x-1) + a_2\Gamma(x-2), \tag{5.27}$$

where $a_1 = 2\cos\omega$ and $a_2 = -1$. The AR coefficients a_1 and a_2 carry the information on the frequency of the sinusoidal signal. If a sufficient number of signal samples are available, we can obtain the least squares estimates of the AR coefficients in accordance with (5.27) followed by the frequency of the sinusoidal signal. This simple example illustrates the capability of the AR representation of a sinusoidal signal in the estimation of its frequencies.

Depending on the setup in use, the fringe signal at the output can either be a real valued or complex valued sinusoidal signal (the EPF) containing single or multiple signal components. The fringe signal can be represented as a *spatially varying* autoregressive (SVAR) model in function of x (along a row) or y (along a column). The SVAR model of a fringe signal in a given row y can be written as

$$\Gamma(x) = \sum_{p=1}^{P} a_p(x)\Gamma(x - p) + \varepsilon(x), \tag{5.28}$$

where $a_p(x)$ are the SVAR coefficients (note the spatial dependence of the coefficients); P is the SVAR model order; $\varepsilon(x)$ represents the additive white Gaussian noise in the fringe signal. The transfer function representation of the SVAR model in (5.28) can be expressed as

$$H(z; x) = \frac{1}{1 - \sum_{p=1}^{P} a_p(x)z^{-p}}. \tag{5.29}$$

Equation (5.29) implies that the one-dimensional fringe signal can be considered as the output of an all-pole infinite-impulse response filter driven by the white noise $\varepsilon(x)$. The roots of the polynomial in the denominator of the transfer function formed by the spatially varying coefficients $a_p(x)$ represent the poles of the transfer function at each pixel x. In the case of a transfer function representation of the AR model of constant frequency sinusoidal signal in (5.27), there are two signal related complex conjugate poles at $\exp(i\omega)$ and $\exp(-i\omega)$ lying on the unit circle. An example of a pole plot for a constant frequency sinusoidal signal $\cos(\omega x)$ with $\omega = 0.4909$ *rad* is shown in figure 5.8. Similarly, in the context of the sinusoidal nature of the fringe signal, the signal related poles are seen to lie on the unit circle in the complex z plane. On account of the non-stationarity of the fringe signal, the angular positions of poles computed at each x correspond to the local spatial frequencies, which are nothing but the required phase derivatives of the respective signal components. In the

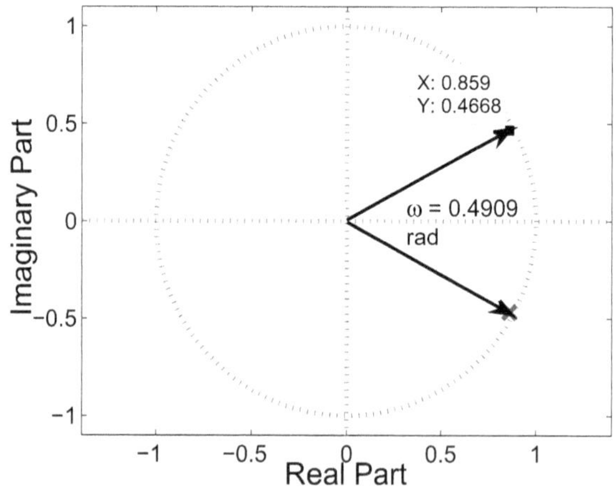

Figure 5.8. Pole plot for the constant frequency sinusoidal signal $\cos(\omega x)$ with $\omega = 0.4909$ *rad*.

noiseless condition, the SVAR model order P depends on the type of the fringe signal, that is, real/complex, and on the number of fringe signal components. It is evident from (5.29) that accurate estimation of the SVAR coefficients is essential for obtaining accurate estimates of the poles. Although in the case of constant AR coefficients, as in (5.27), a linear system of equations is required to be solved to obtain the least squares estimate of the AR coefficients, such a linear system of equations cannot be formulated with the SVAR model of the fringe signal due to the spatially varying characteristics of the AR coefficients.

SVAR coefficient estimation is carried out based on two distinct assumptions imposed on the spatial evolution of the coefficients. Whereas the first assumption is based on the deterministic evolution of the AR coefficients, the second assumption considers the stochastic evolution of the AR coefficients.

5.2.1 Deterministic evolution of SVAR coefficients

First, we consider the SVAR coefficient estimation wherein a deterministic evolution of the AR coefficients is assumed. Each of the SVAR coefficients is thus modeled as a weighted linear combination of the pre-defined linearly independent basis functions. That is

$$a_p(x) = \sum_{k=0}^{K-1} a_{pk}\beta_k(x), \tag{5.30}$$

where $\{\beta_k(x)\}$ is a set of linearly independent basis functions with the basis dimension K and $\beta_0(x) = 1$; $\{a_{pk}\}$ is a set of PK number of constant coefficients, that is, the weights of the basis functions. Essentially, such a representation of the SVAR coefficients has the advantage of converting the difficult problem of their estimation into the relatively simple problem of the estimation of constant coefficients (a_{pk}'s). Substituting the expression of $a_p(x)$ from (5.30) into (5.28) we obtain

$$\Gamma(x) = \sum_{p=1}^{P}\sum_{k=0}^{K-1} a_{pk}\beta_k(x)\Gamma(x-p) + \varepsilon(x). \tag{5.31}$$

Using the preceding P fringe signal samples, the estimate of $\Gamma(x)$ is obtained as

$$\hat{\Gamma}(x) = \sum_{p=1}^{P}\sum_{k=0}^{K-1} a_{pk}\beta_k(x)\Gamma(x-p). \tag{5.32}$$

The noise term $\varepsilon(x)$ is considered to contribute to the error in the estimation of $\Gamma(x)$. This error term can be expressed as

$$\varepsilon(x) = \Gamma(x) - \hat{\Gamma}(x). \tag{5.33}$$

The sum of the squares of the errors can be represented as

$$\|\varepsilon\|^2 = |\varepsilon(1)|^2 + |\varepsilon(2)|^2 + \ldots + |\varepsilon(N)|^2 \,.. \tag{5.34}$$

It is clear from (5.30)–(5.34) that the minimization of $||\varepsilon||^2$ is achieved at the optimum value of a set of constant coefficients $\{a_{pk}\}$. To compute these coefficients, we first define the dot product of one-dimensional vectors u and v, each of length N, as

$$(u, v) = \sum_{n=0}^{N-1} u(x)v^*(x).$$

The sum of the squares of the error in the estimation of $\Gamma(x)$ can now be expressed as

$$||\varepsilon||^2 = (\varepsilon, \varepsilon). \tag{5.35}$$

Substituting the expression of ε from (5.33) into (5.35)

$$||\varepsilon||^2 = ||\Gamma||^2 - 2(\Gamma, \hat{\Gamma}) + ||\hat{\Gamma}||^2 . \tag{5.36}$$

At this juncture, the following definitions are in order [11]:

$$w_{kp} = \{\beta_k(x)\Gamma(x - p)\}, \tag{5.37}$$

$$r_{kl}(p, q) = (w_{kp}, w_{lq}), \tag{5.38}$$

$$\bar{\boldsymbol{r}}_{0l} = [r_{0l}(0, 1), \ldots, r_{0l}(0, P)]^{\mathsf{T}}, \tag{5.39}$$

$$\boldsymbol{r}_0 = \left[\bar{\boldsymbol{r}}_{00}^{\mathsf{T}}, \ldots, \bar{\boldsymbol{r}}_{0(K-1)}^{\mathsf{T}}\right]^{\mathsf{T}}, \tag{5.40}$$

where k, $l = 0, \ldots, (K-1)$ and p, $q = 1, \ldots, P$. Let $\boldsymbol{R}_{kl}$ represent the $P \times P$ matrix with $r_{kl}(p, q)$ as its (pq)-th element. A Hermitian matrix $\boldsymbol{R}$ of size $PK \times PK$ is built with $\boldsymbol{R}_{kl}$ as its (kl)-th sub-block. The constant coefficient vector $\bar{\boldsymbol{a}}$ of size $PK \times 1$ is defined as $\bar{\boldsymbol{a}} = (a_{10}, a_{20}, \ldots, a_{P0}, \ldots, a_{1(K-1)}, \ldots, a_{P(K-1)})^{\mathsf{T}}$. Further, a column vector $\boldsymbol{r}_0$ of size $PK \times 1$ is defined with the ordered vectors $\bar{\boldsymbol{r}}_{0l}$ as its elements. Using the notations in (5.37)–(5.40) into (5.36) we obtain,

$$||\varepsilon||^2 = ||\Gamma||^2 - 2\bar{\boldsymbol{a}}^{\mathsf{T}}\boldsymbol{r}_0 + \bar{\boldsymbol{a}}^{\mathsf{T}}\boldsymbol{R}\bar{\boldsymbol{a}}. \tag{5.41}$$

The optimal value of $\bar{\boldsymbol{a}}$ for which the minimum value of (5.41) is achieved can be computed by differentiating (5.41) with respect to $\bar{\boldsymbol{a}}$ and assigning it to zero. Accordingly, we obtain

$$\boldsymbol{R}\bar{\boldsymbol{a}} = \boldsymbol{r}_0.$$

The vector $\bar{\boldsymbol{a}}$ is obtained as, $\bar{\boldsymbol{a}} = \boldsymbol{R}^{-1}\boldsymbol{r}_0$. The spatially varying coefficients $a_p(x)$s are computed by substituting the estimated constant coefficients, the elements of vector $\bar{\boldsymbol{a}}$, into (5.30).

The SVAR model given in (5.28) is a forward linear prediction model which utilizes the preceding samples of the fringe signal at $x - 1, x - 2, \ldots, x - P$ to compute fringe signal at x. Similarly, the fringe signal at x can also be computed using the backward estimation approach wherein the fringe signal samples at $x + 1$, $x + 2$, ..., etc are utilized. The modified basis function method [12] utilizes both the forward and backward estimation of the fringe signal in order to obtain enhanced accuracy in the estimation of the SVAR coefficients. The model of forward linear estimation of the fringe signal given in (5.28) can thus also be written as

$$\hat{\Gamma}^f(x+1) = \sum_{p=1}^{P} a_p(x)\Gamma(x-p+1),$$

for $x = P, P+1, \ldots, N-1$; and, the error in the forward linear estimation of the fringe signal is

$$\begin{aligned} e^f(x+1) &= \Gamma(x+1) - \hat{\Gamma}^f(x+1) \\ &= \Gamma(x+1) - \sum_{p=1}^{P}\sum_{k=0}^{K-1} a_{pk}\beta_k(x)\Gamma(x-p+1). \end{aligned}$$

On the other hand, the model of backward linear estimation of the fringe signal can be represented as

$$\begin{aligned} \hat{\Gamma}^b(x-1) &= \sum_{p=1}^{P} b_p(x)\Gamma(x+p-1) \\ &= \sum_{p=1}^{P} a_p^*(x)\Gamma(x+p-1), \end{aligned}$$

for $x = 2, \ldots, N-P+1$. In the backward linear estimation of the fringe signal, the spatially varying coefficients $b_p(x)$ are constrained such that $b_p(x) \cong a_p^*(x)$ [12]. The error in the backward linear estimation of the fringe signal is

$$\begin{aligned} e^b(x-1) &= \Gamma(x-1) - \hat{\Gamma}^b(x-1) \\ &= \Gamma(x-1) - \sum_{p=1}^{P}\left(\sum_{k=0}^{K-1} a_{pk}\beta_k(x)\right)^* \Gamma(x+p-1). \end{aligned}$$

The mean square error ε in the forward–backward linear estimation of the fringe signal is defined as

$$\varepsilon = \frac{1}{2(N-P)}\left\{\sum_{x=P}^{N-1} |e^f(x+1)|^2 + \sum_{x=2}^{N-P+1} |e^b(x-1)|^2\right\}.$$

The set of constant coefficients $\{a_{pk}\}$ is now computed by solving the set of linear equations such that ε is minimized. Hence, $r_{kl}(p, q)$ is obtained as

$$\begin{aligned} r_{kl}(p, q) = &\sum_{x=P}^{N-1} \beta_k(x)\beta_l^*(x)\Gamma(x-p+1)\Gamma^*(x-q+1) \\ &+ \sum_{x=2}^{N-P+1} \beta_k(x)\beta_l^*(x)\Gamma^*(x+p-1)\Gamma(x+q-1). \end{aligned}$$

Once the SVAR coefficients are estimated, the poles and subsequently the phase derivatives of signal components are computed by plugging them into (5.29). It is important to note that the estimation of SVAR coefficients depends on the selection of the basis functions and the basis dimension K. Different types of basis functions are reported in the literature. A few of the popularly used ones are given below:

- Polynomial basis

$$\beta_k(x) = \left(\frac{x}{N}\right)^k \tag{5.42}$$

- Fourier basis

$$\beta_k(x) = \begin{cases} \cos\dfrac{k\pi x}{2N}, & k \text{ even} \\ \sin\dfrac{(k+1)\pi x}{2N}, & k \text{ odd} \end{cases} \tag{5.43}$$

- Legendre basis

$$\begin{aligned} \beta_0(x) &= 1 \\ \beta_1(x) &= \frac{x}{N} \\ \beta_{k+1}(x) &= \frac{1}{(k+1)}\left((2k+1)\left(\frac{x}{N}\right)\beta_k(x) - k\beta_{k-1}(x)\right) \end{aligned} \tag{5.44}$$

- Chebyshev basis

$$\begin{aligned} \beta_0(x) &= 1 \\ \beta_1(x) &= \frac{x}{N} \\ \beta_{k+1}(x) &= 2\left(\frac{x}{N}\right)\beta_k(x) - \beta_{k-1}(x) \end{aligned} \tag{5.45}$$

The appropriate selection of SVAR model order P and basis dimension K depends on the available signal length N. As a rule of thumb, the values of K and P should be selected such that $KP \ll N$ [12].

5.2.2 Stochastic evolution of SVAR coefficients

The implicit assumption in the SVAR coefficient estimation based on the deterministic evolution model is that the phases of the fringe signal are spatially continuous. This assumption limits us from extending the estimation of phase derivatives based on the deterministic evolution of SVAR coefficients to, for example, more general material testing situations where object related discontinuities such as the presence of cracks or holes could be expected to be an integral part of the fringe signal. A generic model of the spatial evolution of the SVAR coefficients has been proposed [13] to overcome this restriction.

Considering a stochastic evolution of the SVAR coefficients, the κth order difference equation representation for the pth AR coefficient is given as

$$\nabla^{\kappa} a_p(x) = \xi_p(x),$$

where $\xi_p(x)$ is the coefficient perturbing noise and the difference operator ∇ is defined as

$$\nabla a_p(x) = a_p(x) - a_p(x-1).$$

Based on the difference equation representation for the SVAR coefficient, the SVAR process can be formulated as a state space model. The state vector containing the SVAR coefficients as its elements is defined as

$$\boldsymbol{\mu}(x) = (a_1(x), a_2(x), \ldots, a_P(x), \ldots, a_1(x-\kappa+1), \ldots, a_P(x-\kappa+1))^T.$$

The state transition equation can be written as

$$\boldsymbol{\mu}(x) = \boldsymbol{F}\boldsymbol{\mu}(x-1) + \boldsymbol{G}\boldsymbol{\xi}(x),$$

where

$$\boldsymbol{F} = \boldsymbol{I}_P, \qquad \boldsymbol{G} = \boldsymbol{I}_P, \qquad \text{for } \kappa = 1$$

$$\boldsymbol{F} = \begin{bmatrix} 2\boldsymbol{I}_P & -\boldsymbol{I}_P \\ \boldsymbol{I}_P & 0 \end{bmatrix}, \quad \boldsymbol{G} = \begin{bmatrix} \boldsymbol{I}_P \\ 0 \end{bmatrix}, \quad \text{for } \kappa = 2$$

with,

$$\boldsymbol{\xi}(x) = (\xi_1(x), \xi_2(x), \ldots, \xi_P(x))^T.$$

Here, $\boldsymbol{I}_P$ is defined as a $P \times P$ identity matrix. The fringe signal at x is represented as

$$\Gamma(x) = \boldsymbol{H}(x)\boldsymbol{\mu}(x) + \varepsilon(x),$$

where $\boldsymbol{H}(x)$ is a $(1 \times \kappa P)$ vector defined as

$$\boldsymbol{H}(x) = [\Gamma(x-1), \Gamma(x-2), \ldots, \Gamma(x-P), 0, \ldots, 0].$$

It is assumed that the coefficient perturbing noise $\boldsymbol{\xi}(x)$ is a zero mean additive white Gaussian noise with a known covariance matrix Ξ. The fringe signal noise $\varepsilon(x)$ is assumed to be of a known variance R_Γ. Note that the values of Ξ and R_Γ depend on the spatial rate of change of the phase derivative and on the signal-to-noise ratio of the fringe signal, respectively. As these parameters are not known *a priori*, the values of these parameters are usually specified by the user.

The state space formulation of the evolution of SVAR coefficients allows us to compute their estimates utilizing the linear Kalman filter algorithm. Since a fixed length (N or M) measurement of the fringe signal is available in a given row or column, the fixed length Kalman smoothing algorithm consisting of forward and backward operations on the fringe signal, such as the computationally efficient Rauch–Tung–Striebel (RTS) smoother algorithm can be implemented [14]. The steps in the RTS algorithm are given as follows:

1. The forward state vector $\boldsymbol{\mu}_f$ and its associated forward state estimation error covariance matrix $\boldsymbol{Q}_f$ are initialized at $x = P$ as

$$\hat{\boldsymbol{\mu}}_f^+(P) = \mathbb{E}(\boldsymbol{\mu}(P)),$$
$$\boldsymbol{Q}_f^+(P) = \mathbb{E}\left[\left(\boldsymbol{\mu}(P) - \hat{\boldsymbol{\mu}}_f^+(P)\right)\left(\boldsymbol{\mu}(P) - \hat{\boldsymbol{\mu}}_f^+(P)\right)^T\right].$$

2. The standard forward Kalman filter is implemented for $x = P + 1, \ldots, N$ as

$$\begin{aligned}
\hat{\boldsymbol{\mu}}_f^-(x) &= \boldsymbol{F}\hat{\boldsymbol{\mu}}_f^+(x-1),\\
\boldsymbol{Q}_f^-(x) &= \boldsymbol{F}\boldsymbol{Q}_f^+(x-1)\boldsymbol{F}^T + \boldsymbol{G}\boldsymbol{\Xi}\boldsymbol{G}^T,\\
\boldsymbol{K}_f(x) &= \boldsymbol{Q}_f^-(x)\boldsymbol{H}(x)^T\left(\boldsymbol{H}(x)\boldsymbol{Q}_f^-(x)\boldsymbol{H}(x)^T + R_\Gamma\right)^{-1},\\
\hat{\boldsymbol{\mu}}_f^+(x) &= \hat{\boldsymbol{\mu}}_f^-(x) + \boldsymbol{K}_f(x)\left(\Gamma(x) - \boldsymbol{H}(x)\hat{\boldsymbol{\mu}}_f^-(x)\right),\\
\boldsymbol{Q}_f^+(x) &= \left(\boldsymbol{I} - \boldsymbol{K}_f(x)\boldsymbol{H}(x)\right)\boldsymbol{Q}_f^-(x)\left(\boldsymbol{I} - \boldsymbol{K}_f(x)\boldsymbol{H}(x)\right)^T\\
&\quad + \boldsymbol{K}_f(x)R_\Gamma\boldsymbol{K}_f^T(x).
\end{aligned}$$

where $\boldsymbol{K}_f$ represents the forward Kalman gain; $\boldsymbol{I}$ is the $(\kappa P \times \kappa P)$ sized identity matrix. The superscripts $-$ and $+$ represent the *a priori* and *a posteriori* estimates of their associated variables.

3. In the backward operation on the fringe signal, the RTS smoother algorithm is initialized at $x = N$ as

$$\hat{\boldsymbol{\mu}}(N) = \hat{\boldsymbol{\mu}}_f^+(N),$$
$$\hat{\boldsymbol{Q}}(N) = \boldsymbol{Q}_f^+(N).$$

4. The backward state estimation is performed at $x = N - 1, \ldots, P$ as follows:

$$\begin{aligned}
\boldsymbol{\Pi}_f^-(x+1) &= \left(\boldsymbol{Q}_f^-(x+1)\right)^{-1},\\
\boldsymbol{K}(x) &= \boldsymbol{Q}_f^+(x)\boldsymbol{F}^T\boldsymbol{\Pi}_f^-(x+1),\\
\hat{\boldsymbol{Q}}(x) &= \boldsymbol{Q}_f^+(x) - \boldsymbol{K}(x)\left(\boldsymbol{Q}_f^-(x+1) - \hat{\boldsymbol{Q}}(x+1)\right)\boldsymbol{K}^T(x),\\
\hat{\boldsymbol{\mu}}(x) &= \hat{\boldsymbol{\mu}}_f^+(x) + \boldsymbol{K}(x)\left(\hat{\boldsymbol{\mu}}(x+1) - \hat{\boldsymbol{\mu}}_f^-(x+1)\right).
\end{aligned} \tag{5.46}$$

Finally, the phase derivative estimates are obtained by substituting the SVAR coefficients obtained from the estimated state vector $\hat{\boldsymbol{\mu}}(x)$ at each x into (5.29).

In sections 5.2.1 and 5.2.2, we have discussed two methods of estimating the SVAR coefficients and the phase derivatives along the rows or the columns of a general fringe signal. We will now look into the possibility of applying these methods to the estimation of the phase derivatives of fringe signals of specific kinds.

5.2.3 Case 1: single phase EPF

First, we deal with the processing of a single phase EPF. The single phase EPF in a given row y can be written as

$$\Gamma(x) = \exp(i\phi(x)). \tag{5.47}$$

In its digital implementation, the phase derivative $\dot{\phi}(x)$ is expressed as

$$\dot{\phi}(x) = \phi(x) - \phi(x - 1). \tag{5.48}$$

Substituting the expression of $\phi(x)$ from (5.48) into (5.47), we obtain

$$\begin{aligned}\Gamma(x) &= \exp[i(\dot{\phi}(x) + \phi(x - 1))] \\ &= \exp(i\dot{\phi}(x))\exp(i\phi(x - 1)).\end{aligned} \tag{5.49}$$

Equation (5.49) implies that in the noiseless condition, the single phase EPF can be exactly represented as the SVAR model with $P = 1$ as [15]

$$\Gamma(x) = a_1(x)\Gamma(x - 1),$$

where $a_1(x) = \exp(i\dot{\phi}(x))$ represents the SVAR coefficient. Since the knowledge of $a_1(x)$ is central to obtaining the estimate of the phase derivative, the accurate estimation of $a_1(x)$ is essential. Considering the deterministic spatial evolution of $a_1(x)$, from (5.30), we have

$$a_1(x) = \sum_{k=0}^{K-1} a_{1k}\beta_k(x).$$

A single SVAR coefficient is associated with a single pole in the complex z-plane. The angular position of this pole at each x provides the estimate of the phase derivative at that pixel. A single phase EPF is generated with the simulated phase derivative shown in figure 5.9(a). The SVAR coefficient estimation is performed with the Fourier basis functions, the basis dimension $K = 15$ and the signal length, $N = 256$. The pole plot corresponding to the transfer function associated with the SVAR model in the complex z-plane is shown in figure 5.9(b). The poles lie on the unit circle with varying angular positions due to the spatial variations of the phase derivative. The SVAR coefficient estimation is performed in function of the basis functions and the basis dimension. The root mean square error in the linear estimation of the fringe signal, shown in figure 5.9(c), indicates that the error magnitude is necessarily dependent on the selection of the basis function and its dimension. For the example at hand, the optimal value of the basis dimension can be derived from figure 5.9(c). The estimated phase derivatives computed with each type of basis functions, for the

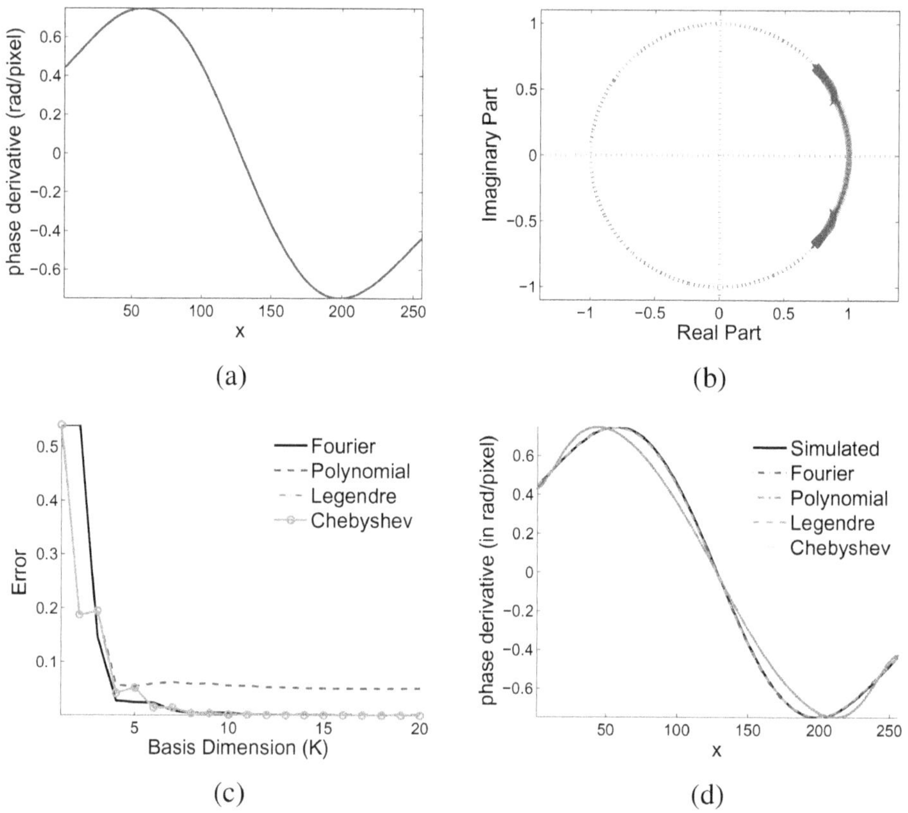

Figure 5.9. (a) One-dimensional simulated phase derivative; (b) pole plot in the complex z-plane computed with Fourier basis functions and $K = 15$; (c) errors in linear estimation of the fringe signal in function of basis functions and basis dimension; (d) estimated phase derivative in function of basis functions with basis dimension $K = 10$.

value of basis dimension, $K = 10$, are shown in figure 5.9(d). The Fourier and the Chebyshev basis functions are seen to provide accurate phase derivative estimates.

It must be noted that the selection of SVAR model order, $P = 1$, provides accurate estimation of the phase derivatives, even in the case of fringe signals with high SNRs. However, in the presence of a significant amount of noise, perturbations are observed in the positions of the poles in the complex z-plane and which may adversely affect the accuracy of the estimated phase derivative. Efforts have been on the way to improve the accuracy of phase derivative estimation at low SNRs [16]. This method selects an SVAR model, order $P > 1$, implying thereby that there would be P spatially varying poles of the transfer function associated with the SVAR model of the single phase EPF. Of these poles, one corresponds to the noiseless single phase EPF and the remaining, $P - 1$, poles are associated with the noise component. Whereas the pole associated with the single phase EPF lies along the unit circle, the poles associated with the noise are randomly distributed inside the unit circle. This characteristic is exploited to identify the pole associated with the

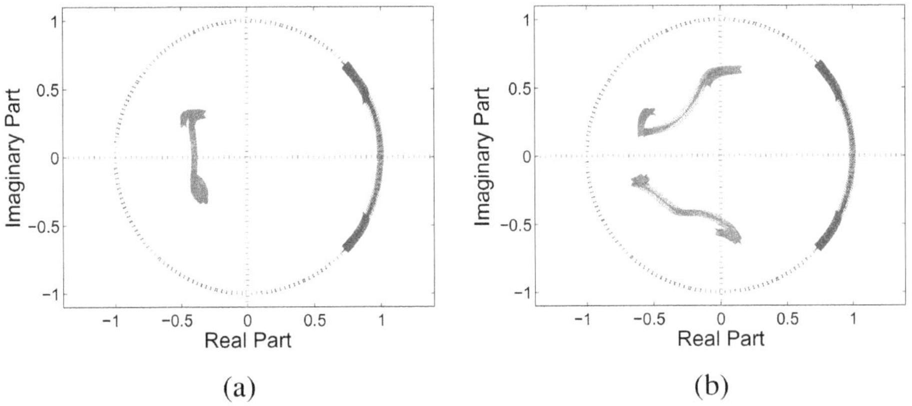

Figure 5.10. Pole plot computed for a noisy EPF using Fourier basis functions and SVAR model order (a) $P = 2$ and (b) $P = 3$.

signal component and subsequently, the phase derivative is computed as the angular position of the pole lying on the unit circle in the complex z-plane. As an example, a noisy single phase EPF is simulated with the phase derivative shown in figure 5.9(a). The poles computed with the SVAR model, orders $P = 2$ and $P = 3$, are plotted in figures 5.10(a) and 5.10(b), respectively. It can be clearly seen that the poles associated with the single phase EPF lie on the unit circle while the poles associated with the noise are randomly distributed inside the unit circle.

The phase derivative estimation is performed based on the assumption of the stochastic evolution of the SVAR coefficients. The simulated phase derivative is shown in figure 5.11(b). The pole plot corresponding to $\kappa = 1$ and $P = 1$ is shown in figure 5.11(a). The selection of the order of difference equation, $\kappa = 1$, is in general sufficient to capture the rapid spatial variations in the phase. However, in the case where the phase is a spatially slowly varying function, the spatial evolution of the SVAR coefficients is seen to keep pace with the variations in phase. Accordingly, $\kappa = 2$ is selected to impose additional constraint on the spatial evolution of the coefficients. The estimated phase derivatives obtained with the pairs $\kappa = 1$, $P = 2$; and $\kappa = 2$ with $P = 2$, are shown in figure 5.11(b). It can be observed that in the regions of rapidly varying phase derivative, the estimated phase derivative with $\kappa = 1$ follows the true phase derivative more closely than the estimated phase derivative with $\kappa = 2$. On the other hand, in the regions of slowly varying phase derivative, the estimated phase derivative with $\kappa = 2$ follows the true phase derivative more closely than the estimated phase derivative with $\kappa = 1$.

Examples of two-dimensional phase fringe patterns of size, 256×256, simulated at SNR = 10 dB are shown in figures 5.12(a) and 5.12(b). The phase derivative estimation performed on the phase fringe pattern shown in figure 5.12(a) is based on the deterministic evolution of SVAR coefficients. The estimated phase derivatives, with respect to x and y, obtained with $K = 15$ and Fourier basis functions, are shown in figures 5.12(c) and 5.12(e), respectively. A simulated phase fringe pattern similar to the one shown in figure 5.12(a) and containing certain fringe-less areas is shown in

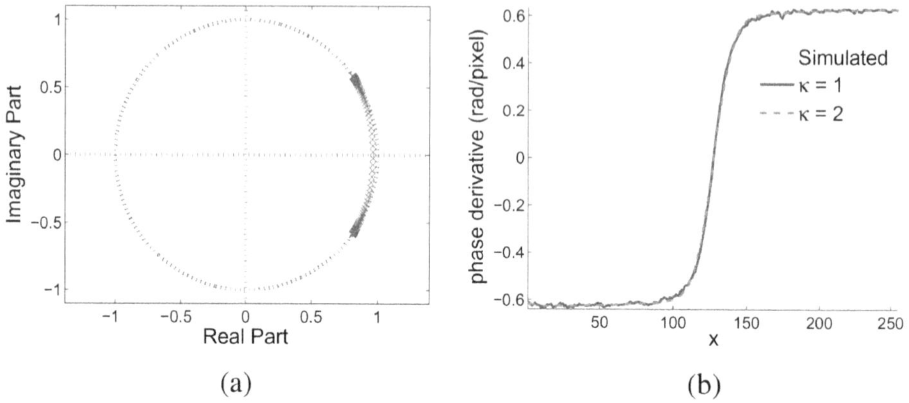

Figure 5.11. (a) Pole plot computed using stochastic coefficient estimation method with $\kappa = 1$ and $P = 1$; (b) estimated phase derivatives computed with $\kappa = 1$ and $\kappa = 2$.

figure 5.12(b). In this case, the phase derivative estimation based on the assumption of the deterministic evolution of SVAR coefficients fails due to the lack of fringe data in certain regions. On the other hand, the phase derivative estimation based on the assumption of the stochastic evolution of SVAR coefficients is capable of handling such a type of fringe patterns. The estimated phase derivatives with respect to x and y using $\kappa = 2$ and $P = 2$ are shown in figures 5.12(d) and 5.12(f), respectively.

The examples of experimentally recorded phase fringe pattern are shown in figures 5.13(a) and 5.13(b). These fringe patterns correspond to the out-of-plane displacement of an aluminum plate recorded in a digital holographic interferometry setup. For the phase fringe pattern shown in figure 5.13(a), the phase derivatives, estimated with respect to x and y, based on the assumption of the deterministic evolution of SVAR coefficients with $K = 10$ and Fourier basis functions, are shown in figures 5.13(c) and 5.13(e), respectively. The phase fringe pattern shown in figure 5.13(b) contains a fringe-less area due to the use of a circular aluminum plate. A binary mask is manually generated for selecting the circular plate as the region of interest. The phase derivatives are estimated based on the assumption of the stochastic evolution of SVAR coefficients with $\kappa = 2$ and $P = 2$. The estimated phase derivatives with respect to x and y are shown in figures 5.13(d) and 5.13(f), respectively.

5.2.4 Case 2: single phase carrier modulated fringe signal

In FT interferometry, a single phase carrier modulated fringe signal can be described by

$$\Gamma_c(x, y) = a(x, y) + b(x, y)\cos(\omega_x x + \omega_y y + \phi(x, y)),$$

where ω_x and ω_y represent the carrier frequencies along the x and y directions, respectively. The background intensity, $a(x, y)$, is eliminated prior to the phase derivative estimation by a high pass filtering of the fringe signal. In general, the fringe amplitude, $b(x, y)$, has a low spatial variation. Although the SVAR model

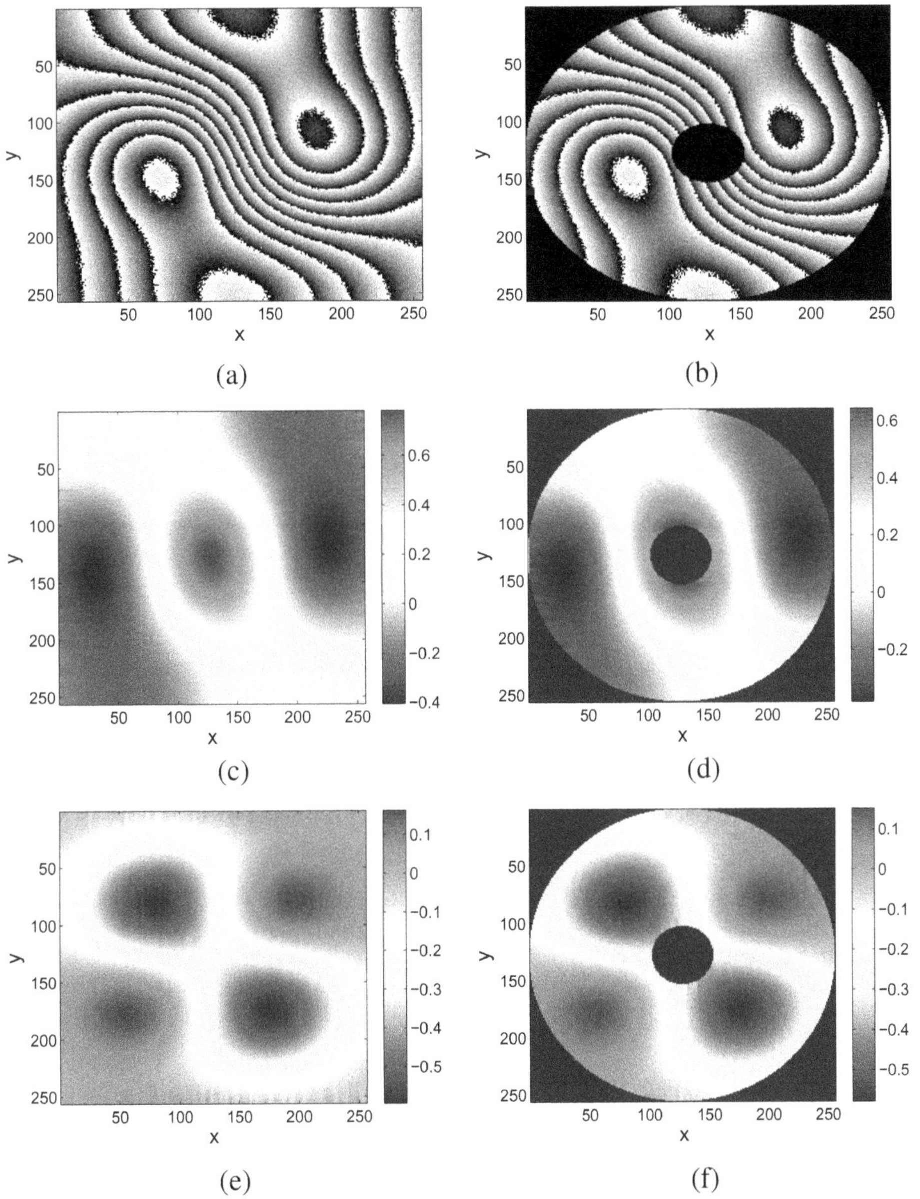

Figure 5.12. (a), (b) Phase fringe patterns simulated at SNR = 10 dB. Estimated phase derivatives with respect to (c) x and (e) y based on deterministic evolution of SVAR coefficients, and with respect to (d) x and (f) y based on stochastic evolution of SVAR coefficients.

based method is applicable in the case of fringe amplitude modulation, we have, for the sake of simplicity, assumed an amplitude normalized fringe signal, that is, $b(x, y) = 1$. Next, we define $\psi(x, y)$ as

$$\psi(x, y) = \omega_x x + \omega_y y + \phi(x, y).$$

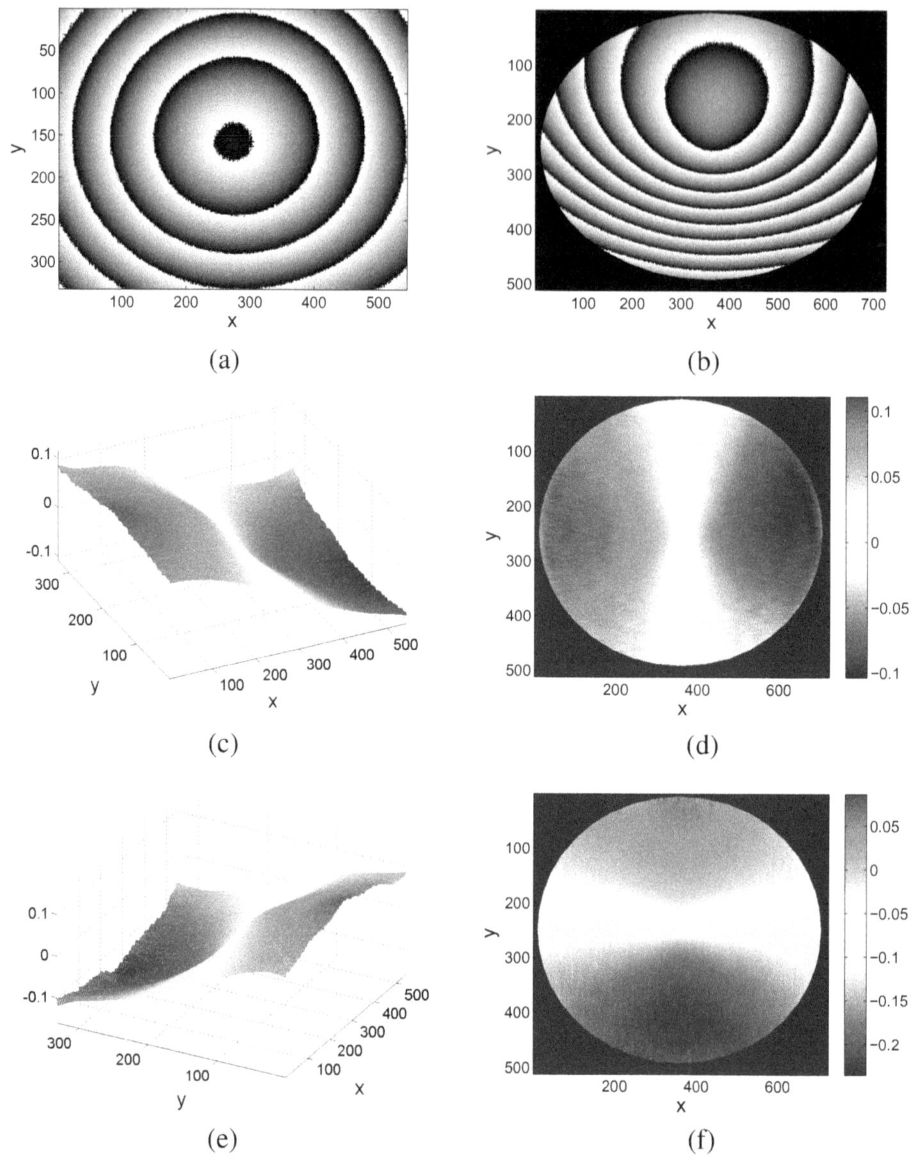

Figure 5.13. (a), (b) Experimentally recorded phase fringe patterns in a digital holographic interferometry setup corresponding to out-of-plane displacement of an aluminum plate. Estimated phase derivatives with respect to (c) x and (e) y based on the assumption of deterministic evolution of SVAR coefficients. Estimated phase derivatives with respect to (d) x and (f) y based on the assumption of stochastic evolution of SVAR coefficients.

$\Gamma(x, y)$ can then be written as

$$\Gamma(x, y) = \cos(\psi(x, y)).$$

Let us consider the fringe intensity in a given row y as

$$\Gamma(x) = \cos(\psi(x)). \tag{5.50}$$

In its digital implementation, the phase derivative, $\dot{\psi}(x)$, can be represented as

$$\dot{\psi}(x) = \psi(x) - \psi(x - 1). \tag{5.51}$$

Substituting the expression of $\psi(x)$ from (5.51) into (5.50), we obtain

$$\begin{aligned}\Gamma(x) &= \cos(\psi(x - 1) + \dot{\psi}(x)) \\ &= \cos(\psi(x - 1))\cos(\dot{\psi}(x)) - \sin(\psi(x - 1))\sin(\dot{\psi}(x)).\end{aligned} \tag{5.52}$$

Expanding the term $\sin(\psi(x - 1))$ in (5.52), we get

$$\begin{aligned}\sin(\psi(x - 1)) &= \sin(\psi(x - 2) + \dot{\psi}(x - 1)) \\ &= \sin(\psi(x - 2))\cos(\dot{\psi}(x - 1)) + \cos(\psi(x - 2))\sin(\dot{\psi}(x - 1)) \\ &= \sin(\psi(x - 1) - \dot{\psi}(x - 1))\cos(\dot{\psi}(x - 1)) \\ &\quad + \cos(\psi(x - 2))\sin(\dot{\psi}(x - 1)) \\ &= (\sin(\psi(x - 1))\cos(\dot{\psi}(x - 1)) - \cos(\psi(x - 1))\sin(\dot{\psi}(x - 1))) \\ &\quad \times \cos(\dot{\psi}(x - 1)) + \cos(\psi(x - 2))\sin(\dot{\psi}(x - 1)).\end{aligned}$$

Arranging the terms containing $\sin(\psi(x - 1))$ on the left side, we have

$$\begin{aligned}(1 - \cos^2(\dot{\psi}(x - 1)))\sin(\psi(x - 1)) &= -\cos(\psi(x - 1))\sin(\dot{\psi}(x - 1))\cos(\dot{\psi}(x - 1)) \\ &\quad + \cos(\psi(x - 2))\sin(\dot{\psi}(x - 1)) \\ \sin(\psi(x - 1)) &= -\cos(\psi(x - 1))\frac{\cos(\dot{\psi}(x - 1))}{\sin(\dot{\psi}(x - 1))} + \frac{\cos(\psi(x - 2))}{\sin(\dot{\psi}(x - 1))}.\end{aligned}$$

Substituting the expression of $\sin(\psi(x - 1))$ into (5.52), we obtain

$$\Gamma(x) = a_1(x)\Gamma(x - 1) - a_2(x)\Gamma(x - 2), \tag{5.53}$$

where

$$a_1(x) = \frac{\sin(\dot{\psi}(x) + \dot{\psi}(x - 1))}{\sin(\dot{\psi}(x - 1))},$$

$$a_2(x) = \frac{\sin(\dot{\psi}(x))}{\sin(\dot{\psi}(x - 1))}.$$

Equation (5.53) essentially representing the SVAR model of the fringe signal with $P = 2$ testifies to two poles of the transfer function associated with it. The angular positions of these two poles in the complex z-plane carry the information on the phase derivative. Since a cosine signal can be considered as the weighted sum of two complex sinusoids, in this case, one pole carries the information on the phase derivative while the other pole carries the information on the negative of the phase derivative. Therefore, tracking the angular position of only a single pole is sufficient. Under the assumption of working with a noisy fringe signal, the selection of $P > 2$ would be more appropriate to achieve the noise robustness in the phase derivative estimation. Based on the reasoning given in the case of a single phase EPF, the extra poles help in capturing the effect of noise in the fringe signal.

A representative validation of this method is shown by considering a one-dimensional fringe signal simulated at SNR = 20 dB. The carrier fringe and phase modulated fringe signals are shown in figures 5.14(a) and (b), respectively. Phase derivative estimation based on the assumption of the deterministic evolution of SVAR coefficients is implemented with $K = 10$, Fourier basis functions and $P = 4$. On the other hand, the phase derivative estimation based on the assumption of the stochastic evolution of SVAR coefficients is implemented with $\kappa = 2$ and $P = 4$. The respective pole plots are shown in figures 5.14(c) and (d). The pole plots clearly show two pole trajectories corresponding to the fringe signal that lie on the unit circle. The remaining two poles corresponding to noise have random trajectories inside the unit circle.

The demonstration is extended to two-dimensional carrier modulated fringe patterns (figures 5.15(a) and (b)) simulated at SNR = 20 dB. Note that the fringe data is missing in certain portions of the frame shown in figure 5.15(b). Since the carrier frequency is introduced along the x direction, the phase derivative estimation

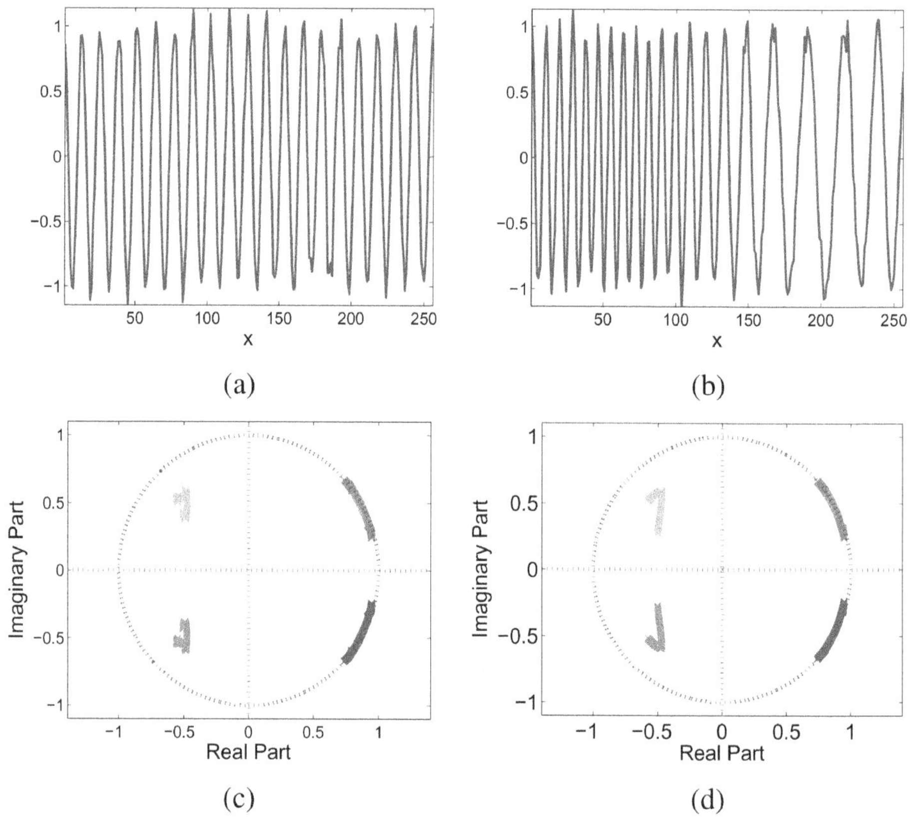

Figure 5.14. (a) One-dimensional simulated carrier signal; (b) modulated carrier fringe signal. Pole plots computed using the method based on (c) deterministic evolution of SVAR coefficients with Fourier basis function, basis dimension $K = 10$ and $P = 4$ and (d) stochastic evolution of SVAR coefficients with $\kappa = 2$ and $P = 4$.

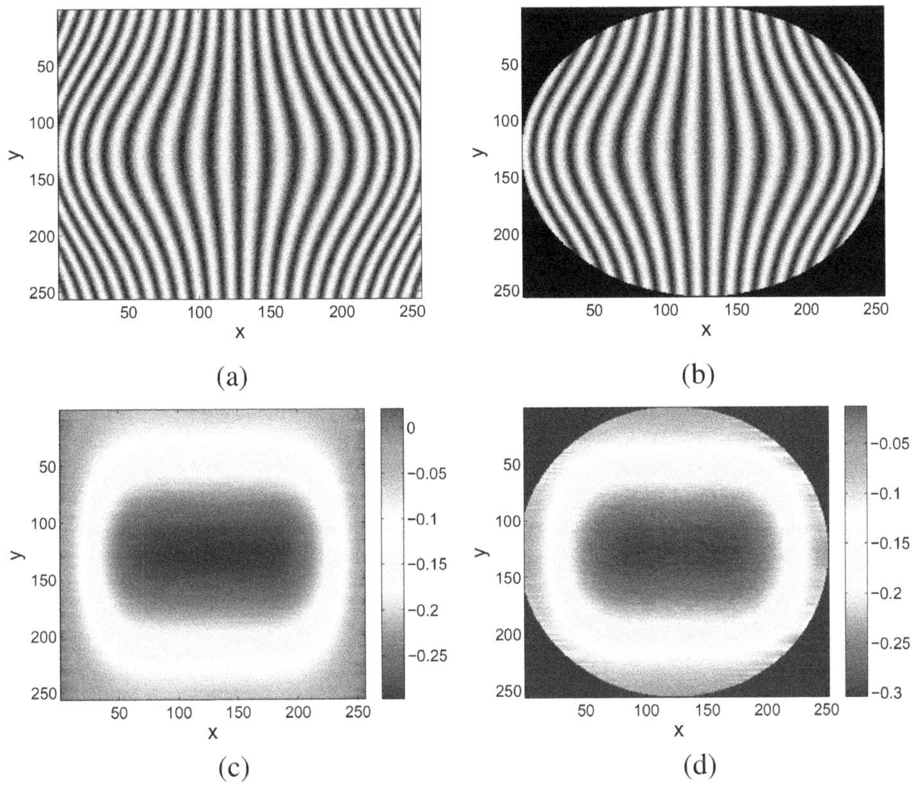

Figure 5.15. (a), (b) Two-dimensional carrier modulated fringe patterns. Estimated phase derivatives with respect to x using (c) deterministic coefficient estimation method and (d) stochastic coefficient estimation method.

is performed with respect to x. The estimated phase derivatives shown in figures 5.15(c) and (d) are obtained by applying the approaches based on the deterministic and stochastic evolutions of SVAR coefficients to the fringe patterns shown in figures 5.15(a) and (b), respectively.

5.2.5 Case 3: multi-phase EPF

Multi-phase EPF is a type of signal typically present in multiple beam digital holographic interferometry. For the sake of simplicity, we consider a dual-phase EPF, a type of signal produced in a two-beam setup and which is sensitive to the displacement components d_x and d_z that are measured simultaneously (see (4.4) and (4.5)). Differentiating both sides of these equations, say, with respect to x, we obtain

$$\frac{\partial d_x}{\partial x} = \frac{\lambda}{4\pi} \frac{\dot{\phi}_1(x, y) - \dot{\phi}_2(x, y)}{\sin(\theta)},$$

$$\frac{\partial d_z}{\partial x} = \frac{\lambda}{4\pi} \frac{\dot{\phi}_1(x, y) + \dot{\phi}_2(x, y)}{(1 + \cos(\theta))},$$

where, $\dot{\phi}_1(x, y)$ and $\dot{\phi}_2(x, y)$ are defined as

$$\dot{\phi}_1(x) = \frac{\partial \phi_1(x)}{\partial x}$$
$$\dot{\phi}_2(x) = \frac{\partial \phi_2(x)}{\partial x}.$$

Thus, the displacement derivatives, that is, the in-plane strain and out-of-plane slopes, can be obtained by estimating the phase derivatives. Let us consider the SVAR based modeling of the EPF for the simultaneous estimation of multiple phase derivatives. The EPF containing two signal components is given in (4.3). For an unambiguous estimation of the phase derivatives, it is essential that the two signal components are separated in the space-frequency plane. This is achieved by introducing auxiliary carrier fringes in one of the two arms of the object illumination beams. A carrier frequency of, say, ω_x in the direction of x is introduced into beam 2 prior to the object deformation. The effective phase term in the signal component related to beam 2 thus becomes $\bar{\phi}_2(x, y) = \phi_2(x, y) + \omega_x x$. Note that the carrier frequency can be similarly introduced along the y direction.

Considering that the phase derivative estimation is performed with respect to x, the EPF in a row y is given by (4.6). Since the EPF in (4.6) can be looked upon as a signal consisting of two complex non-stationary sinusoids embedded in noise, $\Gamma(x)$ can be expressed as the SVAR process of order $P = 2$ [16]. The SVAR model here is similar to that used for carrier modulated fringe signal except that in the latter case both the poles carry the information on the phase derivative associated with a single signal component. On the other hand, the two poles carry the information on the phase derivatives of their respective signal components, that is to say, these two poles provide the estimates of the two-phase derivatives.

The direct phase derivative estimation corresponding to the in-plane strain and out-of-plane slopes of an aluminum plate performed in a DHI experimental setup is illustrated in figure 5.16. Carrier frequency introduced along the y direction by suitably tilting a mirror in the path of beam 2 is shown in figure 5.16(a). A hologram corresponding to the carrier frequency is recorded prior to the plate deformation. The plate is subjected to out-of-plane deformation and in-plane rotation. The recorded moiré fringe pattern after a simple median filtering of the EPF is shown in figure 5.16(b). Phase derivative estimation is performed using the method based on the stochastic evolution of SVAR coefficients with $\kappa = 1$ and $P = 4$. The phase derivatives $\dot{\phi}_1$ and $\dot{\phi}_2$ estimated with respect to y are shown in figures 5.16(c) and 5.16(d), respectively. The sum and the difference of the estimated phase derivatives are shown in figures 5.16(e) and 5.16(f), respectively.

A possible point of contention here could be that since the two signal components are separated in the space–frequency plane, why would we not first filter them out and then analyze them separately? This approach, though considered in the fringe analysis technique proposed in [17], does implicitly assume that in the spectrum of the EPF the signal components maintain their positions in the space–frequency

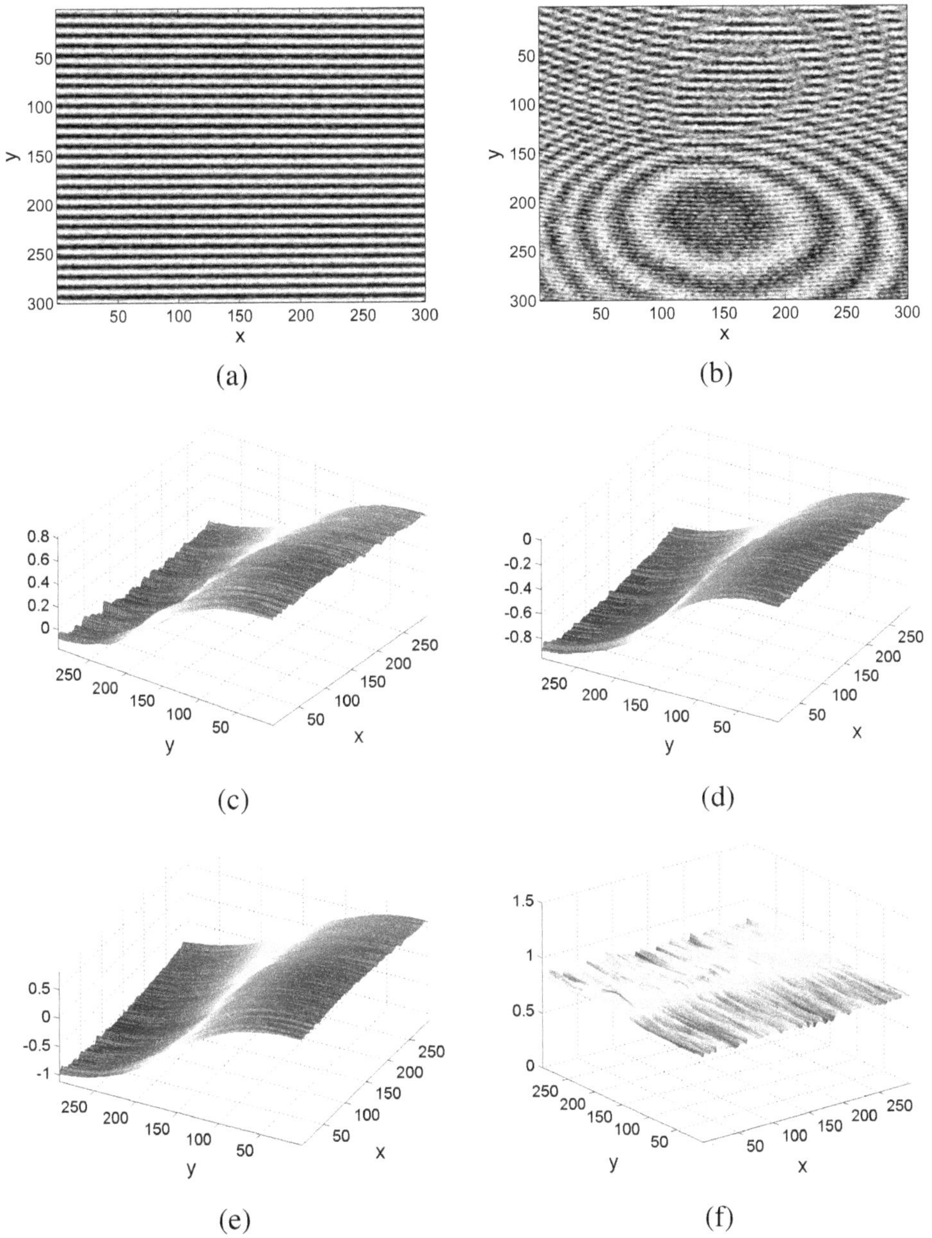

Figure 5.16. (a) Experimentally recorded carrier fringe pattern (b) carrier modulated fringe pattern in a two-beam digital holographic interferometry setup corresponding to out-of-plane and in-plane deformation of an aluminum plate. Estimates of phase derivatives computed with respect to y of (c) first signal and (d) second signal components using the method based on stochastic evolution of SVAR coefficients with $\kappa = 1$ and $P = 4$. (e) Sum and (f) difference of estimated phase derivatives. All the phase derivatives are in radians/pixel.

plane. This assumption may, however, not hold in practice due to the fact that the in-plane displacement of the object surface may introduce a shift in the spectrum of the signal. Hence, the two signal components in the SVAR based multi-phase EPF modeling approach are not filtered in the space–frequency plane prior to the phase derivative estimation.

5.3 Parametric phase derivative estimation: matrix enhancement and matrix pencil

In the one-dimensional signal analysis involved in the SVAR based EPF modeling approach, the phase derivative estimation is obtained with respect to either x or y. It is, however, desirable to obtain simultaneous phase derivative estimation with respect to both x and y. In this section, we describe a method based on matrix pencil and matrix enhancement (MEMP) [18] which considers a generalized model of the EPF containing K signal components recorded, for example, in a multi-beam digital holographic interferometry setup. For an unambiguous estimation of the multiple phase derivatives, the signal components are separated in the space-frequency plane by the addition of auxiliary carrier fringes to each of the beams illuminating the object surface. An EPF recorded after the introduction of carrier and prior to the object deformation can be written as

$$\Gamma(x, y) = \sum_{k=1}^{K} A_k(x, y)\exp[i(\omega_{kx}x + \omega_{ky}y)],$$

where ω_{kx} and ω_{ky} correspond to the carrier frequencies of the kth signal component in the directions of x and y, respectively. The object is next deformed and the corresponding EPF is given by

$$\Gamma(x, y) = \sum_{k=1}^{K} A_k(x, y)\exp[i(\omega_{kx}x + \omega_{ky}y + \phi_k(x, y))].$$

Let us define,

$$\bar{\phi}_k(x, y) = \omega_{kx}x + \omega_{ky}y + \phi_k(x, y).$$

Accordingly, we have

$$\Gamma(x, y) = \sum_{k=1}^{K} A_k(x, y)\exp[i\bar{\phi}_k(x, y)].$$

The derivatives of the phases $\phi_k(x, y)$ with respect to x and y can be represented as

$$\left.\begin{aligned} \dot{\phi}_{kx}(x, y) &= \frac{\partial\phi_k(x, y)}{\partial x} \\ \dot{\phi}_{ky}(x, y) &= \frac{\partial\phi_k(x, y)}{\partial y} \end{aligned}\right\} \text{for } k = 1 \ldots K.$$

In this method, the fringe pattern is analyzed over a rectangular window of size $M_w \times N_w$ around each pixel. Supposing the window size to be sufficiently small, the EPF can be considered as a combination of K complex sinusoids with constant spatial frequencies. Thus, the EPF within the window can be represented as

$$\Gamma_w(x, y) = \sum_{k=1}^{K} A_k(x, y)\exp(i(\hat{\omega}_{kx}x + \hat{\omega}_{ky}y)), \tag{5.54}$$

where $x \in [0, N_w - 1]$ and $y \in [0, M_w - 1]$; $\hat{\omega}_{kx}$ and $\hat{\omega}_{ky}$ represent the frequencies of the kth complex sinusoid with respect to x and y, respectively. The phase derivatives at the pixel (x, y) are obtained as the frequencies, $\hat{\omega}_{kx}$ and $\hat{\omega}_{ky}$, computed at that pixel. Thus, the problem of multiple phase derivative estimation boils down to the problem of accurate estimation of the local fringe frequencies, $\hat{\omega}_{kx}$ and $\hat{\omega}_{ky}$. To accurately capture the local variations in the fringe pattern, we assume the analysis window to be of a small size. As a consequence, the number of EPF samples within the analysis window is limited which adversely affects the resolution in the periodogram based frequency estimation. The negative impact is undercut by a MEMP based high resolution frequency estimation technique described in the following subsection.

5.3.1 Matrix enhancement and related properties

At each pixel (x, y), a data matrix $\boldsymbol{\Gamma}_w$ formed by windowing the EPF signal can be written as

$$\boldsymbol{\Gamma}_w = \begin{bmatrix} \Gamma_w(0, 0) & \Gamma_w(0, 1) & \cdots & \Gamma_w(0, N_w - 1) \\ \Gamma_w(1, 0) & \Gamma_w(1, 1) & \cdots & \Gamma_w(1, N_w - 1) \\ \vdots & \vdots & \ddots & \vdots \\ \Gamma_w(M_w - 1, 0) & \Gamma_w(M_w - 1, 1) & \cdots & \Gamma_w(M_w - 1, N_w - 1) \end{bmatrix}.$$

The amplitude of each signal component is considered to be constant within each analysis window. The data matrix in (5.54) can be represented as

$$\boldsymbol{\Gamma}_w = \sum_{k=1}^{K} A_k p_k^y q_k^x, \tag{5.55}$$

where

$$p_k = \exp(i\hat{\omega}_{ky})$$
$$q_k = \exp(i\hat{\omega}_{kx}).$$

The matrix $\boldsymbol{\Gamma}_w$ can now be written as

$$\boldsymbol{\Gamma}_w = \boldsymbol{PAQ}$$

where

$$\boldsymbol{P} = \begin{bmatrix} 1 & 1 & \cdots & 1 \\ p_1 & p_2 & \cdots & p_K \\ \vdots & & & \\ p_1^{(M_w-1)} & p_2^{(M_w-1)} & \cdots & p_K^{(M_w-1)} \end{bmatrix}, \ \boldsymbol{Q} = \begin{bmatrix} 1 & q_1 & \cdots & q_1^{(N_w-1)} \\ 1 & q_2 & \cdots & q_2^{(N_w-1)} \\ \vdots & & & \\ 1 & q_K & \cdots & q_K^{(N_w-1)} \end{bmatrix}, \ \boldsymbol{A} = \begin{bmatrix} A_1 & 0 & 0 & 0 \\ 0 & A_2 & \cdots & 0 \\ \vdots & & & \\ 0 & 0 & \cdots & A_K \end{bmatrix}.$$

The above decomposition of the data matrix indicates that the singular value decomposition of the data matrix $\boldsymbol{\Gamma}_w$ can provide the estimates of the frequencies, $\hat{\omega}_{kx}$ and $\hat{\omega}_{ky}$. However, depending on the values of frequencies, the rank of matrix $\boldsymbol{\Gamma}_w$ can be less than or equal to K. If any of the frequencies of different signal components along the x or y directions share the same value, then $rank(\boldsymbol{\Gamma}_w)$ will be less than K. In this case, the singular value decomposition of the data matrix $\boldsymbol{\Gamma}_w$ does not provide the complete information on the data matrix and therefore, the frequencies associated with each signal component cannot be estimated accurately and unambiguously.

In [19], a matrix enhancement approach is proposed in order to obtain a matrix which carries all the information on the data matrix $\boldsymbol{\Gamma}_w$ and at the same time possesses an increased rank. Accordingly, an enhanced matrix $\boldsymbol{\Gamma}_e$ generated by the partition and stacking procedure applied to the matrix $\boldsymbol{\Gamma}_w$ can be written as

$$\boldsymbol{\Gamma}_e = \begin{bmatrix} \boldsymbol{\Gamma}_0 & \boldsymbol{\Gamma}_1 & \cdots & \boldsymbol{\Gamma}_{M_w-I} \\ \boldsymbol{\Gamma}_1 & \boldsymbol{\Gamma}_2 & \cdots & \boldsymbol{\Gamma}_{M_w-I+1} \\ \vdots & \vdots & \ddots & \vdots \\ \boldsymbol{\Gamma}_{I-1} & \boldsymbol{\Gamma}_I & \cdots & \boldsymbol{\Gamma}_{M_w-1} \end{bmatrix}, \tag{5.56}$$

where

$$\boldsymbol{\Gamma}_i = \begin{bmatrix} \Gamma_w(i, 0) & \Gamma_w(i, 1) & \cdots & \Gamma_w(i, N_w - J) \\ \Gamma_w(i, 1) & \Gamma_w(i, 2) & \cdots & \Gamma_w(i, N_w - J + 1) \\ \vdots & \vdots & \ddots & \vdots \\ \Gamma_w(i, J-1) & \Gamma_w(i, J) & \cdots & \Gamma_w(i, N_w - 1) \end{bmatrix}. \tag{5.57}$$

Here, $\boldsymbol{\Gamma}_e$ is a Hankel block matrix of size $I \times (M_w - I + 1)$, and $\boldsymbol{\Gamma}_i$ is a Hankel matrix of size $J \times (N_w - J + 1)$. Substituting (5.55) into (5.57), we obtain

$$\boldsymbol{\Gamma}_i = \boldsymbol{Q}_J \boldsymbol{A} \boldsymbol{P}_d^y \boldsymbol{Q}_R, \tag{5.58}$$

where

$$\boldsymbol{Q}_J = \begin{bmatrix} 1 & 1 & \cdots & 1 \\ q_1 & q_2 & \cdots & q_K \\ \vdots & & & \\ q_1^{(J-1)} & q_2^{(J-1)} & \cdots & q_K^{(J-1)} \end{bmatrix}, \ \boldsymbol{Q}_R = \begin{bmatrix} 1 & q_1 & \cdots & q_1^{(N_w-J)} \\ 1 & q_2 & \cdots & q_2^{(N_w-J)} \\ \vdots & & & \\ 1 & q_K & \cdots & q_K^{(N_w-J)} \end{bmatrix},$$

$$\boldsymbol{P}_d = diag\left[p_1, \ldots, p_K\right], \quad \boldsymbol{A} = diag\left[A_1, \ldots, A_K\right].$$

Substituting $\boldsymbol{\Gamma}_i$ from (5.58) into (5.56), we obtain $\boldsymbol{\Gamma}_e$ as

$$\boldsymbol{\Gamma}_e = \boldsymbol{F}_J \boldsymbol{A} \boldsymbol{F}_R, \tag{5.59}$$

where

$$\boldsymbol{F}_J = \begin{bmatrix} \boldsymbol{Q}_J \\ \boldsymbol{Q}_J \boldsymbol{P}_d \\ \vdots \\ \boldsymbol{Q}_J \boldsymbol{P}_d^{I-1} \end{bmatrix}, \ \boldsymbol{F}_R = \left[\boldsymbol{Q}_R, \ \boldsymbol{P}_d \boldsymbol{Q}_R, \ \ldots, \ \boldsymbol{P}_d^{M_w - I} \boldsymbol{Q}_R\right].$$

It is evident from (5.59) that $rank(\boldsymbol{\Gamma}_e) = K$, if and only if $rank(\boldsymbol{F}_J) = rank(\boldsymbol{F}_R) = K$. For this condition to be satisfied, the following criterion on I and J should be fulfilled:

$$\begin{aligned} M_w - I + 1 \geqslant I \geqslant K + 1 \\ N_w - J + 1 \geqslant J \geqslant K + 1 . \end{aligned} \tag{5.60}$$

Once the conditions given in (5.60) are satisfied, we can use the relationship between the decomposition of $\boldsymbol{\Gamma}_e$ given in (5.59) and its singular value decomposition vectors to obtain the required frequency estimates. Let us consider the singular value decomposition (SVD) of $\boldsymbol{\Gamma}_e$ as follows,

$$\boldsymbol{\Gamma}_e = \sum_{j=1}^{minRank} \lambda_j \boldsymbol{u}_j \boldsymbol{v}_j^H = \boldsymbol{U} \boldsymbol{\Lambda} \boldsymbol{V}^H,$$

where $minRank$ is defined as $minRank = \min\left(IJ, \ (M_w - I + 1)(N_w - J + 1)\right)$; λ_j are the singular values, and $\boldsymbol{u}_j$ and $\boldsymbol{v}_j$ are the left and right singular vectors; the superscript H denotes the conjugate transpose operation. Since there are K signal components present in the enhanced data matrix $\boldsymbol{\Gamma}_e$, we can express the SVD of $\boldsymbol{\Gamma}_e$ as

$$\boldsymbol{\Gamma}_e = \boldsymbol{U}_i \boldsymbol{\Lambda}_i \boldsymbol{V}_i^H + \boldsymbol{U}_\eta \boldsymbol{\Lambda}_\eta \boldsymbol{V}_\eta^H,$$

where $\boldsymbol{\Lambda}_i$, $\boldsymbol{U}_i$, $\boldsymbol{V}_i$ correspond to the first K principal components, and $\boldsymbol{\Lambda}_\eta$, $\boldsymbol{U}_\eta$, $\boldsymbol{V}_\eta$ correspond to the remaining $min - K$ non-principal components. Under the assumption of noiseless signal, the singular values in $\boldsymbol{\Lambda}_i$ are non-zero, whereas the singular values in $\boldsymbol{\Lambda}_\eta$ are zero. In the presence of noise, $\boldsymbol{\Gamma}_e$ has $rank$ equal to $minRank$. In this case, the singular values in $\boldsymbol{\Lambda}_\eta$ are non-zero. However, the singular values in $\boldsymbol{\Lambda}_i$ are still significantly higher than the singular values in $\boldsymbol{\Lambda}_\eta$. This leads us to compare the vector space spanned by the column vectors in $\boldsymbol{U}_i$ with $\boldsymbol{F}_J$, and column vectors in $\boldsymbol{V}_i$ with $\boldsymbol{F}_R^H$. Note that if $rank(\boldsymbol{\Gamma}_e) = K$, we have

$$\begin{aligned} range(\boldsymbol{U}_i) = range(\boldsymbol{F}_J) = range(\boldsymbol{\Gamma}_e) \\ range(\boldsymbol{V}_i) = range(\boldsymbol{F}_R^H) = range(\boldsymbol{\Gamma}_e^H). \end{aligned}$$

Now, since $\boldsymbol{U}_i \perp \boldsymbol{U}_\eta$, we have $\boldsymbol{F}_J \perp \boldsymbol{U}_\eta$. Similarly, we have $\boldsymbol{F}_R^H \perp \boldsymbol{V}_\eta$. This property can be utilized to compute the required frequency estimates.

5.3.2 Matrix pencil method for multiple frequency estimation

Following the development in the preceding subsection, we can write

$$\boldsymbol{U}_i = \boldsymbol{F}_J \boldsymbol{W}, \tag{5.61}$$

where $\boldsymbol{W}$ is a unique $K \times K$ non-singular matrix. With this formulation, we generate two matrices

$$\begin{aligned} \boldsymbol{U}_1 &= \boldsymbol{F}_1 \boldsymbol{W} \\ \boldsymbol{U}_2 &= \boldsymbol{F}_1 \boldsymbol{P}_d \boldsymbol{W}, \end{aligned}$$

where $\boldsymbol{U}_1$ and $\boldsymbol{U}_2$ are equal to $\boldsymbol{U}_i$ with its last and the first J rows deleted, respectively; $\boldsymbol{F}_1$ is equal to the matrix $\boldsymbol{F}_J$ with its last J rows deleted. Thus, we obtain the matrix pencil which we define as

$$\boldsymbol{U}_2 - \rho_y \boldsymbol{U}_1 = \boldsymbol{F}_1\left(\boldsymbol{P}_d - \rho_y \boldsymbol{I}'\right)\boldsymbol{W},$$

where $\boldsymbol{I}'$ is the identity matrix. The above equation indicates that the generalized eigenvalues of the matrix pencil provide the estimates of the poles located along the diagonal of $\boldsymbol{P}_d$. The argument of these poles provides the estimates of $\hat{\omega}_{ky}$ for $k = 1, \ldots, K$. On the other hand, to compute the estimates of $\hat{\omega}_{kx}$, we consider the formulation of a matrix $\boldsymbol{F}_{JC}$ as

$$\boldsymbol{F}_{JC} = \boldsymbol{C}\boldsymbol{F}_J = \begin{bmatrix} \boldsymbol{P}_I \\ \boldsymbol{P}_I \boldsymbol{Q}_d \\ \vdots \\ \boldsymbol{P}_I \boldsymbol{Q}_d^{J-1} \end{bmatrix},$$

where

$$\boldsymbol{P}_I = \begin{bmatrix} 1 & 1 & \cdots & 1 \\ p_1 & p_2 & \cdots & p_K \\ \vdots & & & \\ p_1^{(I-1)} & p_2^{(I-1)} & \cdots & p_K^{(I-1)} \end{bmatrix}, \ \boldsymbol{Q}_d = diag[q_1, \ldots, q_K],$$

Here, $\boldsymbol{C}$ is a shuffling matrix defined as

$$\boldsymbol{C} = \begin{bmatrix} \boldsymbol{c}^T(1) \\ \boldsymbol{c}^T(1 + J) \\ \cdots \\ \boldsymbol{c}^T(1 + (I - 1)J) \\ \cdots \\ \boldsymbol{c}^T(J) \\ \boldsymbol{c}^T(J + J) \\ \cdots \\ \boldsymbol{c}^T(J + (I - 1)J) \end{bmatrix},$$

where $\boldsymbol{c}(j)$ is a $IJ \times 1$ sized row vector with 1 at the jth position and zeros elsewhere. That there is a similarity between the structures of matrices $\boldsymbol{F}_J$ and $\boldsymbol{F}_{JC}$ can be readily noted. Now, let us define a matrix $\boldsymbol{U}_{iC}$ as

$$\boldsymbol{U}_{iC} = \boldsymbol{C}\boldsymbol{U}_i.$$

Similar to the case with $\boldsymbol{U}_i$, we generate two matrices $\boldsymbol{U}_{1C}$ and $\boldsymbol{U}_{2C}$ as equal to the matrices $\boldsymbol{U}_{iC}$ with its last and first I rows deleted, respectively. Following (5.61), we have, $\boldsymbol{U}_{iC} = \boldsymbol{C}\boldsymbol{F}_J\boldsymbol{W} = \boldsymbol{F}_{JC}\boldsymbol{W}$. This gives

$$\begin{aligned}\boldsymbol{U}_{1C} &= \boldsymbol{F}_{1C}\boldsymbol{W}\\ \boldsymbol{U}_{2C} &= \boldsymbol{F}_{1C}\boldsymbol{Q}_d\boldsymbol{W}.\end{aligned}$$

where $\boldsymbol{F}_{1C}$ is equal to $\boldsymbol{F}_{JC}$ with its last I rows deleted. Now, we define the matrix pencil $\boldsymbol{U}_{2C} - \rho_x\boldsymbol{U}_{1C}$ as

$$\boldsymbol{U}_{2C} - \rho_x\boldsymbol{U}_{1C} = \boldsymbol{F}_{1C}\left(\boldsymbol{Q}_d - \rho_x\boldsymbol{I}'\right)\boldsymbol{W}. \tag{5.62}$$

Equation (5.62) indicates that the generalized eigenvalues of the matrix pencil $\boldsymbol{U}_{2C} - \rho_x\boldsymbol{U}_{1C}$ provide the estimates of the poles located along the diagonal of $\boldsymbol{Q}_d$. The angular positions of these poles in the complex plane provide the estimates of $\hat{\omega}_{kx}$ for $k = 1, \ldots, K$.

Since in the case of a single signal component EPF the matrix enhancement step is not required to be implemented, as $K = 1$, the SVD computation of the data matrix $\boldsymbol{\Gamma}_w$, instead of $\boldsymbol{\Gamma}_e$, together with the matrix pencil based technique provides the phase derivative estimates. Given that the frequency estimation is performed using the data matrix $\boldsymbol{\Gamma}_w$, instead of $\boldsymbol{\Gamma}_e$, the time for computing the phase derivatives is quite low. On the other hand, the matrix enhancement offers improved noise robustness in the phase derivative estimation. One such example of single component complex EPF signal simulated at SNR = 10 dB is shown in figure 5.17(a). The MEMP based phase derivative estimation is performed with $M_w = N_w = 17$. The estimated phase derivatives with respect to x and y are shown in figures 5.17(b) and (c), respectively. For the purpose of illustration, the fringe density map computed as $\sqrt{\hat{\omega}_x^2 + \hat{\omega}_y^2}$ is provided in figure 5.17(d).

The MEMP based phase derivative estimation performed with an experimentally recorded phase fringe pattern is shown in figure 5.18(a). The estimated phase derivatives with respect to x and y are shown in figures 5.18(b) and 5.18(c), respectively. It can be clearly observed that this method is capable of providing phase derivative estimates even in the case of fringe patterns containing fringe-less areas. For the sake of illustration, the fringe density map is shown in figure 5.18(d).

5.3.3 Pairing of multiple frequencies in the case of a multicomponent signal

Since for an EPF containing multiple signal components the frequencies, $\hat{\omega}_{kx}$ and $\hat{\omega}_{ky}$ are estimated separately in the MEMP based method, these would also

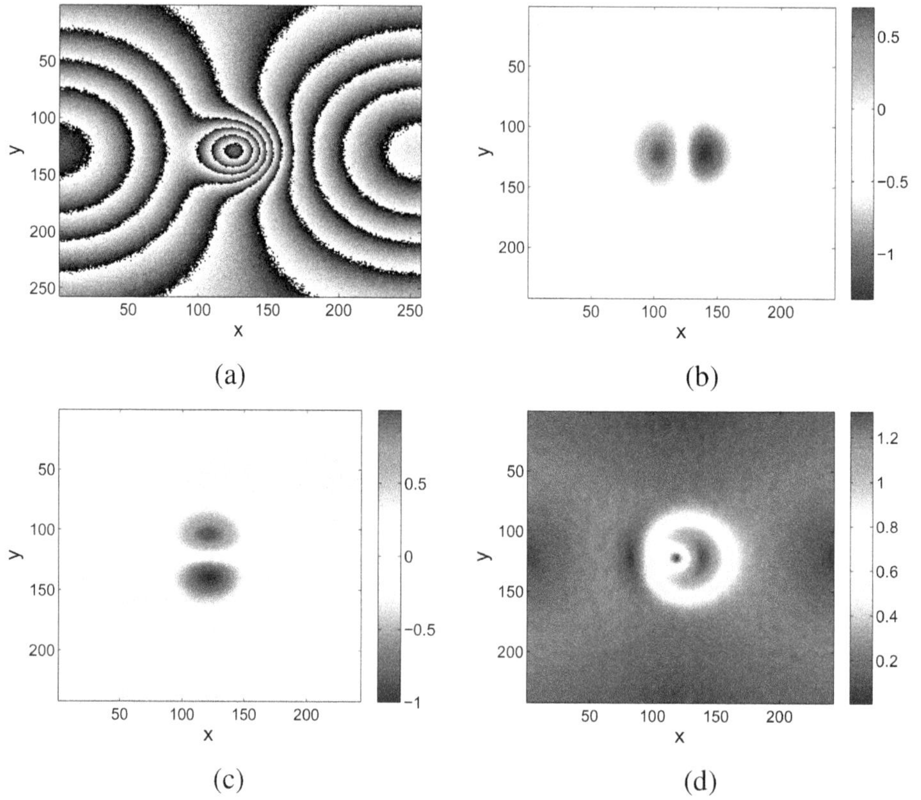

Figure 5.17. (a) Two-dimensional simulated phase fringe pattern. Estimated phase derivatives with respect to (b) x and (c) y using the MEMP method. (d) Fringe density map.

need to be appropriately paired using a pairing mechanism. Let us define a vector $\boldsymbol{f}_J$ as,

$$\boldsymbol{f}_J(\omega_x, \omega_y) = \boldsymbol{p}_J \otimes \boldsymbol{q}_J,$$

where $\otimes$ represents the Kronecker product and,

$$\boldsymbol{p}_J = [1, p, \ldots, p^{(I-1)}]^T, \; \boldsymbol{q}_J = [1, q, \ldots, q^{(J-1)}]^T$$

$$p = \exp\left(i\omega_y\right), \; q = \exp(i\omega_x).$$

If $\omega_x = \hat{\omega}_{kx}$ and $\omega_y = \hat{\omega}_{ky}$ for the kth signal component, then $\boldsymbol{f}_J \perp \boldsymbol{U}_\eta$. Utilizing this property, we implement the steps given below in order to pair the estimated frequencies together:

1. Set $h = 1$.
2. Compute the function $G(h, l)$ for $l = 1 \ldots K$ as follows,

$$G(h, l) = \sum_{k=1}^{K} \left\| \boldsymbol{u}_k^H \boldsymbol{f}_J\left(\hat{\omega}_{hx}, \hat{\omega}_{ly}\right)\right\|^2 \tag{5.63}$$

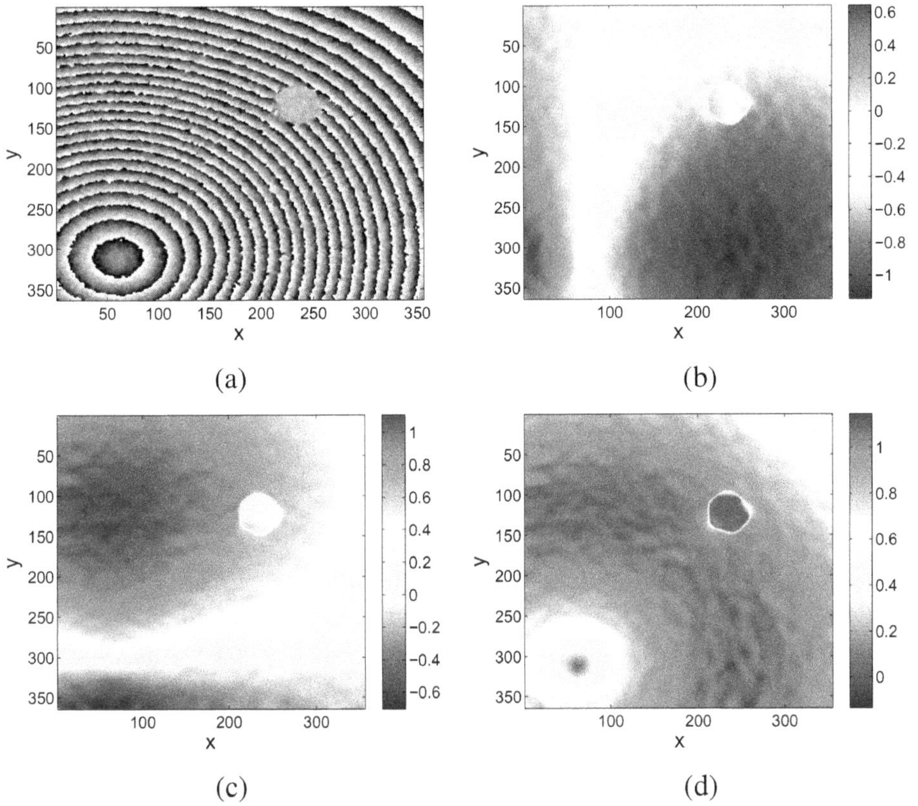

Figure 5.18. (a) Experimentally recorded phase fringe pattern. Estimated phase derivatives with respect to (b) x and (c) y (in radians/pixel) using the MEMP method. (d) Fringe density map.

3. Obtain a frequency pair $(\hat{\omega}_{hx}, \hat{\omega}_{l'y})$ using the value $l = l'$ for which $G(h, l)$ is maximum.
4. Set $h = h + 1$.
5. Compute the function $G(h, l)$ as per (5.63) for all values of l except for the previously computed set of l'.
6. Repeat the procedure in step (4) till $h = K - 1$.

The frequency estimation steps are repeated at all pixels. In the case of a multi-component EPF, although multiple frequency estimates are obtained at each pixel, they are not ordered the same for all the pixels. That is, the frequency estimates corresponding to a particular signal component do not appear at the same location in the estimated frequency vector at all the pixels. A criterion is therefore required to be able to appropriately assign the estimated frequencies to their respective signal components. The use of amplitude discrimination criterion has been proposed in

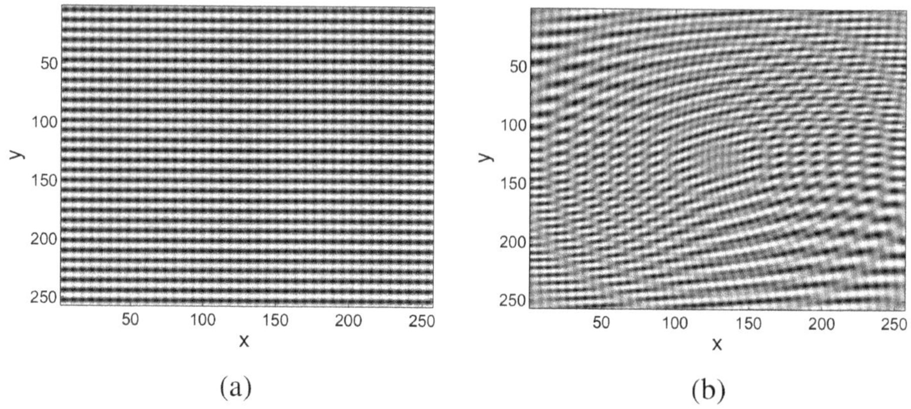

Figure 5.19. (a) Simulated and (b) deformation modulated carrier fringe patterns.

which the amplitudes of the signal components are set to different levels by adjusting appropriately the intensities in the object illumination beams. Equation (5.54) indicates that the EPF $\Gamma_w(x, y)$ is linearly dependent on the amplitudes A_k. A least squares method is implemented using the estimated frequencies to compute the amplitudes A_k. The order of the computed amplitudes associated with the signal components allows for assigning the associated frequency pairs to their respective signal components.

To demonstrate the utility of the procedure, a numerical experiment is performed using a simulated multicomponent EPF containing three signal components, i.e., $K = 3$. The amplitudes of the signal components are set such that $A_1 > A_2 > A_3$. A linear fringe pattern corresponding to the carrier frequencies added to it is shown in figure 5.19(a). The same fringe pattern modulated by object deformation is shown in figure 5.19(b). The MEMP method is implemented with an analysis window of size 16×16. The estimated phase derivatives of the three signal components with respect to x and y, respectively, obtained at SNR = 20 dB, are shown in figure 5.20.

A DHI setup consisting of two object illumination beams similar to that shown in figure (4.1) is used to measure the in-plane and out-of-plane displacement derivatives with respect to x and y simultaneously. A carrier fringe pattern added into beam 2 is recorded. This is shown in figure 5.21(a). The object is then deformed and the corresponding moiré fringe fringe pattern is shown in figure 5.21(b). The implementation of the MEMP based phase derivative estimation with an analysis window of size 20×20, provides the sums and differences of the phases derivatives with respect to x and y, as shown in figure 5.22.

The capability of multiple displacement derivative measurements from a single frame of EPF makes the MEMP based method highly suitable for dynamic deformation analysis. On the one hand, the equipment requirement mainly involving

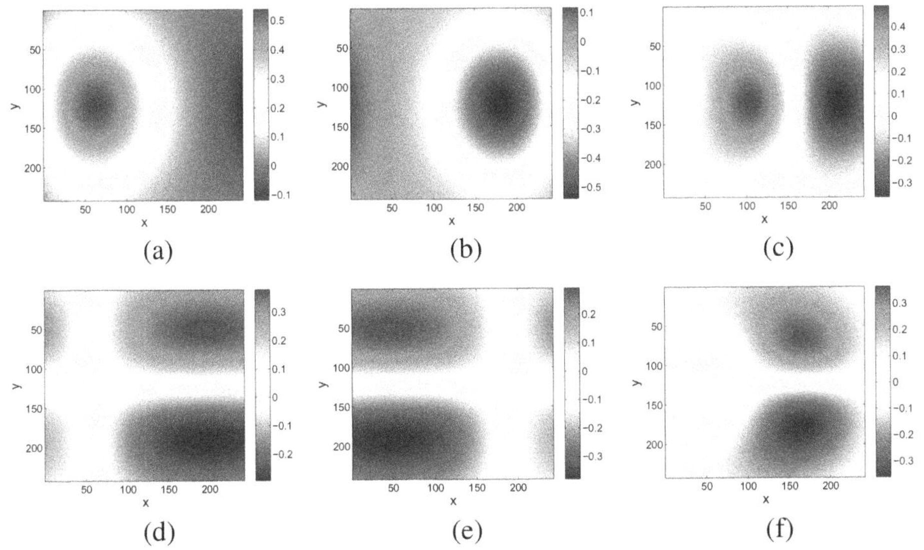

Figure 5.20. Estimated phase derivatives (in rad/pixel) with respect to x (a) $\dot{\phi}_{1x}$ (b) $\dot{\phi}_{2x}$ (c) $\dot{\phi}_{3x}$. Estimated phase derivatives (in rad/pixel) with respect to y (d) $\dot{\phi}_{1y}$ (e) $\dot{\phi}_{2y}$ and (f) $\dot{\phi}_{3y}$.

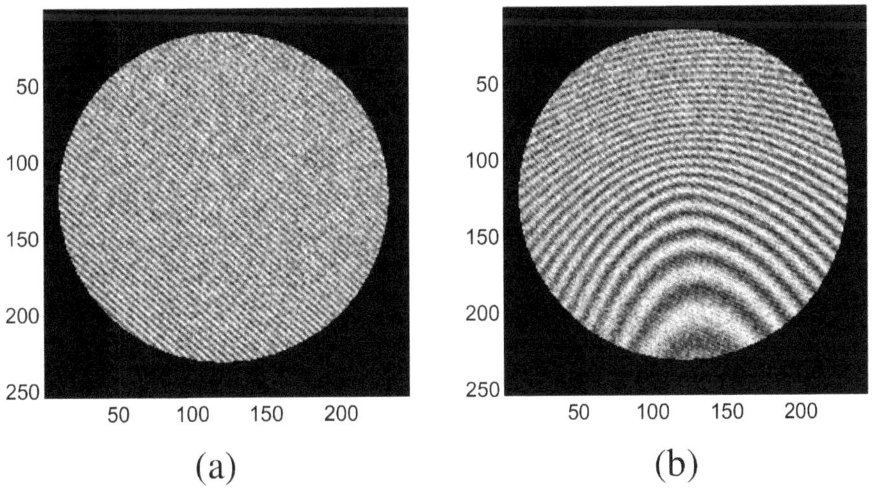

Figure 5.21. (a) Experimental linear fringe pattern corresponding to the carrier; (b) deformation modulated carrier fringe pattern.

a single laser light source and a single CCD camera highlights the simplicity of the method, and on the other hand, the ability of computing the fringe density in terms of spatial fringe frequencies makes it a valuable tool to detect and locate possible defects in an object in a non-destructive manner.

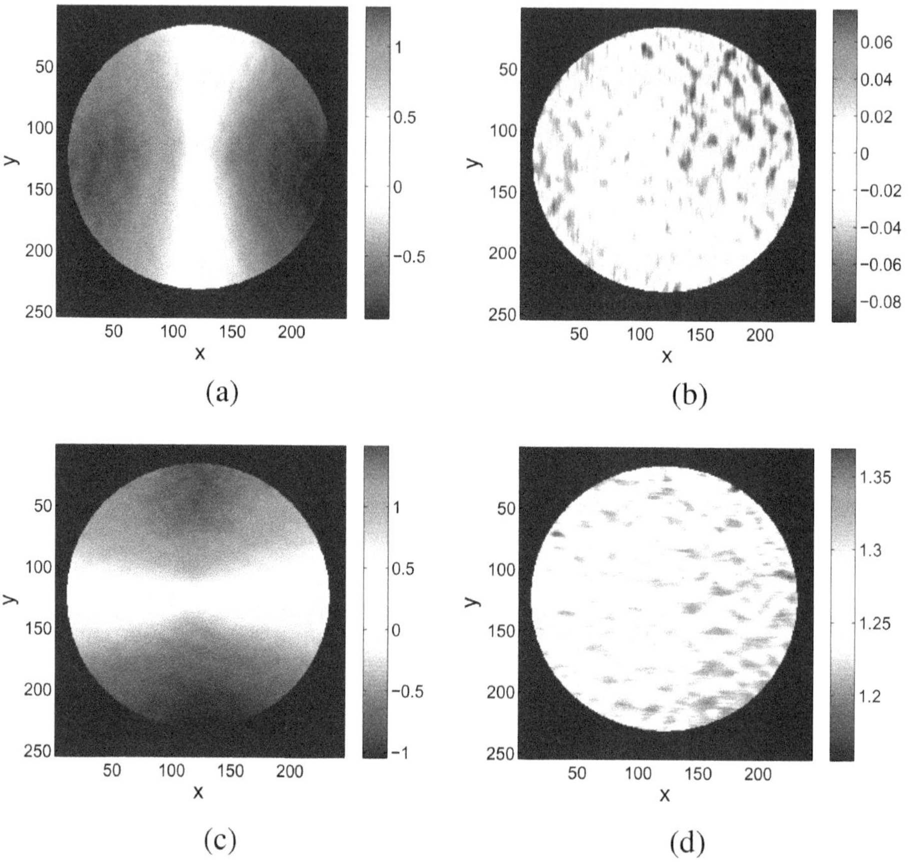

Figure 5.22. (a) $\dot{\phi}_{1x} + \dot{\phi}_{2x}$ (b) $\dot{\phi}_{1x} - \dot{\phi}_{2x}$ (c) $\dot{\phi}_{1y} + \dot{\phi}_{2y}$ (d) $\dot{\phi}_{1y} - \dot{\phi}_{2y}$. All values are in rad/pixel.

References

[1] Boashash B 2003 *Time Frequency Analysis: A Comprehensive Reference in Time Frequency Analysis: A Comprehensive Reference* ed B Boashash (Amsterdam: Elsevier) pp 1–743

[2] Rajshekhar G, Gorthi S S and Rastogi P 2009 Adaptive window Wigner–Ville-distribution-based method to estimate phase derivative from optical fringes *Opt. Lett.* **34** 3151–3

[3] Rajshekhar G and Gorthi S S 2010 Estimation of the phase derivative using an adaptive window spectrogram *J. Opt. Soc. Am.* **27** 69–75

[4] Katkovnik V and Stankovic L 1997 Instantaneous frequency estimation using the Wigner distribution with varying and data-driven window length *IEEE Trans. Sig. Proc.* **45** 2147

[5] Katkovnik V and Stanković L J 1998 Periodogram with varying and data-driven window length *Sig. Proc.* **67** 345–58

[6] Cornu C, Stankovic S, Ioana C, Quinquis A and Stankovic L 2007 Generalized representation of phase derivatives for regular signals *IEEE Trans. Sig. Proc.* **55** 4831–8

[7] Kulkarni R and Rastogi P 2015 Multiple phase derivative estimation using autoregressive modeling in holographic interferometry *Meas. Sci. Tech.* **26** 035202

[8] Kulkarni R and Rastogi P 2014 Direct phase derivative estimation using difference equation modeling in holographic interferometry *J. Opt.* **16** 105708

[9] Kulkarni R and Rastogi P 2014 Digital holographic moiré for the direct and simultaneous estimation of strain and slope fields *Opt. Exp.* **22** 23192–201

[10] Kulkarni R and Rastogi P 2015 Phase derivative estimation from a single interferogram using a Kalman smoothing algorithm *Opt. Lett.* **40** 3794–7

[11] Liporace L A 1975 Linear estimation of nonstationary signals *J. Acous. Soc. Am.* **58** 1288–95

[12] Sodsri C 2003 Time-varying autoregressive modelling for nonstationary acoustic signal and its frequency analysis *PhD thesis* (Pennsylvania: The Pennsylvania State University)

[13] Kitagawa G 1983 Changing spectrum estimation *J. Sound Vib.* **89** 433–45

[14] Simon D 2006 *Optimal State Estimation: Kalman, H Infinity, and Nonlinear Approaches* (Hoboken, NJ: Wiley)

[15] Kayhan A S 1998 Representation and analysis of complex chirp signals *Sig. Proc.* **66** 111–6

[16] Beex A A and Shan P 1999 Time-varying Prony method for instantaneous frequency estimation at low SNR *Proc. IEEE Int. Sym. Cir. Sys.* **3** III-5–8

[17] Rajshekhar G, Gorthi S S and Rastogi P 2011 Simultaneous multidimensional deformation measurements using digital holographic moiré *Appl. Opt.* **50** 4189–97

[18] Kulkarni R and Rastogi P 2015 Iterative signal separation based multiple phase estimation in digital holographic interferometry *Opt. Exp.* **23** 26842–52

[19] Hua Y 1992 Estimating two-dimensional frequencies by matrix enhancement and matrix pencil *IEEE Trans. Sig. Proc.* **40** 2267–80

IOP Publishing

Single and Multicomponent Digital Optical Signal Analysis
Estimation of phase and its derivatives
Rishikesh Kulkarni and Pramod Rastogi

Chapter 6

Simultaneous estimation of unwrapped phase and its derivatives

The fringe analysis techniques discussed so far are focused on either the phase or phase derivative estimation. However, for the phase estimates which are obtained in a wrapped form to be useful, they need to be unwrapped. It is also important to note here that most of the phase derivative estimation methods described in chaper 5 are limited to obtaining the estimates of first order phase derivatives only. The higher order phase derivatives are known to provide a better localization of the damage in a deformed object [1]. Digital signal processing, as we will see in this chapter, is ideally suited to the simultaneous estimation of unwrapped phase and its pth order derivatives from an EPF. Of the two methods described in this chapter, the first method is based on the premise of modeling the true phase as a weighted linear combination of linearly independent basis functions. Two Kalman filtering algorithms are described to accurately estimate the weights of the basis functions. The unwrapped phase and phase derivative of any given order can be derived using the estimated weights. In the second method, the estimated phase derivative based on the stochastic evolution of the SVAR modeling of the EPF, described in chapter 5, is plugged into the phase evolution model and a nonlinear Kalman filter based algorithm is designed to obtain the unwrapped phase estimate.

6.1 Simultaneous estimation of unwrapped phase and arbitrary order phase derivatives

The basic system of obtaining a simultaneous estimation of unwrapped phase and arbitrary order phase derivatives from an EPF assumes the phase to be spatially continuous. A noisy EPF is described as

$$\Gamma(x, y) = \exp[i\phi(x, y)] + \varepsilon(x, y).$$

doi:10.1088/978-0-7503-1469-5ch6

The fringe analysis is performed either in a row-wise or column-wise manner. Considering that the fringe analysis is performed in a row-wise manner, the EPF in a given row y can be written as

$$\Gamma(x) = \exp[i\phi(x)] + \varepsilon(x). \tag{6.1}$$

Based on the assumption of the spatial continuity of the phase, $\phi(x)$ is modeled as a weighted linear combination of spatially continuous linearly independent basis functions of x as [2–4],

$$\phi(x) = \sum_{k=0}^{K-1} a_k \beta_k(x), \tag{6.2}$$

where $\beta_k(x)$ represents the kth basis function with $\beta_0(x) = 1$; a_k is the weight associated with the basis function β_k; K is the basis dimension. Considering this phase model, the pth order phase derivative with respect to x can be represented as

$$\begin{aligned} \phi^{(p)}(x) &= \frac{d^p\phi(x)}{dx^p} \\ &= \sum_{k=0}^{K-1} a_k \frac{d^p\beta_k(x)}{dx^p}. \end{aligned} \tag{6.3}$$

Although different types of continuous basis functions are reported in the literature, we will, in the present context of the estimation of phase and its derivatives, limit ourselves to considering those described in (5.42)–(5.45) and

- Gaussian radial basis functions

$$\beta_k(x) = \exp\left(-\frac{(x - x_k)^2}{2\sigma^2}\right).$$

In the case of Gaussian radial basis functions, x_k represents the location of the peak of the Gaussian function; σ represents the standard deviation which controls the width of the Gaussian. The selection of x_k and σ depends on the signal length N; and the centers of the Gaussian functions are assumed to be distributed uniformly over x for a preset value of σ.

The maximum value of the basis dimension K for which a reliable estimation of weights can be obtained is decided by the signal length N. The basis functions with a given basis dimension can be computed *a priori* and stored for later use. During the estimation of the weights of the basis functions, the values of the basis function at each pixel are retrieved from the stored arrays. This procedure greatly helps in achieving a high computational efficiency. With the basis functions computed, we may attack the problem of computing accurately the weights of the basis functions such that the error in the phase approximation is minimized. Substituting the phase in (6.2) into (6.1), we obtain

$$\Gamma(x) = \exp\left(i \sum_{k=0}^{K-1} a_k \beta_k(x)\right) + \varepsilon(x). \tag{6.4}$$

We observe from (6.4) that due to the phase approximation in (6.2), the EPF becomes a nonlinear function of the weights of the basis functions. The estimates of the weights of the basis functions can thus be obtained by applying a nonlinear parameter estimation technique. To this end, a state space formulation, wherein the weights of the basis functions are defined to be the elements of the state vector, has been developed which allows for utilizing the nonlinear Kalman filter to efficiently estimate the weights of the basis functions. Let us consider the state space formulation as

$$\boldsymbol{\mu}_x = \boldsymbol{F}\boldsymbol{\mu}_{x-1}$$
$$\boldsymbol{\Gamma}_x = \boldsymbol{h}_x(\boldsymbol{\mu}_x) + \boldsymbol{\epsilon}_x.$$

The state vector $\boldsymbol{\mu}_x$ at x is defined as a column vector containing the weights as its elements. Thus

$$\boldsymbol{\mu}_x = [a_0, a_1, \dots, a_{K-1}]^T.$$

Since the weights are constant in a given row y, the state transition matrix $\boldsymbol{F}$ can be defined as $\boldsymbol{F} = \boldsymbol{I}_K$, where $\boldsymbol{I}_K$ is an identity matrix of size $K \times K$. The measurement vector $\boldsymbol{\Gamma}_x$ is defined as a column vector

$$\boldsymbol{\Gamma}_x = [\Re(\Gamma(x)), \Im(\Gamma(x))]^T.$$

The measurement noise $\boldsymbol{\epsilon}_x$ is a column vector containing the real and imaginary parts of the complex noise $\varepsilon(x)$. The estimate of the measurement vector, which according to (6.4) is a nonlinear function of the state vector, is

$$\boldsymbol{h}_x(\boldsymbol{\mu}_x) = \left[\cos\left(\boldsymbol{\mu}_x^T \boldsymbol{\beta}\right), \sin\left(\boldsymbol{\mu}_x^T \boldsymbol{\beta}\right)\right]^T, \tag{6.5}$$

where $\boldsymbol{\beta}$ is defined as

$$\boldsymbol{\beta} = [\beta_0(x), \beta_1(x), \dots, \beta_{K-1}(x)]^T.$$

Two most commonly used nonlinear Kalman filters are the extended Kalman filter (EKF) and the unscented Kalman filter (UKF). The implementation of both of these filters is discussed in subsections 6.1.1 and 6.1.2.

6.1.1 Extended Kalman filter

The algorithm for the implementation of the EKF has the form [5]:

1. Initialize the state vector and the covariance of the error in the state estimate as

$$\hat{\boldsymbol{\mu}}_{0|0} = \mathbb{E}(\boldsymbol{\mu}_0), \tag{6.6}$$

$$\boldsymbol{P}_{0|0} = \mathbb{E}\left[(\boldsymbol{\mu}_0 - \hat{\boldsymbol{\mu}}_{0|0})(\boldsymbol{\mu}_0 - \hat{\boldsymbol{\mu}}_{0|0})^T\right]. \tag{6.7}$$

The process and measurement updates of the state vector at $x = 1, 2, \ldots, N$ are computed as in steps (2) and (3):

2. Process update

$$\hat{\boldsymbol{\mu}}_{x|x-1} = \boldsymbol{F}\hat{\boldsymbol{\mu}}_{x-1|x-1},$$
$$\boldsymbol{P}_{x|x-1} = \boldsymbol{F}\boldsymbol{P}_{x-1|x-1}\boldsymbol{F}^T.$$

3. The nonlinear function representing the state vector $\boldsymbol{h}_x(\boldsymbol{\mu}_x)$, which depicts the estimate of the measurement vector, is linearized around the current state estimate $\hat{\boldsymbol{\mu}}_{x|x-1}$ using its first order Taylor series expansion as

$$\boldsymbol{h}_x(\boldsymbol{\mu}_x) = \boldsymbol{h}_x(\hat{\boldsymbol{\mu}}_{x|x-1}) + \left.\frac{\partial \boldsymbol{h}_x}{\partial \boldsymbol{\mu}}\right|_{\hat{\boldsymbol{\mu}}_{x|x-1}} (\boldsymbol{\mu}_x - \hat{\boldsymbol{\mu}}_{x|x-1}).$$

The partial derivative matrix $\boldsymbol{H}_x = \left.\frac{\partial \boldsymbol{h}_x}{\partial \boldsymbol{\mu}}\right|_{\hat{\boldsymbol{\mu}}_{x|x-1}}$ is computed using (6.5) and the current state estimate.

4. Measurement update

$$\boldsymbol{G}_x = \boldsymbol{P}_{x|x-1}\boldsymbol{H}_x^T(\boldsymbol{H}_x\boldsymbol{P}_{x|x-1}\boldsymbol{H}_x^T + \boldsymbol{R}_\Gamma)^{-1},$$
$$\hat{\boldsymbol{\mu}}_{x|x} = \hat{\boldsymbol{\mu}}_{x|x-1} + \boldsymbol{G}_x\left[\boldsymbol{\Gamma}_x - \boldsymbol{h}_x(\hat{\boldsymbol{\mu}}_{x|x-1})\right],$$
$$\boldsymbol{P}_{x|x} = (\boldsymbol{I} - \boldsymbol{G}_x\boldsymbol{H}_x)\boldsymbol{P}_{x|x-1}(\boldsymbol{I} - \boldsymbol{G}_x\boldsymbol{H}_x)^T + \boldsymbol{G}_x\boldsymbol{R}_\Gamma\boldsymbol{G}_x^T,$$

where $\boldsymbol{R}_\Gamma$ is the EPF noise covariance matrix; $\boldsymbol{G}_x$ is the optimal Kalman gain at x; $\hat{\boldsymbol{\mu}}_{x|x-1}$ and $\hat{\boldsymbol{\mu}}_{x|x}$ represent the state estimates at the xth pixel obtained using the measurement of the EPF up to and including the $(x-1)$ th and xth pixel, respectively. Since, in this particular state space formulation, the state transition matrix $\boldsymbol{F}$ is an identity matrix and the state space model is assumed to be accurate, the process update equations do not introduce any changes in the state estimate $\hat{\boldsymbol{\mu}}_x$ and its error covariance matrix $\boldsymbol{P}_x$. Consequently, only the measurement update equations need to be implemented.

5. The estimated weights in the state vector $\hat{\boldsymbol{\mu}}_{x|x}$ at $x = N$ are substituted into (6.2) and (6.3) to obtain the respective unwrapped phase and its derivative estimates with respect to x.
6. The state estimation procedure is implemented in all the rows to obtain two-dimensional phase and phase derivative estimates. The state vector estimated at the current row is used as an initial estimate in the subsequent row. As a result, accurate state estimates are obtained and the resulting two-dimensional phase estimate is obtained in an unwrapped form.

In the EKF algorithm, the first order linearization of the estimate of the measurement vector is performed around the estimate of the state vector at each x. On the other hand, in the UKF algorithm, the underlying Gaussian distribution associated with the state vector estimate is represented with deterministically sampled points carrying complete information on the mean and covariance of the distribution. The estimate of the measurement vector computed at these sampled points provides the estimate of the mean and covariance of the distribution associated with the actual measurement vector with an accuracy reaching up to the third order of its Taylor series expansion.

6.1.2 Unscented Kalman filter

The implementation details of the UKF algorithm are provided below:

1. Initialize the mean and the covariance of the state estimate and estimation error as in (6.6) and (6.7), respectively.
2. Since the state vector and error covariance matrix are considered to remain constant during the process update operation, the state transition matrix $\boldsymbol{F}$ is an identity matrix. Thus

$$\hat{\boldsymbol{\mu}}_{x|x-1} = \hat{\boldsymbol{\mu}}_{x-1|x-1},$$
$$\boldsymbol{P}_{x|x-1} = \boldsymbol{P}_{x-1|x-1}.$$

3. Using the state estimate $\hat{\boldsymbol{\mu}}_{x|x-1}$, a set of deterministically sampled sigma points $\bar{\boldsymbol{\mu}}_x$ is computed as

$$\begin{aligned}
\bar{\boldsymbol{\mu}}_x^{(0)} &= \hat{\boldsymbol{\mu}}_{x|x-1} \\
\bar{\boldsymbol{\mu}}_x^{(k)} &= \hat{\boldsymbol{\mu}}_{x|x-1} + \sqrt{(k+\lambda)\boldsymbol{P}_{x|x-1}} \quad \forall\ k = 1, \ldots, K \\
\bar{\boldsymbol{\mu}}_x^{(k)} &= \hat{\boldsymbol{\mu}}_{x|x-1} - \sqrt{(k+\lambda)\boldsymbol{P}_{x|x-1}} \quad \forall\ k = K+1, \ldots, 2K.
\end{aligned} \tag{6.8}$$

4. The weights required in the computation of the non-linearly transformed sigma points are given as

$$w_\mu^{(0)} = \frac{\lambda}{K+\lambda},$$
$$w_P^{(0)} = \frac{\lambda}{K+\lambda} + (1 - \alpha^2 + \gamma),$$
$$w_\mu^{(k)} = w_P^{(k)} = \frac{\lambda}{2(K+\lambda)} \quad \forall\ k = 1, \ldots, 2K.$$

   In these equations, $\lambda = \alpha^2(K+\rho) - K$ is the scaling parameter; ρ is the secondary scaling parameter; the value of α decides the amount of the spread of sigma points; γ value carries prior knowledge on the distribution of $\boldsymbol{\mu}$. Reference [6] recommends to set the values of these variables as $\alpha = 10^{-3}$, $\rho = 0$ and $\gamma = 2$.
5. The sigma points in (6.8) are substituted into (6.5) to obtain the respective estimates of the measurement vectors as

$$\hat{\mathbf{\Gamma}}_x^{(k)} = \begin{bmatrix} \cos\left(\bar{\boldsymbol{\mu}}_x^{(k)}\boldsymbol{\beta}\right) \\ \sin\left(\bar{\boldsymbol{\mu}}_x^{(k)}\boldsymbol{\beta}\right) \end{bmatrix} \quad \forall\ k = 0, \ldots, 2K.$$

6. The effective estimate of the measurement vector $\hat{\mathbf{\Gamma}}_x$ is computed as the weighted sum of $\hat{\mathbf{\Gamma}}_x^{(k)}$

$$\hat{\mathbf{\Gamma}}_x = \sum_{k=0}^{2K} w_\mu^{(k)} \hat{\mathbf{\Gamma}}_x^{(k)}.$$

7. The covariance matrix $\boldsymbol{P}_\Gamma(x)$ associated with the effective estimate of the measurement vector is computed as

$$\boldsymbol{P}_\Gamma(x) = \sum_{k=0}^{2K} w_P^{(k)} \left(\hat{\mathbf{\Gamma}}_x^{(k)} - \hat{\mathbf{\Gamma}}_x\right)\left(\hat{\mathbf{\Gamma}}_x^{(k)} - \hat{\mathbf{\Gamma}}_x\right)^T + \boldsymbol{R}_\Gamma.$$

8. The matrix $\boldsymbol{P}_{\mu\Gamma}(x)$, a cross covariance between the state vector and the effective estimate of the measurement vector, is computed as

$$\boldsymbol{P}_{\mu\Gamma}(x) = \sum_{k=0}^{2K} w_P^{(k)} \left(\bar{\boldsymbol{\mu}}_x^{(k)} - \hat{\boldsymbol{\mu}}_{x|x-1}\right)\left(\hat{\mathbf{\Gamma}}_x^{(k)} - \hat{\mathbf{\Gamma}}_x\right)^T.$$

9. The measurement updates of the state and the covariance matrix are computed as

$$\begin{aligned} \boldsymbol{G}_x &= \boldsymbol{P}_{\mu\Gamma}(x)\boldsymbol{P}_\Gamma^{-1}(x), \\ \hat{\boldsymbol{\mu}}_{x|x} &= \hat{\boldsymbol{\mu}}_{x|x-1} + \boldsymbol{G}_x(\mathbf{\Gamma}_x - \hat{\mathbf{\Gamma}}_x), \\ \boldsymbol{P}_{x|x} &= \boldsymbol{P}_{x|x-1} - \boldsymbol{G}_x\boldsymbol{P}_\Gamma(x)\boldsymbol{G}_x^T. \end{aligned}$$

10. The state vector estimate $\hat{\boldsymbol{\mu}}_{x|x}$ computed at $x = N$ provides the estimates of the weights of the basis functions.

A one-dimensional example of an EPF of length $N = 512$ with a simulated phase is illustrated in figure 6.1(a). An appropriate selection of the basis dimension is crucial to the accurate estimation of the unwrapped phase and its derivatives. In general, for an accurate and efficient estimation of the weights of the basis functions, the basis dimension should be sufficiently small compared to the signal length. Accordingly, the implementation of the EKF and UKF based state estimation is performed with the basis dimension ranging between $K = 15$ and $K = 35$. A plot of the residual signal energy corresponding to the error in the estimation of the EPF computed in function

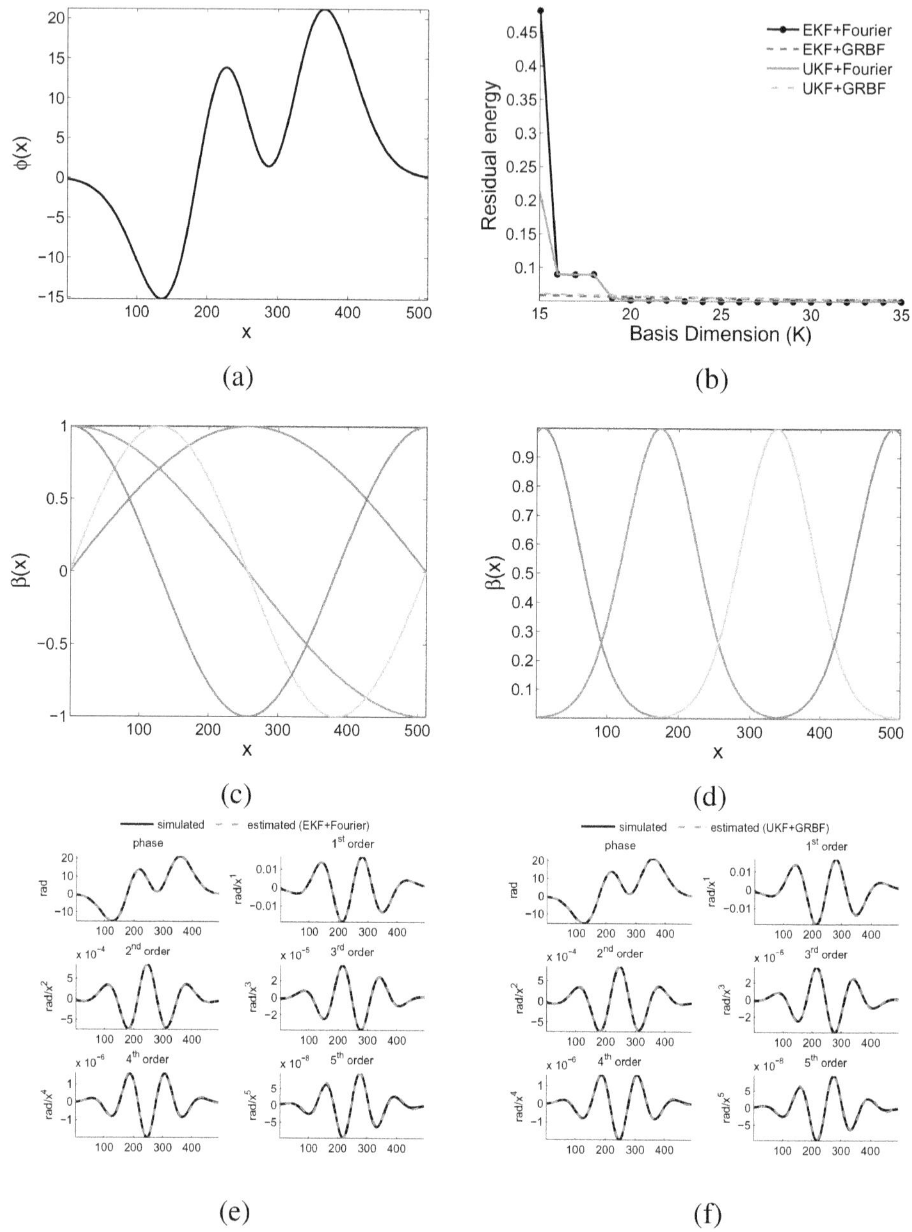

Figure 6.1. (a) Simulated one-dimensional phase; (b) residual signal energy in the estimation of EPF signal in function of basis dimension; (c) Fourier basis functions (FBFs); (d) Gaussian radian basis functions (GRBFs). Estimated phase and phase derivatives with $K = 25$ using (e) EKF + FBFs and (f) UKF + GRBFs.

of the basis dimension is shown in figure 6.1(b). The plot indicates that for the given EPF, the basis dimension can be selected in the range of 20 to 30 to achieve the required estimation accuracy using either the EKF or UKF algorithm. For the purposes of illustration, the Fourier and Gaussian radial basis functions with the basis dimension of $K = 5$ are shown in figures 6.1(c) and (d), respectively. The EKF and UKF based state estimation performed to compute the estimates of the unwrapped phase and its derivatives up to the sixth order with $K = 25$ are shown in figures 6.1(e) and (f), respectively.

Two-dimensional examples of the phase fringe patterns simulated at SNR = 10 dB are shown in figures 6.2(a) and (b), respectively. For the phase fringe patterns given in figures 6.2(a) and 6.2(b), the estimated unwrapped phase, and the first and second order phase derivatives with respect to x computed using the EKF algorithm with Fourier basis functions are shown in figures 6.2(c), (e) and (g), respectively. Idem, using the UKF algorithm with Gaussian radial basis functions, are shown in figures 6.2(d), (f) and (h), respectively. The basis dimension $K = 25$ is selected for both the examples.

Figures 6.3(a) and (b) show the examples of experimentally recorded phase fringe patterns. For the phase fringe patterns in figures 6.3(a) and 6.3(b), the sets of estimated unwrapped phase, and the first and second order phase derivatives with respect to x computed using the EKF algorithm with Fourier basis functions, and using the UKF algorithm with Gaussian radial basis functions are shown in figures 6.3(c), (e) and (g), and figures 6.3(d), (f) and (h), respectively. The basis dimension $K = 20$ is selected in both the examples.

6.2 Closed fringe demodulation

Although the problem of closed fringe demodulation for phase estimation is discussed in section 2.2, the phase estimation method described in the previous section is also applicable in the case of close fringe pattern demodulation [7, 8]. Representing a normalized closed fringe signal as

$$\Gamma(x, y) = \cos(\phi(x, y)),$$

the fringe analysis can be performed either along the rows or columns depending upon the amount of fringe density along the respective directions. Let us suppose that the analysis is performed in a row-wise manner. The fringe signal in a given row y can be represented as

$$\Gamma(x) = \cos(\phi(x)).$$

Similar to the approach described in section 6.1, the phase $\phi(x)$ is modeled as a weighted linear combination of a certain number of linearly independent basis functions. Say, we consider the Fourier basis functions of x. As any linear combination of Fourier basis functions is a continuous function, a sign-unambiguous unwrapped estimate of $\phi(x)$ is obtained even in the presence of *critical points*, the points with zero phase slopes. Based on the phase representation in (6.2), the approximation of the fringe signal can be written as

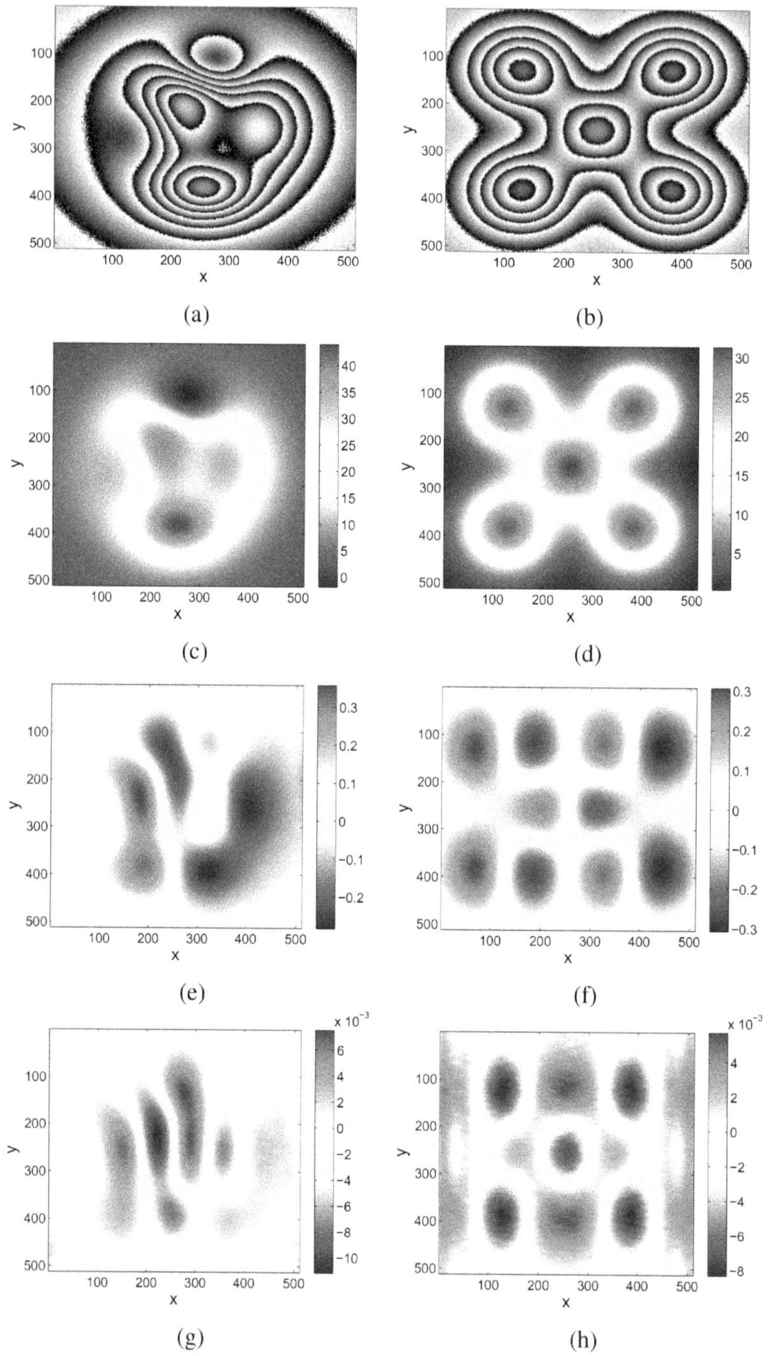

Figure 6.2. (a), (b) Simulated phase fringe patterns. Estimated (c) phase, and (e) first and (g) second order phase derivative using EKF and FBFs with $K = 25$. Idem for (d), (f) and (h) using UKF and GRBFs with $K = 25$.

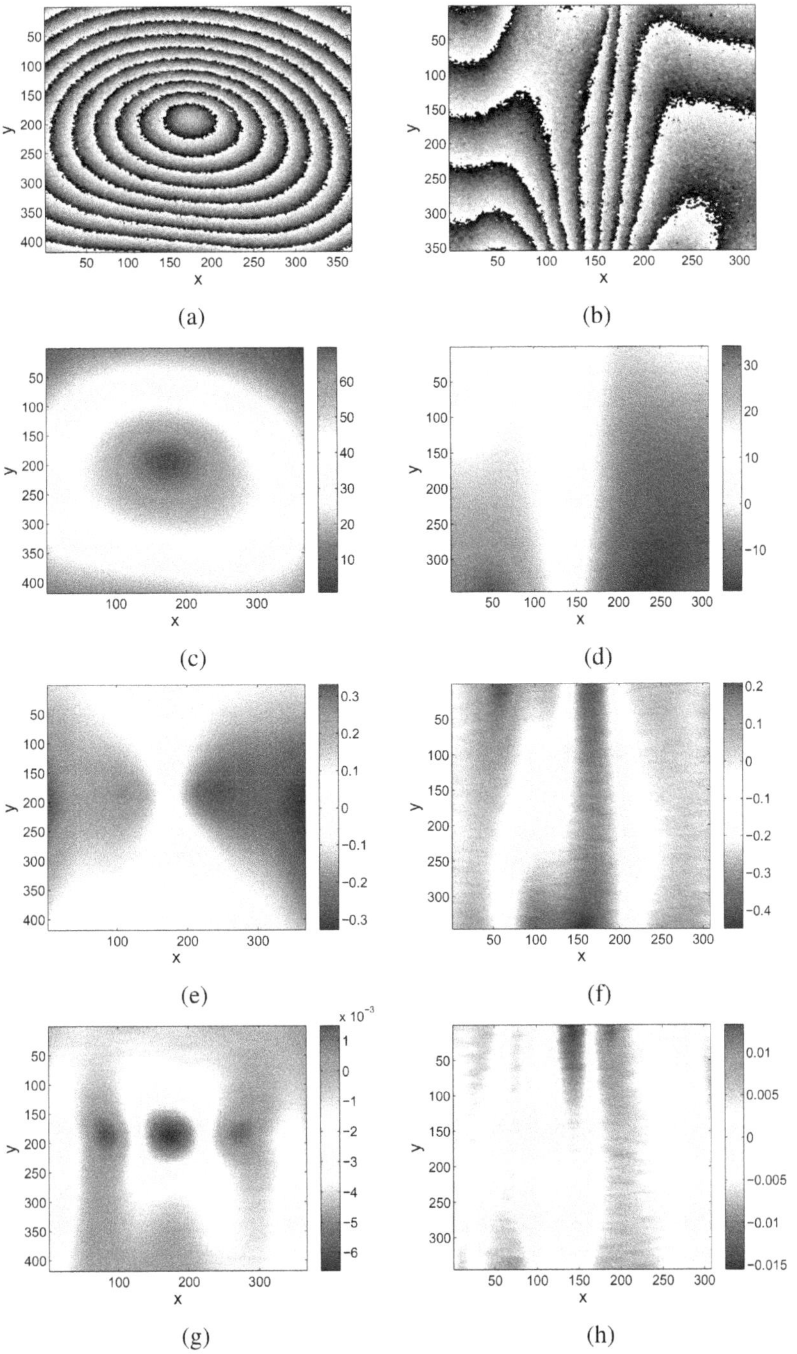

Figure 6.3. Experimentaly recorded phase fringe patterns in (a) digital holographic interferometry setup corresponding to the out-of-plane displacement of an aluminum plate (b) digital speckle pattern interferometry setup corresponding to the in-plane displacement of a fiber-reinforced wood specimen. Estimated (c) phase, and (e) first and (g) second order phase derivatives using UKF and GRBFs with $K = 20$. respectively, and (d), (f) and (h) using EKF and FBFs with $K = 20$, respectively.

$$\hat{\Gamma}(x) = \cos\left(\sum_{k=0}^{K-1} a_k \beta_k(x)\right).$$

In order to accurately estimate the weights of the basis functions, the weights are represented as the elements of a state vector in a state space model. The state estimation is performed in each row for obtaining a complete two-dimensional estimate of the phase. Note that the estimate of the state vector in the current row is used in the initialization step of the EKF or the UKF algorithm in the subsequent row. As a result, the two-dimensional estimate of the phase is obtained in an unwrapped form. Since the state estimation is obtained from a closed fringe signal, the state estimate may not converge to the true solution due to a lack of sufficient information and existence of more than one possible solution. Accordingly, iterative state estimation is performed in each row to obtain robust estimates of the state vector. However, the iterative operation also results in an increase in the computation time which could be notably undesirable in the case of large size fringe patterns. Since the basis dimension K is in general much smaller than either M or N (the length of the fringe signal), the fringe signal information at pixels spaced one or more pixels apart are utilized in the state estimation. This results in a significant reduction in the computation time without affecting the accuracy of the state estimation or for that matter that of the phase estimation.

Selection of seed row/column and optimal basis dimension

Information on the fringe frequency is essential in order to select the seed row/column to begin the demodulation of a closed fringe pattern. The total local fringe frequency (TLF) has been defined in [9] as

$$\omega(x, y) = \sqrt{\omega_x^2(x, y) + \omega_y^2(x, y)},$$

where $\omega(x, y)$ represents the total fringe frequency at pixel (x, y); $\omega_x(x, y)$ and $\omega_y(x, y)$ represent the local fringe frequencies along the x and y directions, respectively. First, the sum of $\omega(x, y)$ along the x and y directions are computed as $\Sigma_\omega(x)$ and $\Sigma_\omega(y)$, respectively. Next, the column/row corresponding to the maximum value of $\Sigma_\omega(x)$ and $\Sigma_\omega(y)$ is selected as the seed column/row. Note that only a rough estimation of the total fringe frequency is required for the selection of the seed column/row. The windowed Fourier ridges (WFR) algorithm based technique [10] is utilized for the estimation of the local fringe frequencies. Considering that y_{seed} is selected as the seed row based on this criterion, the implementation of the proposed method is done starting this row. The optimal basis dimension is computed in the seed row based on the residual signal energy minimization approach for which at first a fixed set of basis dimensions is defined. The fringe signal in y_{seed} is estimated using the weights estimated for each selection of the basis dimension. Subsequently, the optimal basis dimension is selected as

$$K_{opt} = \min_{K}\left\{\left[\sum_{x=1}^{N}(\Gamma(x) - \hat{\Gamma}(x))^2\right] < E_{thr}\right\},$$

where E_{thr} is the maximum pre-defined residual signal energy. For example, E_{thr} can be selected as 0.01, 0.02, 0.03, etc, depending on the noise strength in the fringe signal.

Low density closed fringe demodulation

In general, the closed fringe demodulation techniques described in section 2.2 are mainly based on the accurate local frequency estimation of the closed fringes, providing accurate phase estimation for fringe patterns of sufficiently high density, and unreliable phase estimation for fringe patterns of low fringe densities. Having said that, the method described in section 6.1 is still capable of demodulating low density closed fringe patterns. However, in the case of decidedly low density fringe patterns, the Fourier basis functions may find it difficult to map the slowly varying spatial phase. Such scenarios call for the use of the polynomial basis functions while keeping the procedure for selecting optimal basis dimension and seed column/row evaluation the same as described in the previous subheading.

A simulation example of a closed fringe pattern of size 256 × 256 is shown in figure 6.4(a). The fringe pattern is corrupted by the additive white Gaussian noise at SNR of 20 dB. The total local frequency computed using the windowed Fourier ridges algorithm is shown in figure 6.4(b). The seed column and the optimal basis dimension are found as $x_{seed} = 203$ and $K_{opt} = 10$, respectively. The simulated and estimated phases in the seed column are shown in figure 6.4(c). First, as shown in figure 6.4(d), we obtain the unwrapped phase estimate starting from seed column $x_{seed} = 203$ till $x = 256$. The phase is next estimated starting from $x = x_{seed} - 1$ till $x = 1$ to obtain the complete phase distribution as shown in figure 6.4(e). The re-wrapped phase fringe pattern is shown in figure 6.4(f) to help relate back to the corresponding phase fringe pattern.

An experimentally recorded closed fringe pattern in a classical holographic interferometry setup is shown in figure 6.5(a). The fringe pattern normalized using the adaptive monogenic filter described in section 3.1.4 is shown in figure 6.5(b). The unwrapped phase computed using the Fourier basis functions with basis dimension $K = 10$ and its corresponding re-wrapped phase maps are shown in figures 6.5(c) and (d), respectively.

6.3 Simultaneous estimation of unwrapped phase and its first order phase derivative

In this section, we will describe an approach for the simultaneous estimation of phase and its first order derivative based on the parallel implementation of two Kalman filter algorithms. While the linear Kalman filter is dedicated to the phase derivative estimation, the nonlinear Kalman filter is utilized for the unwrapped phase estimation. The phase derivative estimation is performed using the SVAR

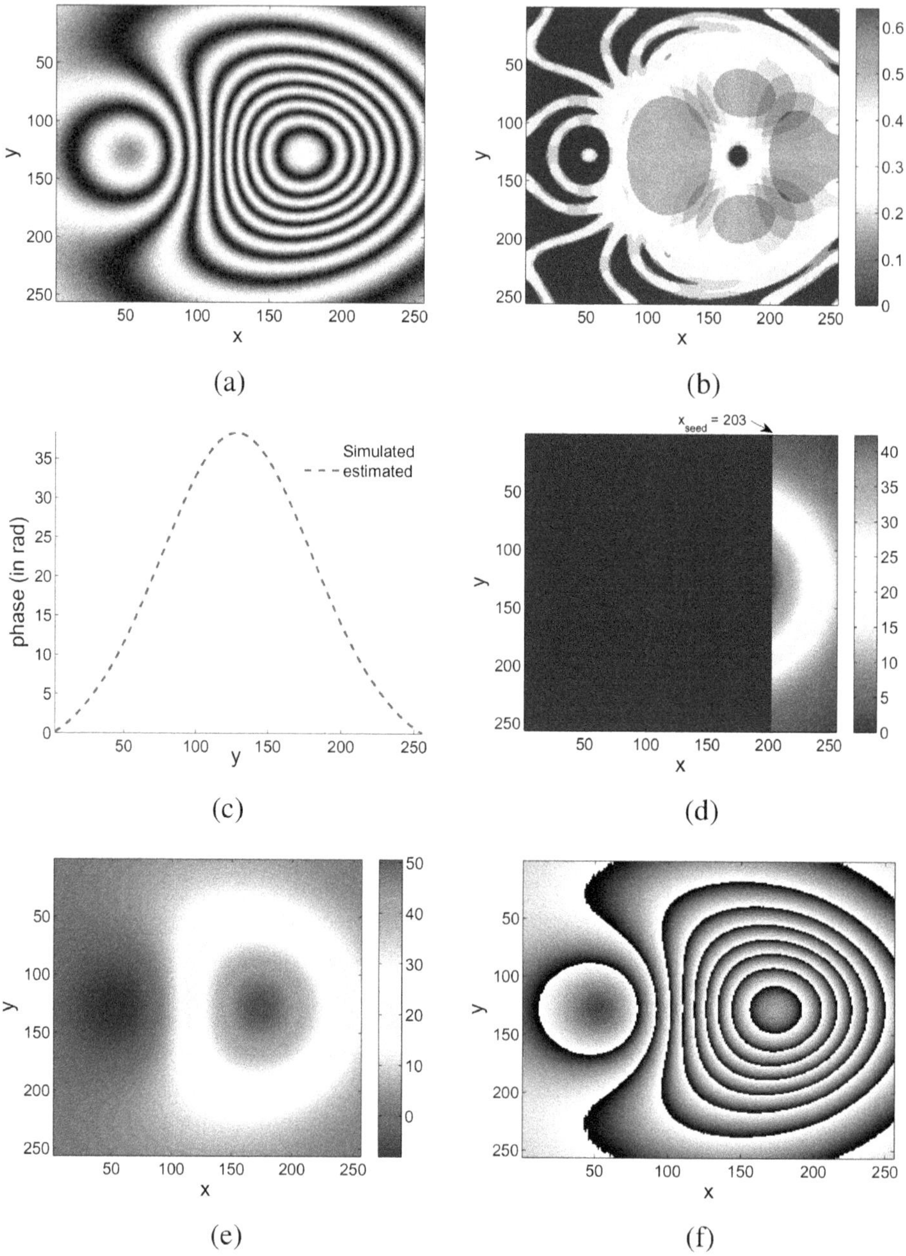

Figure 6.4. (a) Simulated closed fringe pattern; (b) total local fringe frequency (TLF); (c) simulated and estimated phase in the seed column $x = 203$; (d) intermediate estimated unwrapped phase; (e) complete estimated unwrapped phase; (f) re-wrapped estimated phase. All phase values are in radians.

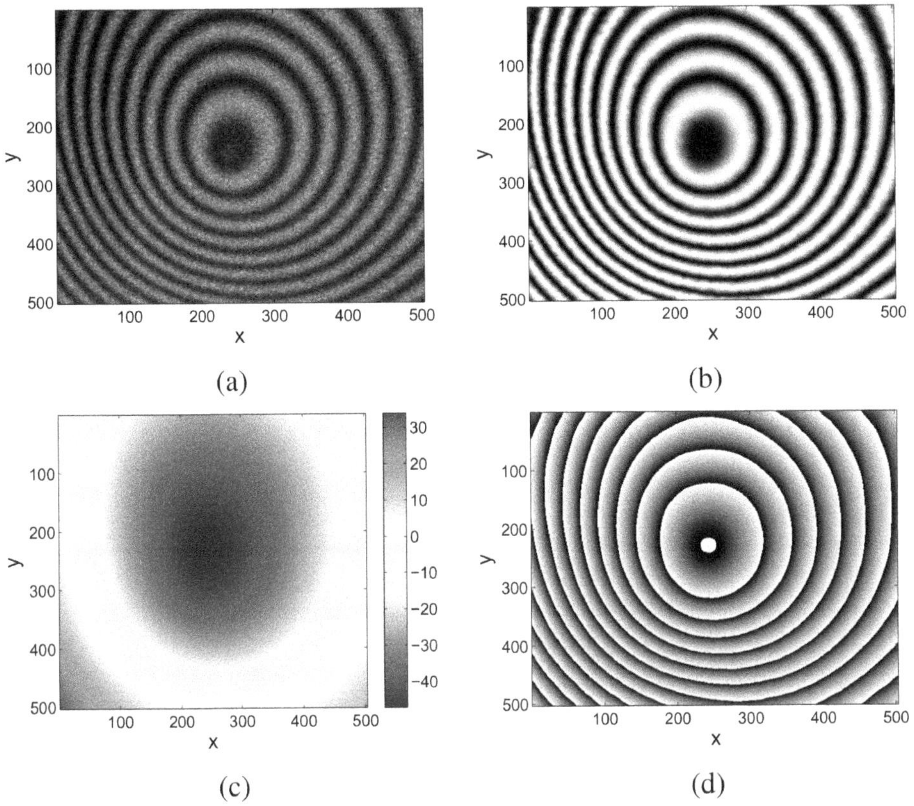

Figure 6.5. (a) Experimentally recorded closed fringe pattern in a classical holographic interferometry setup corresponding to the out-of-plane displacement of an aluminum plate (b) normalized fringe pattern; (c) estimated unwrapped phase (in radians); (d) re-wrapped phase estimate.

modeling of the EPF under the assumption of the stochastic spatial evolution of the AR coefficients, as explained in section 5.2.2. We have seen that during the backward filtering operation in the stochastic coefficient estimation procedure, the state estimate $\boldsymbol{\mu}(x)$ obtained at each pixel provides the required AR coefficients' estimates as in (5.46). It thus follows that at each pixel the phase derivative estimate is derived using the estimated SVAR coefficients at that pixel.

The unwrapped phase estimate at each pixel is subsequently obtained by utilizing the estimated phase derivative. Consider the relationship between two successive phase values [11]

$$\phi(x+1) = \phi(x) + \dot{\phi}(x) + \eta(x), \tag{6.9}$$

where the phase at the $(x+1)$ th pixel is computed as the sum of the phase at the xth pixel and the phase derivative at the xth pixel; $\eta(x)$ is the possible error in this phase evolution model. Equation (6.9) can be represented as a state space model

$$\mu_{x+1} = \mu_x + u(x) + \eta(x),$$

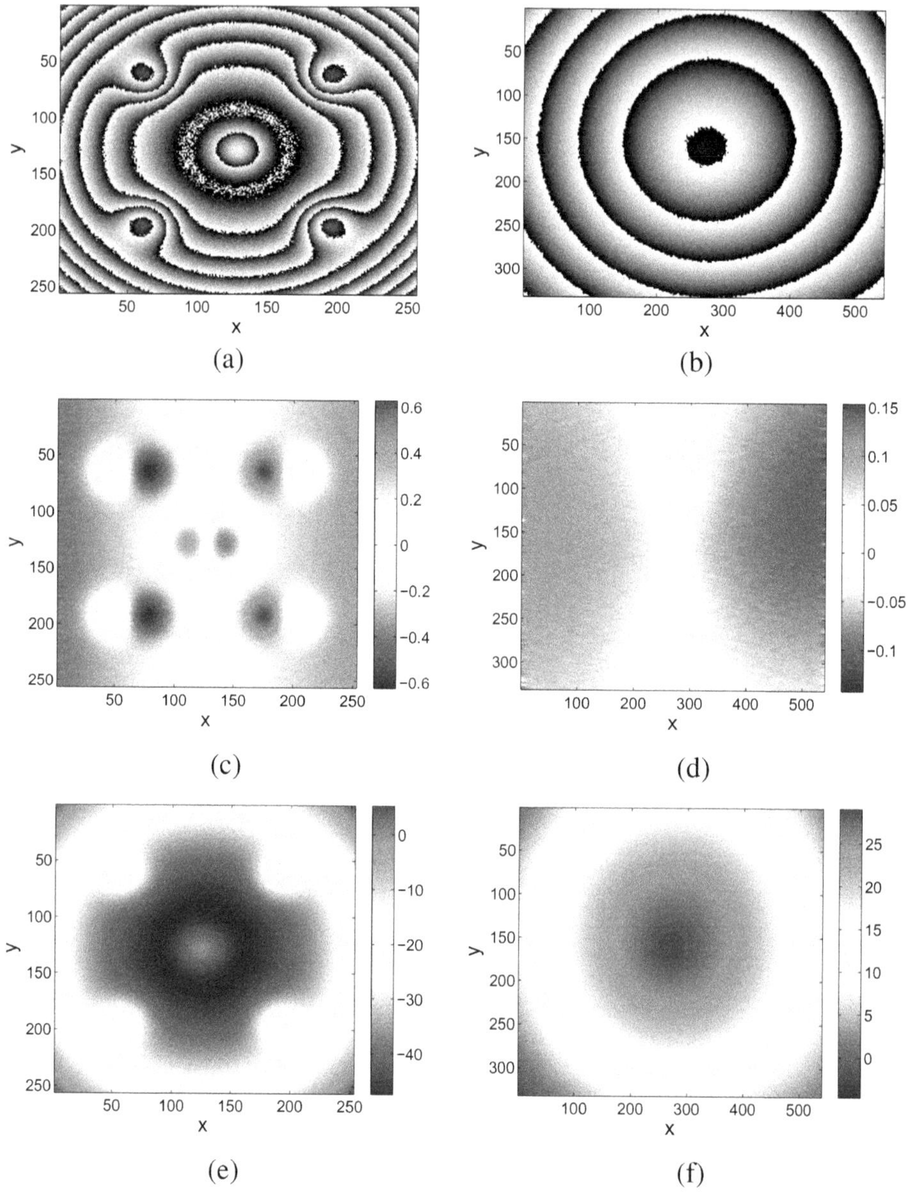

Figure 6.6. (a) Simulated and (b) experimentally recorded phase fringe patterns. (c) and (d) Estimates of first order phase derivatives with respect to x; (e) and (f) unwrapped phase estimates.

where the state vector μ_x represents the unwrapped phase at x; $u(x)$ is the control input vector which in this case is the estimated phase derivative. $\eta(x)$ can be treated as the process noise which is modeled as a zero mean white Gaussian noise with variance ζ^2. Since the phase derivative is estimated during the backward filtering, the state transition equation can be modified as

$$\mu_x = \mu_{x+1} - u(x+1) + \eta(x). \tag{6.10}$$

For the estimation of the unwrapped phase, the real and imaginary parts, that is, the in-phase and quadrature components of the EPF, are accounted for in the measurement vector $\boldsymbol{\Gamma}(x)$ as

$$\boldsymbol{\Gamma}(x) = \begin{bmatrix} \cos(\mu_x) \\ \sin(\mu_x) \end{bmatrix} + \boldsymbol{\varepsilon}(x). \tag{6.11}$$

In the procedure of unwrapped phase estimation, the state space model is defined by (6.10) and (6.11). Note that the measurement vector is a nonlinear function of the state vector. Hence, the EKF or UKF algorithm can be used to estimate the state vector at each pixel location. In the row-wise implementation of this method, the unwrapped phase estimates are obtained along the rows. In order to obtain unwrapped phase estimation along the columns, one can either perform a simple phase unwrapping based on an appropriate addition of integer multiples of 2π to the estimated phase, or, first estimate the unwrapped phase along any given column x and then add appropriate offsets to the estimated unwrapped phase along the respective rows.

Two-dimensional examples of simulated and experimentally recorded phase fringe patterns are shown in figures 6.6(a) and 6.6(b), respectively. The stochastic coefficient estimation based method is implemented with $\kappa = 1$ and AR model order $P = 2$. The phase derivative estimates computed with respect to x are shown in figures 6.6(c) and 6.6(d), respectively. The unwrapped phase estimates are shown in figures 6.6(e) and 6.6(f), respectively.

References

[1] Lopes H, Ferreira F, Araújo dos Santos J V and Moreno-García P 2014 Localization of damage with speckle shearography and higher order spatial derivatives *Mech. Sys. Sig. Proc.* **49** 24–38

[2] Kulkarni R and Rastogi P 2015 Phase estimation from noisy phase fringe patterns using linearly independent basis functions *J. Opt.* **17** 125611

[3] Kulkarni R and Rastogi P 2016 Simultaneous estimation of phase and its *p*th order derivatives *J. Opt.* **18** 055603

[4] Kulkarni R and Rastogi P 2016 Estimation of phase and its higher order derivatives from a single complex interferogram *Exp. Mech.* **2016** 1029–38

[5] Simon D 2006 *Optimal State Estimation: Kalman, H Infinity, and Nonlinear Approaches* (Hoboken, NJ: Wiley) pp 395–433

[6] Wan E A and Van Der Merwe R 2000 The unscented Kalman filter for nonlinear estimation *IEEE Adaptive Systems for Signal Processing, Communications, and Control Symp. 2000 (AS-SPCC)* pp 153–158

[7] Kulkarni R and Rastogi P 2016 Simultaneous estimation of unwrapped phase and phase derivative from a closed fringe pattern *Opt. Lasers Eng.* **87** 168–75

[8] Kulkarni R and Rastogi P 2016 Closed fringe demodulation using phase decomposition by Fourier basis functions *J. Opt. Soc. Am.* A **33** 1120–5

[9] Kemao Q and Soon S H 2007 Sequential demodulation of a single fringe pattern guided by local frequencies *Opt. Lett.* **32** 127–9
[10] Kemao Q 2007 Two-dimensional windowed Fourier transform for fringe pattern analysis: principles, applications and implementations *Opt. Lasers Eng.* **45** 304–17
[11] Loffeld O, Nies H, Knedlik S and Yu W 2008 Phase unwrapping for sar interferometry–a data fusion approach by Kalman filtering *IEEE Trans. Geosci. Remote Sens.* **46** 47–58

Chapter 7

Signal separation techniques: multicomponent spatial fringe analysis

In the context of multi-beam digital holographic interferometry, the problem of estimation of multiple phases or their derivatives from a single frame of the EPF can be looked upon from a perspective different than that offered by the parametric estimation methods discussed in chapters 4 and 5. Driven by this consideration, this chapter is based on the separation of the multicomponent EPF into its constituent signal components. At this point in time, this approach enables the signal component fringe analysis techniques to be applied to these individually separated signal components, and thereby contribute to providing the multiple phase and/or phase derivative estimates. However, the combination of these procedures assumes the ability of the methods developed in this context to discern between the single component signals making up the multicomponent EPF. The signal separation techniques are therefore broadly classified based on their ability to discriminate the individual signal components from one another depending on their characteristics of *frequency and amplitude discrimination.*

7.1 Signal separation based on frequency discrimination

For the sake of simplicity, let us consider a multicomponent EPF containing two signal components. Practically, such a situation is generated in a two-beam digital holographic interferometric setup whose schematic is shown in figure 4.1. The EPF corresponding to the object deformation recorded in this setup is

$$\Gamma(x, y) = A_1(x, y)\exp(i\phi_1(x, y)) + A_2(x, y)\exp(i\phi_2(x, y)) + \varepsilon(x, y).$$

The presence of two signal components implies that the arctan operation performed on $\Gamma(x, y)$ can no longer provide the estimates of ϕ_1 and ϕ_2. Moreover, since the frequency spectra of the signal components overlap each other, it is not even feasible to analyze these signal components in the frequency domain by computing the FT of the EPF. The processing of multicomponent signals has been one of the most

doi:10.1088/978-0-7503-1469-5ch7

difficult challenges in fringe analysis, and has for long restricted its application to only single component signals. Efforts have been reported with by and large satisfactory results. Procedures have been developed that allow for unambiguously discerning between the single component signals making up the multicomponent signal. In general, the overlapping spectra of the two signal components lie around the zero frequency in the frequency domain. The spectra of these signal components can be separated by adding a carrier frequency to one of the two object beams. Since the setup is sensitive to displacement measurement in the x and z directions, the carrier along the y direction is added in order to ensure that the signal spectrum is unaffected by its introduction. Considering that the carrier frequency is introduced in beam 2, the carrier modified EPF can be represented as

$$\Gamma(x, y) = A_1(x, y)\exp(i\phi_1(x, y)) + A_2(x, y)\exp\big(i(\omega_c y + \phi_2(x, y))\big) + \varepsilon(x, y).$$

where ω_c is the carrier frequency. The two-dimensional FT $\mathscr{F}_\Gamma$ of the modified EPF is

$$\mathscr{F}_\Gamma(\omega_x, \omega_y) = \mathscr{F}_1(\omega_x, \omega_y) + \mathscr{F}_2(\omega_x, \omega_y - \omega_c) + \mathscr{F}_\varepsilon(\omega_x, \omega_y),$$

where $\mathscr{F}_\varepsilon$ represents the FT of the noise component. The Fourier spectra of the signal components are given as

$$\mathscr{F}_1(\omega_x, \omega_y) = \mathscr{F}\left\{A_1(x, y)\exp(i\phi_1(x, y))\right\},$$
$$\mathscr{F}_2(\omega_x, \omega_y) = \mathscr{F}\left\{A_2(x, y)\exp(i\phi_2(x, y))\right\}.$$

Due to the introduction of the carrier, $\mathscr{F}_1(\omega_x, \omega_y)$ and $\mathscr{F}_2(\omega_x, \omega_y)$ are separated in the frequency domain. Application of spectral filtering and then inverse FT allows for reconstructing the two signal components as $\hat{\Gamma}_1(x, y) = \hat{A}_1(x, y)\exp(i\hat{\phi}_1(x, y))$ and $\hat{\Gamma}_2(x, y) = \hat{A}_2(x, y)\exp(i\hat{\phi}_2(x, y))$. The phase estimates are obtained as

$$\hat{\phi}_1(x, y) = \arctan\left[\frac{\Im\left\{\hat{\Gamma}_1(x, y)\right\}}{\Re\left\{\hat{\Gamma}_1(x, y)\right\}}\right], \tag{7.1}$$

$$\hat{\phi}_2(x, y) = \arctan\left[\frac{\Im\left\{\hat{\Gamma}_2(x, y)\right\}}{\Re\left\{\hat{\Gamma}_2(x, y)\right\}}\right]. \tag{7.2}$$

In contrast to the unwrapped phase estimation obtained using the parametric techniques described in chapter 4, the estimated phases in (7.1) and (7.2) are in wrapped forms. The unwrapped phase estimates can be obtained using a suitable phase unwrapping technique described in chapter 3. These unwrapped phases provide the in-plane and out-of-plane displacement measurements in accordance with (4.4) and (4.5). It is important to note that the unwrapped phase estimate $\hat{\phi}_2$ also includes phase related to the carrier. This linear phase can be removed by estimating the carrier frequency. The slope of the estimated unwrapped phase is computed with respect to y and the mean value of slope is treated as the estimate of the carrier frequency.

As an example, a simulated two-dimensional moiré pattern containing two signal components at SNR = 10 dB is shown in figure 7.1(a). The Fourier spectrum of the

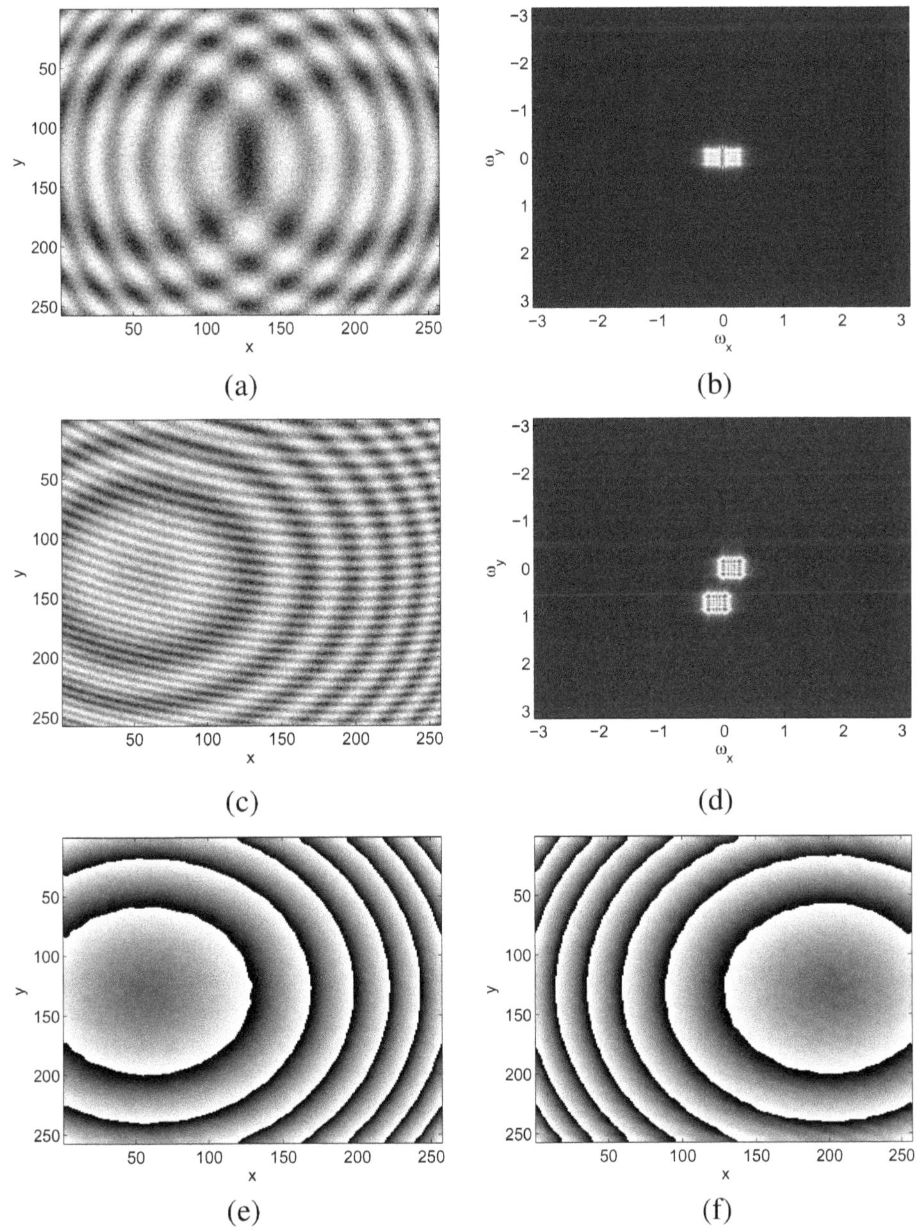

Figure 7.1. (a) Simulated moiré fringe pattern; (b) Fourier spectrum of the EPF; (c) modulated carrier moiré fringe pattern; (d) Fourier spectrum of the modulated carrier EPF. Estimated wrapped phases (e) ϕ_1 and (f) ϕ_2.

EPF signal shown in figure 7.1(b) shows the spectral overlapping of the two signal components. A carrier is added into the second signal component and the modulated carrier moiré pattern is shown in figure 7.1(c). The effect of carrier introduction is visible in the Fourier spectrum of the modulated carrier EPF, figure 7.1(d), wherein the spectra of the two signal components are separated. These two signal components filtered in the frequency domain are reconstructed back in the spatial domain. The wrapped phase estimates related to the separated signal components are shown in figures 7.1(e) and (f), respectively.

While the introduction of carrier allows us to separate the signal components in the frequency domain, it also puts a limit on the dynamic range of the phases that can be reliably estimated using this method. This limitation becomes even more evident in the case of an EPF containing three signal components. Note that, in the presence of three signal components, separate carrier frequencies will need to be added into at least two object beams. These carrier frequencies should be known *a priori* for the purposes of compensating the linear phase terms from the estimated phase maps.

7.2 Signal separation based on amplitude discrimination

Another approach to discriminating the signal components is based on the amplitude discrimination criterion. In this approach, the amplitudes (A_1 and A_2) of the two signal components are adjusted in a manner to make them in a strict sense distinct from each other. Such a criterion is established experimentally by injecting different intensities of light in the two object beams. The signal components can thus be separated in an iterative manner based on the difference in the signal amplitudes. Since, as opposed to the frequency discrimination criterion, no addition of carrier is required, this method allows us to obtain the phase estimates with a comparatively higher dynamic range.

7.2.1 Windowed Fourier transform

The ease of implementation and high noise robustness offered by the windowed Fourier transform (WFT) has made it a versatile tool in the domain of fringe analysis. The principle of WFT has been discussed in chapter 3 in the context of phase estimation from a single component EPF.

In order to extend, the applicability of the WFT to demodulate the EPF containing multiple signal components, we call upon one of its sister techniques, the windowed Fourier filtering (WFF). The WFT of the EPF is defined as

$$WF_{\Gamma}(u, v, \varsigma, \nu) = \sum_{x=0}^{N-1} \sum_{y=0}^{M-1} \Gamma(x, y) g^{*}_{u,v,\varsigma,\nu}(x, y), \tag{7.3}$$

and the inverse windowed Fourier transform (IWFT) is

$$\Gamma(x, y) = \frac{1}{4\pi^2} \sum_{\varsigma=-\pi}^{\pi} \sum_{\nu=-\pi}^{\pi} \sum_{u=0}^{N-1} \sum_{v=0}^{M-1} WF_{\Gamma}(u, v, \varsigma, \nu) g_{u,v,\varsigma,\nu}(x, y) \Delta\varsigma \Delta\nu. \tag{7.4}$$

The WFT basis function is defined as,

$$g_{u,v,\varsigma,\nu}(x, y) = g(x - u, y - v)\exp(i[\varsigma x + \nu y]),$$

with the Gaussian window function $g(x, y) = \exp\left[-\frac{x^2}{2\sigma_x^2} - \frac{y^2}{2\sigma_y^2}\right]$. σ_x and σ_y represent the standard deviations of the Gaussian function along x and y, respectively. In the WFF based fringe denoising, the WFT spectrum of the EPF is *thresholded* such that the noise contribution is removed from the fringe pattern. The filtered EPF can be represented as

$$\bar{\Gamma}(x, y) = \frac{1}{4\pi^2}\sum_{\varsigma=\varsigma_1}^{\varsigma_2}\sum_{\nu=\nu_1}^{\nu_2}\sum_{u=0}^{N-1}\sum_{y=0}^{M-1}\overline{WF_\Gamma}(u, v, \varsigma, \nu)g_{u,v,\varsigma,\nu}(x, y)\Delta\varsigma\Delta\nu,$$

with

$$\overline{WF_\Gamma}(u, v, \varsigma, \nu) = \begin{cases} WF_\Gamma(u, v, \varsigma, \nu), & \text{if } |WF_\Gamma(u, v, \varsigma, \nu)| \geqslant Thr \\ 0, & \text{otherwise} \end{cases}.$$

where $\overline{WF_\Gamma}(u, v, \varsigma, \nu)$ represents the thresholded WFT spectrum of the EPF. The pre-defined threshold value *Thr* is selected based on the noise power in the fringe pattern. High value of *Thr* is selected in the case of high noise and vice versa. This value is selected empirically by the user. While studying the effect of the selection of *Thr*, Kemao [1] emphasized the importance that the size of the Gaussian window function holds in deciding the maximum value of the modulus of the WFT spectrum given in (7.3). Hence, the value of the modulus is also fixed for a fixed window size. Based on the fact that the noise is spread across the whole frequency range while the fringe signal is mainly concentrated in the low frequency region, the spectrum modulus corresponding to the signal component is, in general, sufficiently higher than the spectrum modulus corresponding to the noise component. For the analysis window of size, $\sigma_x = \sigma_y$, the maximum value of the spectrum modulus suggested is $2\sqrt{\pi}\sigma_x$ [1] . Typically, the maximum value of the spectrum modulus reduces with the increase in the noise level. The ability to control the selection of the threshold finally allows us to separate the two signal components present in the EPF by performing the WFF operation in an iterative manner and selecting appropriate threshold values at each iteration. The appropriate selection of this threshold value is crucial in the reliable implementation of the signal separation. The steps for signal separation based on the amplitude discrimination criterion take the form:

1. Without loss of generality, the intensities of beams 1 and 2 are set such that $A_1 > A_2$. The object illumination is supposed to be uniform over the entire object surface for both the beams.
2. The WFT of the EPF is computed, as in (7.3), with the Gaussian window of size $\sigma_x = \sigma_y$. As per the WFF algorithm, the spectrum of the WFT is thresholded with a pre-defined value of $Thr = Thr_1$. Thr_1 is selected such that the spectral contribution due to the signal component with amplitude A_2 is removed along with that of the noise component.

3. After this filtering operation, the signal component with amplitude A_1 is reconstructed as $\hat{\Gamma}_1(x, y)$, as in (7.4). The wrapped phase estimate $\hat{\phi}_{1w}$ of ϕ_1 is computed as

$$\hat{\phi}_{1w} = \arctan\left[\frac{\Im\left\{\hat{\Gamma}_1(x, y)\right\}}{\Re\left\{\hat{\Gamma}_1(x, y)\right\}}\right].$$

4. In order to separate the signal component with amplitude A_1 from the EPF, the estimate of amplitude A_1 is computed as

$$\hat{A}_1 = \left\langle \Re\left\{\Gamma(x, y)\exp(-i\hat{\phi}_{1w}(x, y))\right\}\right\rangle.$$

5. The signal remaining after the separation of the signal component with amplitude A_1 is computed as

$$\hat{\Gamma}_2(x, y) = \Gamma(x, y) - \hat{A}_1\exp(i\hat{\phi}_{1w}(x, y)).$$

   Now, the signal component with amplitude A_2 is the dominant signal in $\hat{\Gamma}_2(x, y)$. A small amount of residue of the signal component with amplitude A_1, in addition to the noise component, is present in $\hat{\Gamma}_2(x, y)$.
6. The signal $\hat{\Gamma}_2(x, y)$ turns out to be basically a single component signal. Consequently, the WFF algorithm applied to it with the threshold value Thr_2 set such that $Thr_2 < Thr_1$, provides the denoised spectrum associated with the signal component with amplitude A_2 (of the EPF). Based on the fact that $A_1 > A_2$, the threshold selection is set as $Thr_2 < Thr_1$.
7. After this filtering operation, the signal component with amplitude A_2 is reconstructed as $\hat{\Gamma}_2(x, y)$. The wrapped phase estimate $\hat{\phi}_{2w}$ of ϕ_2 is computed as

$$\hat{\phi}_{2w} = \arctan\left[\frac{\Im\left\{\hat{\Gamma}_2(x, y)\right\}}{\Re\left\{\hat{\Gamma}_2(x, y)\right\}}\right].$$

8. The amplitude estimate $\hat{A}_2$ associated with the second signal component of the EPF is computed as

$$\hat{A}_2 = \left\langle \Re\left\{\hat{\Gamma}_2(x, y)\exp(-i\hat{\phi}_{2w}(x, y))\right\}\right\rangle.$$

9. The procedure of separation of the signal component with amplitude A_2 from the EPF is similar to that described in Step 5 for the separation of the first signal component. The accuracy of the reconstruction of individual signal components and in consequence that of the respective phase estimations is significantly improved by performing the reconstruction of the individual signal components followed by their removal from the EPF in an iterative manner.

10. Phase unwrapping is applied to the wrapped phase estimates, $\hat{\phi}_{1w}$ and $\hat{\phi}_{2w}$, to obtain the unwrapped estimates, ϕ_1 and ϕ_2, respectively. A simple unwrapping algorithm suffices for this purpose as the wrapped phase estimates are in denoised forms.

Since both the phases together carry the information on the in-plane and out-of-plane displacement components (equations (4.4) and (4.5)), the two signal components are correlated to each other. We also note that the ridge values of the individual signal components have the same relationship to the signal amplitudes [1]. The ratio of the ridge values of the individual signal components is thus equal to the ratio of the amplitudes of the two signal components at each pixel. This observation is conducive in the appropriate selection of the threshold values. Thus we have,

$$\frac{Thr_1}{Thr_2} = \frac{A_1}{A_2}. \tag{7.5}$$

It is, however, of interest to note that although a strict adherence to the relationship given in (7.5) is not essential, this relationship is still helpful because while providing the flexibility of threshold selection to the user it makes the method less sensitive to its variations.

Figure 7.2(a) shows a 256 × 256 sized simulated moiré pattern containing two signal components at SNR = 10 dB. The spectral overlap of the two signal components can be observed from the Fourier spectrum of the EPF shown in figure 7.2(b). We chose a Gaussian window of size $\sigma_x = 10$ and $\sigma_y = 10$. In the first iteration of the algorithm, the threshold values are selected as $Thr_1 = 20$ and $Thr_2 = 10$. Note that the selection of threshold values is performed empirically. The wrapped estimates of ϕ_1 and ϕ_2 are shown in figures 7.2(c) and 7.2(d), respectively. Since in each phase map, some traces of the residues of the other phase can still be observed, we perform a second iteration of the WFT based signal separation algorithm. The improved estimates of ϕ_1 and ϕ_2 are shown in figures 7.2(e) and 7.2(f), respectively.

7.2.2 Noise subspace inflation

A second category of method is based on the noise subspace inflation property of an enhanced matrix. How this property can be utilized for efficient fringe denoising is described in section 3.2.3. We will now see how this property can also be put to use for signal separation from a multicomponent EPF.

We suppose that the EPF is divided into a number of blocks, as illustrated in figure 3.11. The signal separation is performed within each block in an iterative manner. Let us first consider the block of the multicomponent EPF as a data matrix. The procedure of data matrix enhancement and the noise subspace inflation of the enhanced matrix are described in section 3.2.3.

The principle of signal separation is based on the supposition that the size of each block is sufficiently small to allow for the assumption that the phase within any block has a simple form, say, linear. In ideal conditions, that is, in the absence of

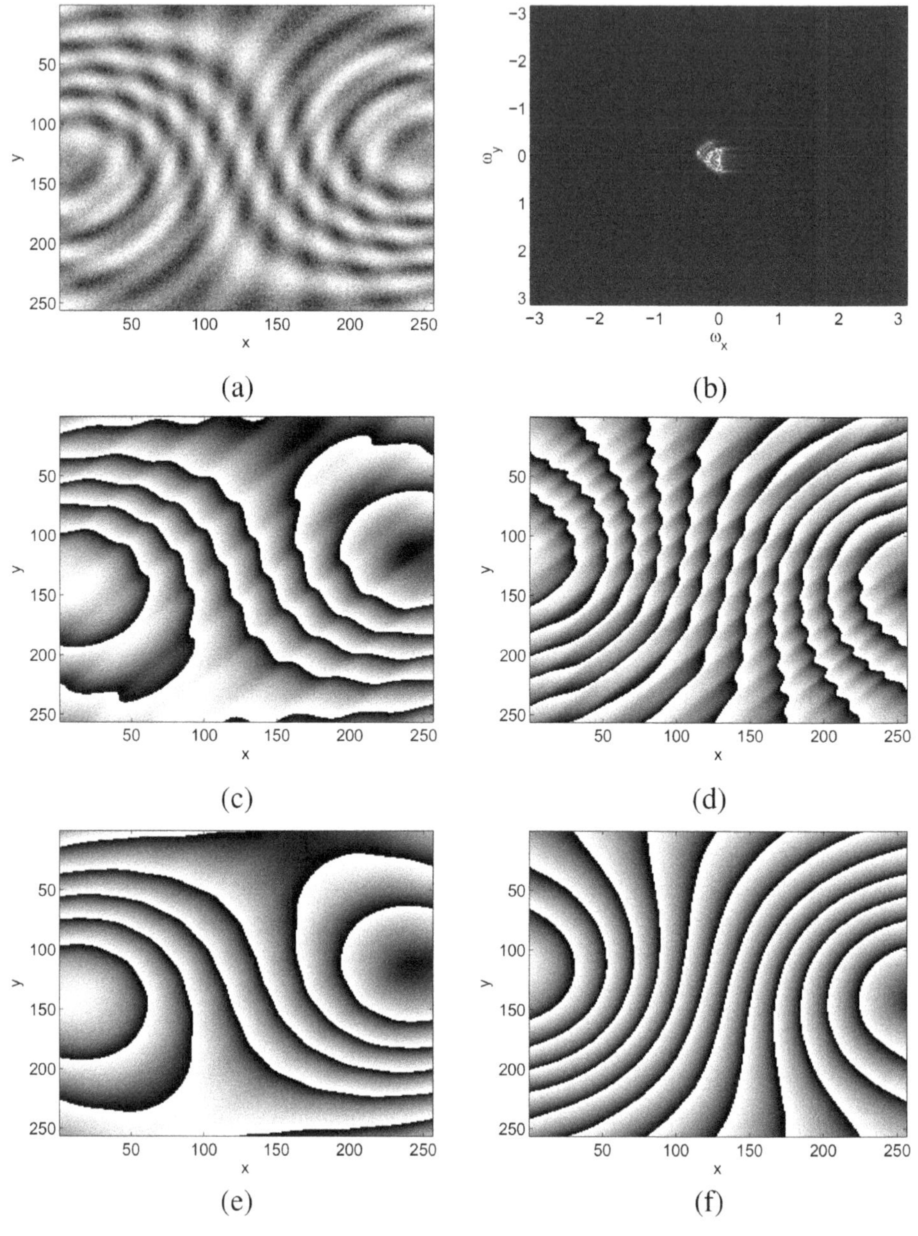

Figure 7.2. (a) Simulated moiré fringe pattern; (b) Fourier spectrum of the EPF. Wrapped estimate of (c) ϕ_1 and (d) ϕ_2 after the first iteration. Wrapped estimates of (e) ϕ_1 and (f) ϕ_2 after second iteration.

noise, the SVD of the data matrix reveals the presence of signals, in the form of distinct singular values. Considering a data matrix $\mathbf{\Gamma}_w$ of size $L \times L$, the SVD of $\mathbf{\Gamma}_w$ can be written as

$$\mathbf{\Gamma}_w = \boldsymbol{U}_w \mathbf{\Lambda}_w \boldsymbol{V}_w^*,$$

where $\boldsymbol{U}_w$ and $\boldsymbol{V}_w$ are matrices containing the left and right singular vectors, $\boldsymbol{u}_w$ and $\boldsymbol{v}_w$, respectively; $\boldsymbol{\Lambda}_w$ is a rectangular diagonal matrix with non-negative real singular values arranged in a descending order on its diagonal. Due to the presence of two signal components in the EPF, we note the presence of two dominant singular values with the remaining singular values being either zero, or very small. This distinction among the singular values offers a way of discerning signal components in the EPF. In the case of a noise corrupted EPF, the presence of noise related non-zero singular values is equally observed. Since due to the random nature of noise, its energy is prone to spreading throughout the frequency spectrum, the singular values associated with the noise are significantly smaller than the singular values related to the signal components. Thus, the signal components can still be identified in a reliable manner in the presence of noise.

An example of signal component identification in the presence of noise is shown in figure 7.3. A 100 × 100 size moiré pattern simulated at SNR = 10 dB and containing two signal components with linear phases is shown in figure 7.3(a). The signal amplitudes are set such that $A_1 = A_2$. The SVD of the EPF is computed and

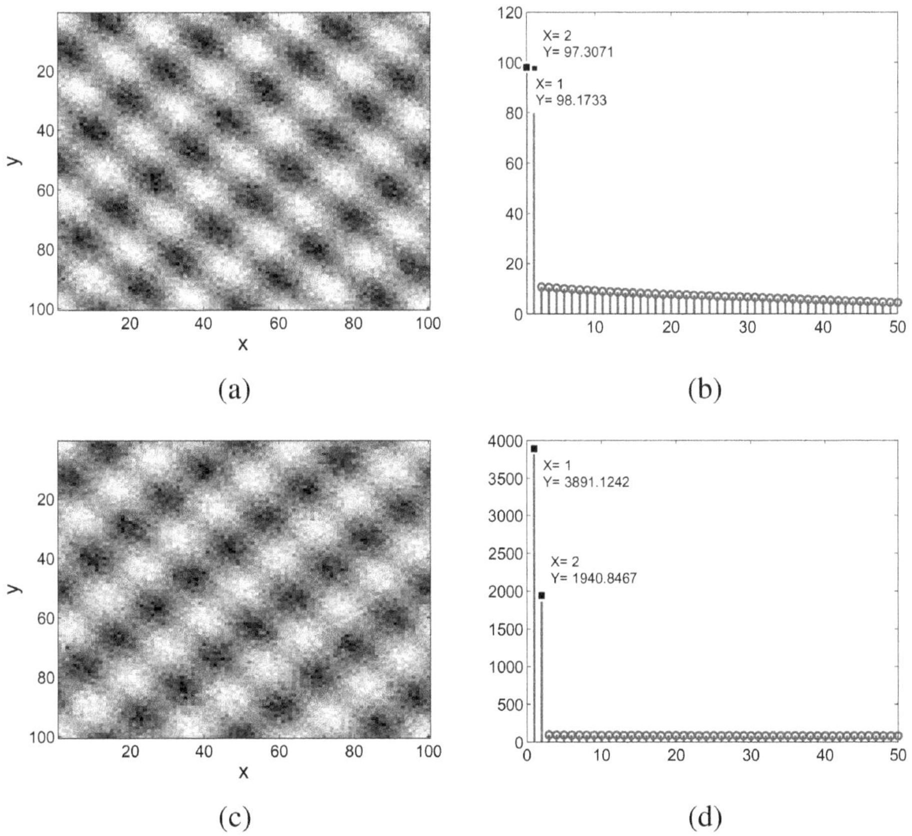

Figure 7.3. (a) Moiré pattern simulated with $A_1 = A_2$; (b) diagonal singular values in the SVD of $\boldsymbol{\Gamma}_w$; (c) moiré pattern simulated with $A_1 > A_2$; (d) diagonal singular values in the SVD of $\boldsymbol{\Gamma}_e$.

the plot of the diagonal values of the singular matrix is given in figure 7.3(b). Two dominant singular values corresponding to the two signal components are observed as the first and second singular values. The remaining singular values are related to noise. Note that the singular values of the two signal components are almost equal due to the setting, $A_1 = A_2$.

In the absence of any pertinent information on the relationship between the singular values and their respective signal components, the presence of ambiguity in the assignment of singular values to their respective signal components is clearly observed in figure 7.3(b). The amplitude discrimination criterion based on setting different intensities to the object illumination beams is utilized to remove the ambiguity in the assignment of singular values to their respective signal components. The amplitude discrimination criterion combined with the noise subspace inflation property of the enhanced matrix contribute to providing a noise robust and reliable signal separation.

The enhanced matrix $\mathbf{\Gamma}_e$ is generated by partitioning and stacking of the rows and columns of $\mathbf{\Gamma}_w$ as given in (3.9). The SVD of the enhanced matrix is

$$\mathbf{\Gamma}_e = \boldsymbol{U}_e \mathbf{\Lambda}_e \boldsymbol{V}_e^*.$$

In the case of the enhanced matrix too, the first two dominant singular values correspond to the two signal components of the EPF and the remaining singular values are related to the noise. The robustness against the noise is achieved by its effective decomposition in the inflated noise subspace whose dimension depends on the selection of the parameters K and P in (3.9) and (3.10). The values of K and P are selected [2] depending on the number of signal components (n_c) present in the signal as

$$n_c + 1 \leqslant K,\ P \leqslant \frac{1}{2}(L + 1).$$

In the case of two-beam digital holographic interferometry, $n_c = 2$. If we set, say, $A_1 > A_2$, the first and the second singular values and their associated singular vectors are related to the signal components with amplitudes A_1 and A_2, respectively. The remaining singular values represent the noise contribution. In the absence of noise, these singular values are zero. In the presence of noise, these singular values are still significantly smaller than those associated with the signal components. These facts are re-emphasized as follows:

1. In general, due to the whiteness of noise, the singular values associated with the signal components are greater than those associated with the noise.
2. Establishment of the amplitude discrimination criterion ($A_1 > A_2$) sets the first and second singular values as the singular values associated with the signal components with amplitudes A_1 and A_2, respectively.

If we reconstruct a signal $\mathbf{\Gamma}^{(1)}$ using the first singular value and its associated singular vectors, it corresponds to the signal component with amplitude A_1. Thus we have

$$\mathbf{\Gamma}^{(1)} = \boldsymbol{u}_{1c} \lambda_1 \boldsymbol{v}_{1c}^*, \tag{7.6}$$

where $\boldsymbol{u}_{1c}$ and $\boldsymbol{v}_{1c}$ represent the first columns of $\boldsymbol{U}_e$ and $\boldsymbol{V}_e$, respectively; λ_1 is the first diagonal element of $\boldsymbol{\Lambda}_e$. Since $\boldsymbol{\Gamma}^{(1)}$ is of the same size as $\boldsymbol{\Gamma}_e$, it has to be resized to the size of $\boldsymbol{\Gamma}_w$ prior to subsequent processing. This reconstructed and resized signal $\boldsymbol{\Gamma}^{(1)}$ is stored and its contribution from $\boldsymbol{\Gamma}_w$ is removed as follows:

$$\boldsymbol{\Gamma}^{(2)} = \boldsymbol{\Gamma}_w - \boldsymbol{\Gamma}^{(1)}. \tag{7.7}$$

The resulting signal $\boldsymbol{\Gamma}^{(2)}$ primarily contains the signal component with amplitude A_2 along with some residues of the signal component with amplitude A_1 and the noise. An enhanced matrix is generated using this signal and, similar to the method described in section 3.2.3, a noise filtering based on the selection of single highest singular value is performed. The filtered signal thus carries the information on the signal component with amplitude A_2.

In the example of signal component identification in figure 7.3, a moiré pattern simulated with the amplitude discrimination criterion, $A_1 > A_2$, is shown in figure 7.3(c). An enhanced matrix of the EPF is generated and the SVD of this enhanced matrix is computed. The plot of the diagonal values of the singular matrix is shown in figure 7.3(d). Two *dominant* and *distinct* singular values corresponding to the two signal components appear as the first and second singular values. The remaining singular values correspond to the noise. Observe the distinct nature of the first and second singular values appearing on account of the amplitude discrimination in figure 7.3(d). The noise related singular values are seen to be greatly reduced in comparison to the signal related singular values. This indicates an improved noise robustness resulting from the inflated noise subspace. Note that only first 50 singular values are considered for the purposes of illustration. The iterative method of signal separation from a multicomponent EPF takes the form:

1. The multicomponent EPF is divided into a number of blocks $\boldsymbol{\Gamma}_w$'s, each of size $L \times L$.
2. Within each block, the first signal component $\boldsymbol{\Gamma}^{(1)}$ is reconstructed as in (7.6).
3. A signal $\boldsymbol{\Gamma}^{(2)}$ containing the signal component with amplitude A_2 is obtained in accordance with (7.7).
4. The noise subspace inflation based denoising of $\boldsymbol{\Gamma}^{(2)}$ is performed.
5. For a more robust signal reconstruction, the first signal component is again reconstructed by re-applying the signal reconstruction method to the signal $\boldsymbol{\Gamma}^{(1)} = \boldsymbol{\Gamma}_w - \boldsymbol{\Gamma}^{(2)}$.
6. Steps 4 and 5 for extracting a single signal component at a time are implemented *iteratively* to obtain a reliable separation of the signal components.

In the WFT based signal separation approach, a frequency band whose value depends on the fringe density is provided as an input parameter to the algorithm. A larger frequency band is selected for high density fringe patterns and vice versa. The performance of signal separation also depends on the appropriate selection of the threshold values (Thr_1 and Thr_2). The noise subspace inflation based method is particularly well suited for applications involving fringe patterns with widely varying fringe densities, as its performance is independent of the fringe density (see section 3.2.3). In addition, no threshold selection is required in the signal identification and reconstruction as the singular values automatically appear in descending order in the SVD operation.

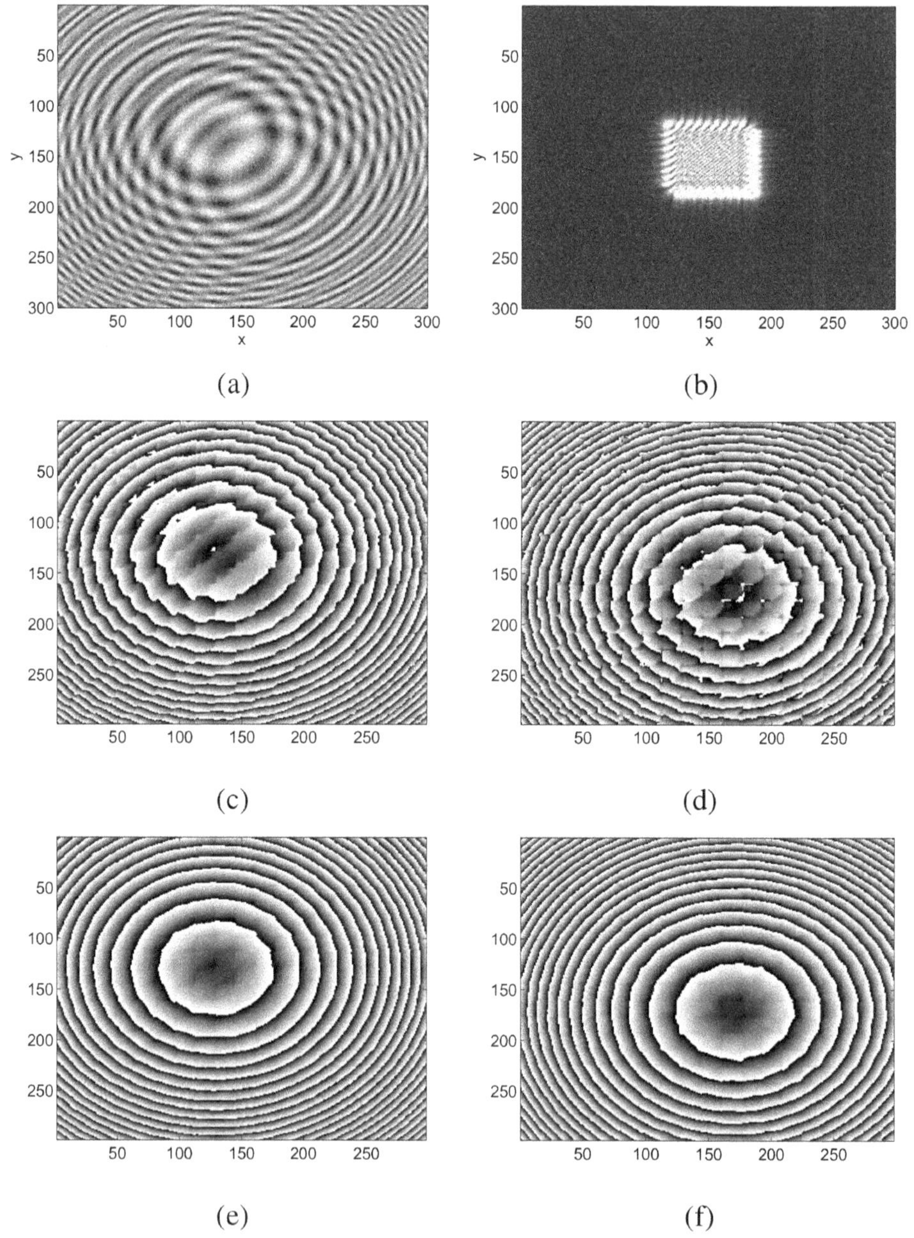

Figure 7.4. (a) Moiré fringe pattern simulated at SNR = 10 dB; (b) Fourier spectrum of the EPF; wrapped estimates of separated signal components with phases (c) ϕ_1 and (d) ϕ_2 after first iteration, and (e) ϕ_1 and (f) ϕ_2 after eight iterations.

To illustrate this point, a two-dimensional example of a moiré fringe pattern simulated with $A_1 > A_2$ at SNR = 10 dB is shown in figure 7.4(a). The EPF is divided into a number of blocks each of size 17 × 17. The Fourier spectrum of the EPF shown in figure 7.4(b) is indicative of a high spectral overlap of the two signal

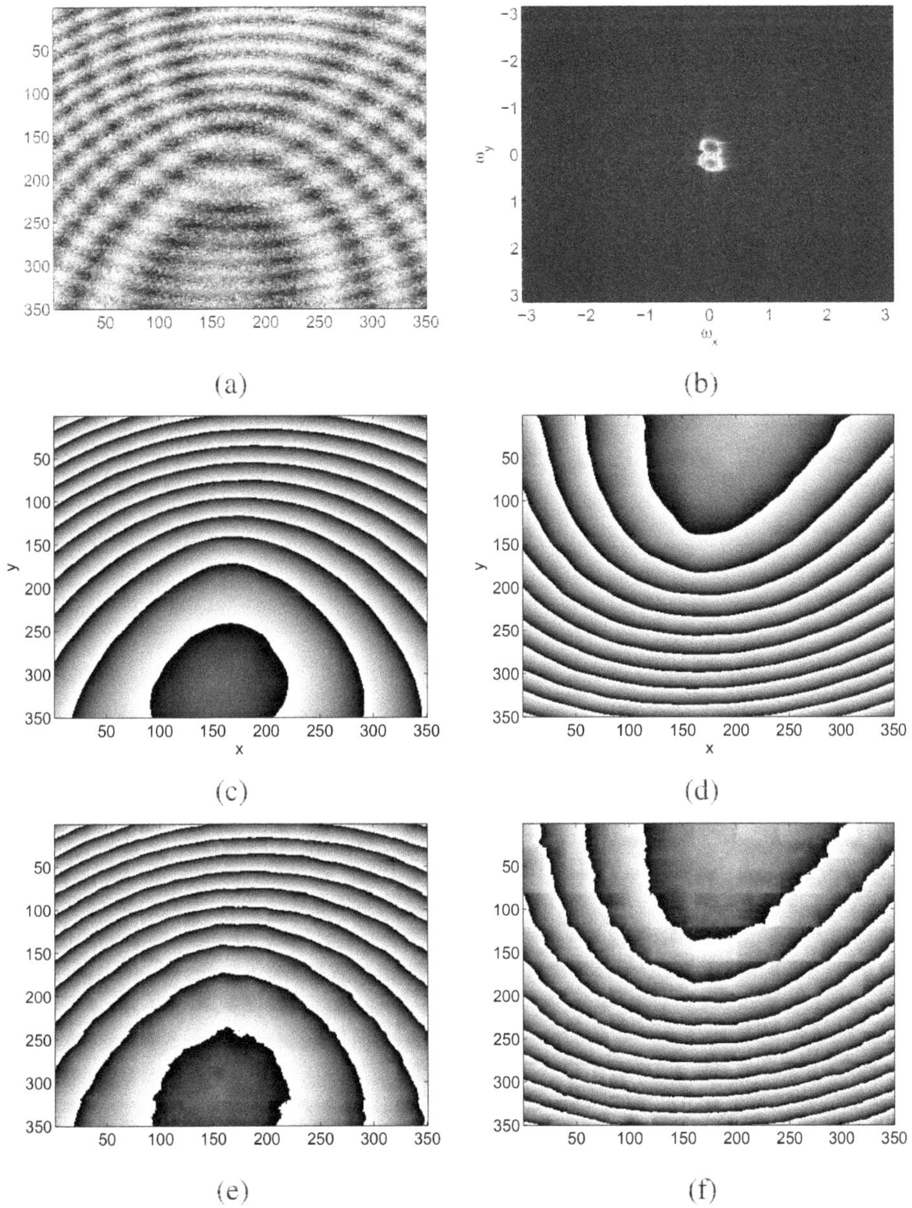

Figure 7.5. (a) Experimentally recorded moiré pattern in a two-beam digital holographic interferometry setup; (b) Fourier spectrum of the EPF signal. Estimated wrapped phases (c) ϕ_1 and (d) ϕ_2 using the WFT based method. Estimated wrapped phases (e) ϕ_1 and (f) ϕ_2 using the noise subspace inflation based method.

components. The wrapped phases ϕ_1 and ϕ_2 estimated after the first iteration are shown in figures 7.4(c) and (d), respectively. It is obvious from the figures that a reliable signal separation is not obtained in this case. However, after eight iterations of the application of the signal separation algorithm, a reliable signal separation is achieved. The wrapped phases of the separated signal components with amplitudes A_1 and A_2 are shown in figures 7.4(e) and 7.4(f), respectively. It is important to point out that in spite of the fact that the EPF contains high density fringes, the method provides a reliable and noise robust signal separation.

An experimentally recorded moiré pattern recorded in a two-beam digital holographic interferometry setup and corresponding to the out-of-plane and in-plane displacements of an aluminum plate is shown in figure 7.5(a). The Fourier spectrum of the EPF shown in figure 7.5(b) indicates the presence of two signal components. The WFT based signal separation implemented with a window size of $\sigma_x = 10$ and $\sigma_y = 10$ to obtain the wrapped estimates of ϕ_1 and ϕ_2 is illustrated in figures 7.5(c) and (d), respectively. The thresholds are selected as $Thr_1 = 10$ and $Thr_2 = 5$ in the first iteration and as $Thr_1 = 5$ and $Thr_2 = 2.5$ in the second iteration. The wrapped phase estimates of ϕ_1 and ϕ_2 obtained using the noise subspace inflation based method are shown in figures 7.5(e) and (f), respectively. The block size is selected as 41×41.

References

[1] Kemao Q 2008 A simple phase unwrapping approach based on filtering by windowed Fourier transform: a note on the threshold selection *Opt. Laser Tech.* **40** 1091–8

[2] Hua Y 1992 Estimating two-dimensional frequencies by matrix enhancement and matrix pencil *IEEE Trans. Sig. Proc.* **40** 2267–80

Chapter 8

Phase estimation: temporal fringe analysis

The previous chapters have been mainly devoted to spatial fringe analysis for extracting the phase and/or phase derivative estimates from a single fringe pattern. An alternate method of fringe analysis is based on temporal fringe processing in which the phase estimates are derived from a stack of multiple phase shifted fringe patterns. The class of interferometric techniques involving the recording of such phase shifted interferograms is generally referred to as phase shifting interferometry (PSI). The PSI setup consists of a phase shifting device placed along the path of a reference or object beam. A stack of fringe patterns is recorded through applications of controlled phase shifts to each consecutive fringe pattern recorded during the build-up of the stack.

PSI is by far the most commonly used interferometric technique across a wide range of metrology applications probably because of the features it has on offer, which include high accuracy in phase estimation, improved noise immunity in comparison to the spatial fringe analysis and localized fringe analysis for measurements with high spatial resolution. A number of algorithms have been developed over the years for extracting the phase from a stack of fringe patterns. Most of these algorithms are based on the concept of fringe signal modulation and demodulation similar to that performed in spatial fringe analysis. The basic difference resides in the way the auxillary information, in the form of a carrier signal, is introduced into the optical setup during the fringe modulation process. Whereas the signal modulation and demodulation operations are performed in the spatial domain in the spatial fringe analysis, they are performed in the temporal domain in temporal fringe analysis. To illustrate this, we consider the fringe patterns recorded in a PSI setup

$$\Gamma_n(x, y) = a(x, y) + b(x, y)\cos(\phi(x, y) + \alpha n). \tag{8.1}$$

The subscript n in $\Gamma_n(x, y)$ indicates the discretized time index with $n \in [0, N_f - 1]$, where N_f represents the number of recorded fringe patterns. The phase shift, α, is

doi:10.1088/978-0-7503-1469-5ch8

applied between two consecutive frames. Basically, α represents the frequency of the temporal carrier fringe signal. In (8.1), the background intensity, the fringe amplitude and the phase are considered to be functions of spatial coordinates x and y and independent of n. This makes the PSI associated setups more suited to static measurements. The fringe signal at a pixel (x,y) along the temporal direction can be expressed as

$$\Gamma(n) = a + b\cos(\phi + \alpha n). \tag{8.2}$$

The temporal fringe signal model in (8.2) is similar to the one-dimensional carrier modulated fringe signal model in (2.3) in spatial fringe analysis. The only difference is that whereas in the PSI the background intensity, a, the fringe amplitude, b, and the phase, ϕ, are constant with respect to n, these quantities are functions of x in the spatial fringe analysis. Similar to the spatial fringe analysis, the aim of many PSI algorithms is to derive the complex fringe signal $\Gamma_c = \frac{1}{2}b\exp(i\phi)$ from the real temporal fringe signal given in (8.2). The *arctan* operator applied on Γ_c provides the phase estimate in a wrapped form.

Transform based techniques discussed in chapter 2 in the context of spatial fringe analysis are equally applicable for analyzing the temporal fringe signal in (8.2). This being the case, a question that may arise is why should one need to develop and use dedicated algorithms for processing the fringe signals in PSI? It is simply because of the experimental limitation on the total number of fringe patterns (N_f) that can be recorded. The application of transform based spatial fringe analysis techniques would seem feasible if one could record a sufficiently large number of fringe samples temporally under the assumption of phase constancy. However, such an assumption being largely restrictive may not hold validity under normal circumstances. Therefore, the transform based techniques developed in the context of spatial fringe analysis cannot be applied directly to process the temporal fringe signal. In addition, there is also a physical limitation on the phase shifting device to record a very high number of fringe patterns with consistently constant phase shifts. These reasons have motivated researchers to develop dedicated PSI algorithms for retrieving the phase from a limited number of fringe patterns. An alternative representation of (8.2) is

$$\Gamma(n) = a + \frac{b}{2}\exp[i(\phi + \alpha n)] + \frac{b}{2}\exp[-i(\phi + \alpha n)]. \tag{8.3}$$

For the time being, considering a temporal fringe signal to be of infinite length, the FT of (8.3) can be written as

$$\Gamma(\omega) = a\delta(\omega) + \frac{b}{2}\exp(i\phi)\delta(\omega - \alpha) + \frac{b}{2}\exp(-i\phi)\delta(\omega + \alpha). \tag{8.4}$$

The representation of the temporal fringe signal in (8.4) is similar to the one-dimensional modulated carrier fringe signal given in (2.7). This chapter is an attempt to summarize different types of PSI algorithms for extracting phase from a stack of fringe patterns.

8.1 PSI algorithms with known phase shifts

Based on the *a priori* knowledge of the phase shift α, the PSI algorithms with known phase shifts, usually referred to as conventional N_f-step algorithms, involve the recording of N_f equally phase shifted interferograms. At any given pixel, a complex variable carrying the phase as its argument is computed as the weighted linear combination of the phase shifted intensities at that pixel. Considering the temporal fringe signal $\Gamma(n)$ at a pixel location (x,y), this complex variable can be written as

$$A_c \exp(i\hat{\phi}) = \sum_{n=0}^{N_f-1} C_n \Gamma(n), \tag{8.5}$$

where the amplitude A_c is a function of the local fringe amplitude, $b(x, y)$, and the complex valued weights, C_n. The phase estimate is computed as

$$\hat{\phi} = \arctan\left(\frac{\sum_{n=0}^{N_f-1} \Im\{C_n\Gamma(n)\}}{\sum_{n=0}^{N_f-1} \Re\{C_n\Gamma(n)\}}\right). \tag{8.6}$$

For a given number of frames (N_f), the weights C_n decide the performance of the PSI algorithm. A number of algorithms in relation with different values of N_f are reported in the literature [1]. The least squares phase shifting algorithm (LS-PSA), based on the pre-defined phase step of $\alpha = 2\pi/N_f$ [2, 3], provides a good example of such algorithms. In this case, the weights of the linear combination in (8.5) are given as $C_{LS-PSA} = \exp(i\alpha n)$ where $n \in [0, N_f - 1]$. And, the phase is estimated as

$$\hat{\phi} = \arctan\left(\frac{\sum_{n=0}^{N_f-1} \Gamma(n)\sin(\alpha n)}{\sum_{n=0}^{N_f-1} \Gamma(n)\cos(\alpha n)}\right), \quad \alpha = 2\pi/N_f.$$

Apart from the LS-PSA, the other PSI algorithms that we would like to cite (based on their popularity) in this category are 3-step, 4-step and 5-step algorithms [1, 4]:

$$\hat{\phi}_{3-\text{step}} = \arctan\left\{\left(\frac{1-\cos\alpha}{\sin\alpha}\right)\frac{\Gamma(0)-\Gamma(2)}{2\Gamma(1)-\Gamma(0)-\Gamma(2)}\right\}, \tag{8.7}$$

$$\hat{\phi}_{4-\text{step}} = \arctan\left\{\frac{\Gamma(3)-\Gamma(1)}{\Gamma(0)-\Gamma(2)}\right\}, \tag{8.8}$$

$$\hat{\phi}_{5-\text{step}} = \arctan\left\{\frac{2(\Gamma(3)-\Gamma(1))}{\Gamma(0)-2\Gamma(2)+\Gamma(4)}\right\}. \tag{8.9}$$

Whereas the 3-step PSI algorithm can be applied for any specified value of phase step α, the 4-step and 5-step algorithms are designed specifically for $\alpha = \pi/2$. Comparing (8.6) and (8.7)–(8.9), we obtain the set of weights that are associated with these algorithms:

$$C_{3\text{-step}} = [-\sin\alpha + (1 - \cos\alpha)i,\ 2\sin\alpha,\ -\sin\alpha - (1 - \cos\alpha)i],$$

$$C_{4\text{-step}} = [1,\ -i,\ -1,\ i],$$

$$C_{5\text{-step}} = [1,\ -2i,\ 2,\ 2i,\ 1].$$

At this point, let us consider a simple example of phase estimation. A simulated one-dimensional fringe signal and its associated phase are shown in figure 8.1(a). Phase shifted fringe signals are simulated with phase steps of $\alpha = \pi/2$. The 3-step algorithm in (8.7) is implemented with $\alpha = \pi/2$ to obtain the wrapped phase estimate shown in figure 8.1(b). In the ideal conditions, that is, under the assumption of pure sinusoidal temporal fringe signal (equation (8.2)) and exact phase step value of $\alpha = \pi/2$, all the three PSI algorithms described in (8.7)–(8.9) provide equivalent phase estimates. However, such ideal conditions are never met in practice. As a result, systematic and random errors appear in the thus computed estimated phase. These errors are caused by the presence of phase step miscalibration, non-linearity in the phase-shifting device, multiple beam interference, electronic noise, and nonlinearity in the imaging device [5]. Of the different sources of errors observed in practice, the two most important ones are the presence of *phase step miscalibration* and *harmonics* in the fringe signal. Much effort has been devoted towards the development of algorithms that allow for the accurate estimation of phase in the presence of such sources of error. In the following, we discuss two powerful methods of analyzing the performance of these algorithms designed in the case of known phase steps.

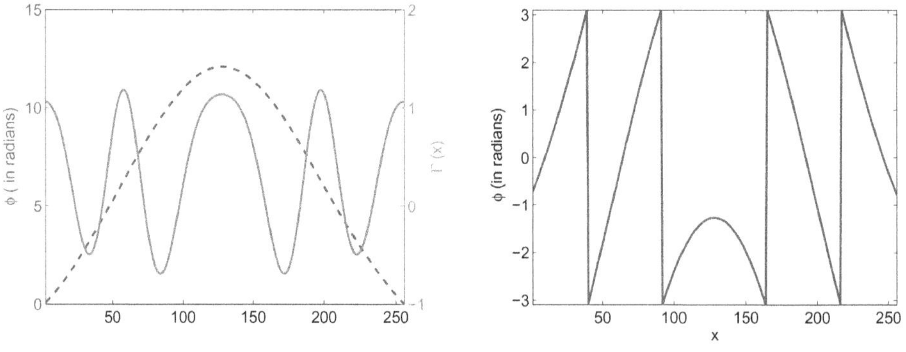

(a) (b)

Figure 8.1. (a) Simulated one-dimensional fringe signal and phase; (b) estimated wrapped phase using the 3-step PSI algorithm.

In the first method, an approach based on its characteristic polynomial (CP) representation of the PSI algorithms is proposed [6]. The temporal fringe signal in (8.3) can be represented as

$$\Gamma(n) = \sum_{l=-1}^{1} A_l \exp(i\phi l) \exp(i\alpha l n), \tag{8.10}$$

where $A_{-1} = b/2$, $A_0 = a$ and $A_1 = b/2$. Although the author in his analysis has considered the infinite order Fourier series expansion of the fringe signal, we will for the sake of simplicity take into account the Fourier series expansion of the fringe signal only up to the first order. Substitution of (8.10) into (8.5) gives,

$$\begin{aligned} A_c \exp(i\hat{\phi}) &= \sum_{n=0}^{N_f-1} C_n \sum_{l=-1}^{1} A_l \exp(i\phi l) \exp(i\alpha l n) \\ &= \sum_{l=-1}^{1} A_l \exp(i\phi l) \sum_{n=0}^{N_f-1} C_n [\exp(i\alpha l)]^n \\ &= \sum_{l=-1}^{1} A_l \exp(i\phi l) P[\exp(i\alpha l)], \end{aligned} \tag{8.11}$$

where the polynomial $P(t)$ of degree $N_f - 1$ is defined as

$$P(t) = \sum_{n=0}^{N_f-1} C_n t^n. \tag{8.12}$$

We can deduce accurate phase estimation from (8.11) by filtering out the zero order and the first order Fourier components associated with amplitudes A_0 and A_{-1}, respectively, and preserving only the first order Fourier component with amplitude A_1. The necessary and sufficient condition for achieving this is

$$P[\exp(i\alpha l)] = 0 \text{ for } l = -1, 0.$$

Note that although the above condition is given in the context of filtering the zero order and the first order Fourier components of the fringe signal, a similar condition holds too for filtering out higher order Fourier components ($l = \ldots, -3, -2, 2, 3, \ldots$). An important aspect of the above analysis is that the performance of any PSI algorithm can be characterized in the form of a polynomial, aptly named as CP and defined in (8.12). Basically, it is the location of the roots of the polynomial that decides upon the Fourier components of the fringe signal whose contributions are to be filtered out from the complex output signal.

The algorithms developed for the case of known phase shifts perform as quadrature filters. Similar to the discussion based on robust quadrature filtering, given in the context of spatial fringe analysis in section 2.1.6, the aim of a typical PSI algorithm is usually to filter out the zero order spectral component ($a\delta(\omega)$) and one of the first order spectral components $\left\{\frac{b}{2}\exp(-i\phi)\delta(\omega + \alpha)\right\}$ in (8.4). Indeed, the general theory of PSI algorithms [7] developed along these lines provides, as well, a common platform for the analysis and synthesis of these algorithms. This method is also a very useful tool in the design of digital finite impulse response (FIR) quadrature filter with the user defined frequency response characteristics. The complex fringe signal associated with the temporal fringe signal can be retrieved by passing the temporal fringe signal through a carefully designed digital FIR filter. Let us represent such a digital FIR filter with its impulse response $h(n)$ as

$$h(n) = \sum_{k=0}^{N_f-1} C_k\delta(n - k).$$

Desiring to obtain a complex signal at the output of the FIR filter, we consider the coefficients (C_k) associated with the filter to be complex valued. Representing the output of the FIR filter as the convolution of the temporal fringe signal and the FIR filter response

$$\begin{aligned}\Gamma_c(n) &= \Gamma(n)*h(n)\\ &= \sum_{k=0}^{N_f-1} h(k)\Gamma(n - k).\end{aligned}$$

Since in the filter output, the maximum overlap of the temporal fringe signal and the impulse response occurs at $n = N_f - 1$, the phase estimate can be derived as

$$\hat{\phi} = \mathrm{Arg}\{\Gamma_c(N_f - 1)\}.$$

Using the Fourier transform property of convolution, we obtain

$$\Gamma_c(\omega) = \Gamma(\omega)H(\omega), \tag{8.13}$$

where $\Gamma_c(\omega)$ and $H(\omega)$ represent the FTs of the complex fringe signal and the impulse response function, respectively.

Digital FIR filters can be characterized in terms of either a pole-zero or frequency response representation of the filters. The z-transform of the digital FIR filter can be represented as

$$H(z) = \sum_{k=0}^{N_f-1} C_k z^{-k}. \tag{8.14}$$

Similarly, the frequency response of the digital FIR filter is given by

$$H(\omega) = \sum_{k=0}^{N_f-1} C_k \exp(-i\omega k). \tag{8.15}$$

Basically, the frequency response or the frequency transfer function (FTF) associated with the PSI algorithm of the digital FIR filter is computed as the FT of its impulse response. Based on equations (8.4) and (8.13), the minimum conditions that are required to be satisfied by the PSI algorithm in order to qualify as a quadrature filter can be formally expressed as

$$|H(0)| = 0 \tag{8.16}$$

$$|H(-\alpha)| = 0. \tag{8.17}$$

Based on the relationship $z = \exp(i\omega)$ on the unit circle in complex z-plane, (8.14) and (8.15) are equivalent. It is interesting to note the similarity between the CP of the PSI algorithm in (8.12) and the z-transform of the impulse reponse of the digital FIR filter in (8.14). The equivalance between the two methods of analyzing PSI algorithms has been established in [8].

Now, let us analyze the 3-step, 4-step and 5-step algorithms using the CP and FTF methods. The roots of the CP associated with these algorithms, which are nothing but the zeros of the equivalent z-transform of the filter impulse response, are plotted in the first column of figure 8.2. The roots (zeros) are indicated by circles in the complex z-plane. The angular positions of these zeros indicate the Fourier order of the fringe signal which is filtered out in the process of computation of the associated complex variable. However, information on the performance of the PSI algorithm for signals with frequencies corresponding to locations other than those of the zeros is not available explicitly in these plots. This also makes it difficult to analyze the algorithm in the presence of perturbations around the phase step due to miscalibration. This limitation is overcome in the FTF based analysis of the PSI algorithms. The second column in figure 8.2 shows the plots of the normalized gain of the FTF, $|H(\omega)|$, associated with the PSI algorithms. The gains are plotted as a continuous function of frequency, normalized with respect to the phase step $\alpha = \pi/2$. The locations of zero gain indicated in these plots correspond to the respective location of zeros in the CP representation. Additionally, the algorithm performance can be clearly visualized at each frequency. It is interesting to observe that the gain of the FTF of a 3-step PSI algorithm does not reach its maximum at the selected phase step of $\alpha = \pi/2$. On the other hand, the gain values associated with the 4-step and 5-step algorithms reach their maximum at $\alpha = \pi/2$. As has been pointed out in the above, at least the zero and the first order fringe signal component should be filtered out in keeping with (8.16) and (8.17). Figure 8.2 also puts in perspective that these minimum conditions are met by all the three PSI algorithms. As a result, in ideal conditions, accurate phase estimate can be derived from these algorithms. However, the experimental imperfections and environmental disturbances may violate the very assumptions on which the PSI algorithms are based. This is where the performances of these algorithms differ from one another. The non-ideal conditions involve the sources of systematic and random errors. Numerous algorithms have been developed to compensate for systematic and random errors [5]. In most of these algorithms, the effect of different sources of errors on the accuracy of phase estimation is studied independently. The effect of sources of errors taken in a coupled manner has also been reported [9].

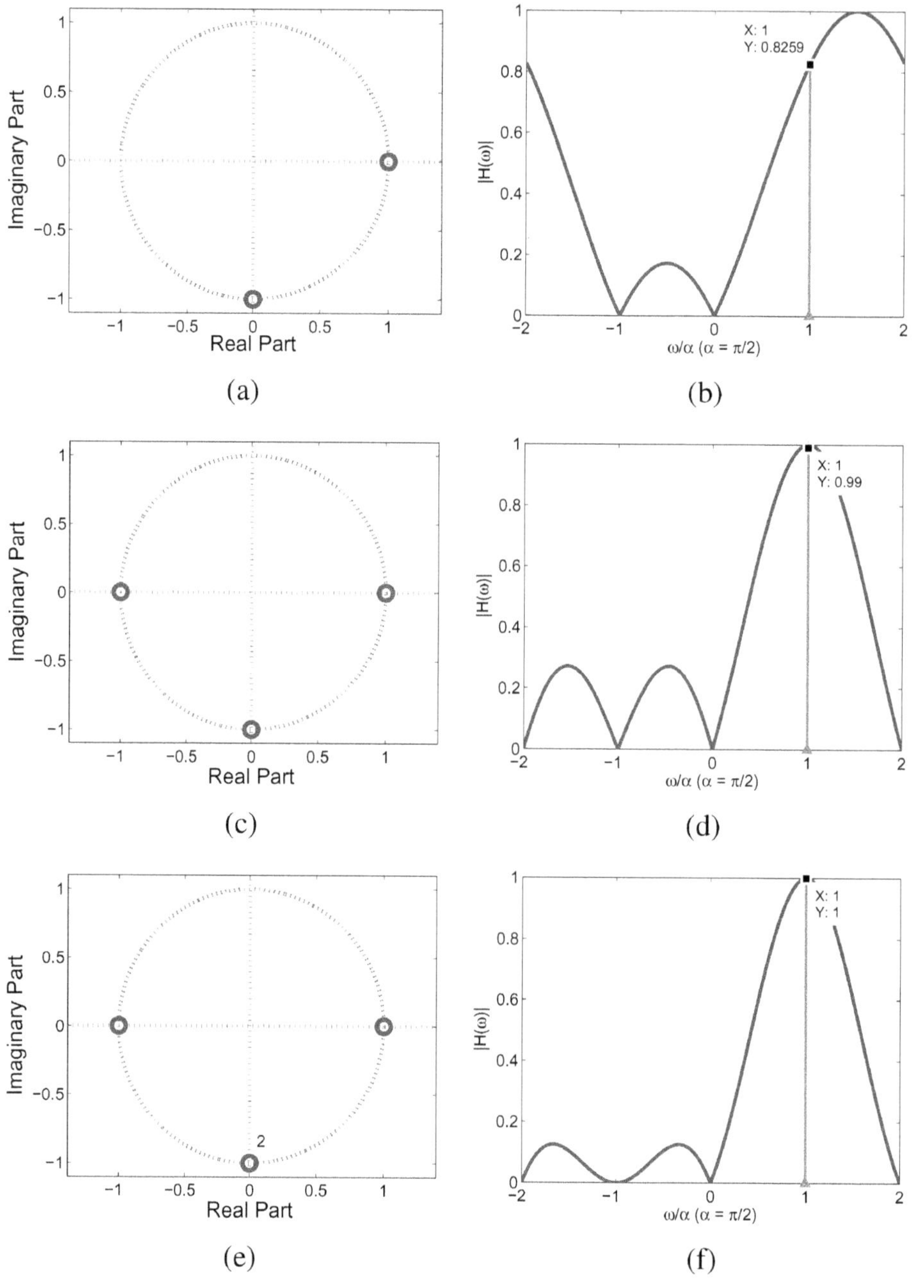

Figure 8.2. Zeros/roots of the characteristic polynomial associated with (a) 3-step, (c) 4-step and (e) 5-step PSI algorithms. Normalized gain of the frequency transfer functions associated with (b) 3-step, (d) 4-step and (f) 5-step PSI algorithms.

One of the most common sources of error in phase estimation is the phase step miscalibration. Although there are different ways of introducing a phase shift between the reference and object beams in an optical interferometry setup [5], the piezoelectric transducer (PZT) is the most common device used for this purpose. The PZT is known to show undesirable characteristics such as nonrepeatability, aging effect and nonlinear response to the applied voltage. In practice, the phase step imparted by the PZT may differ from the desired value. This mismatch between the actual and ideal phase steps is termed phase step miscalibration. Depending on the light beam configuration, the phase step may also vary spatially as well as between the consecutive interferogram pairs.

The phase step miscalibration creates a mismatch between the actual phase step and the phase step for which a PSI algorithm is designed. The effect of this mismatch appears in the form of error in the phase estimation, usually termed the detuning error. For a given level of phase step miscalibration, the detuning error may vary significantly in function of the algorithm used. The actual phase step imparted by the PZT can be written as

$$\alpha' = \alpha(1 + \varepsilon),$$

where ε is the normalized phase step miscalibration usually expressed as the percentage of the desired phase step. Considering a 20% phase step miscalibration with $\alpha = \pi/2$, we calculate the phase estimation errors using 3-step, 4-step and 5-step algorithms for the fringe signal example given in figure 8.1(a). The error plots shown in figure 8.3(a) demonstrate the high amount of robustness offered by the 5-step algorithm in comparion to the 3-step and 4-step algorithms vis à vis the phase step miscalibration. This improvement in the performance results from the presence of two zeros at $\omega = -\pi/2$, as shown in figure 8.2(e). The gain in the FTF associated with the 5-step PSI algorithm is of very low value in the vicinity of $\omega = -\pi/2$ as displayed in figure 8.2(f). Consequently, the fringe signal components with frequencies around $\alpha = -\pi/2$ are effectively filtered out and the detuning error is significantly reduced. In contrast, the gain in the FTFs associated with the 3-step and 4-step algorithms take significant values in the vicinity of $\omega = -\pi/2$. Therefore, in the presence of phase step miscalibration, the traces of fringe signal component at the actual phase step value lying around $\omega = -\pi/2$ appear in the complex output signal. This results in high detuning errors in the estimated phase. The analytical expression of detuning error derived in [10] is

$$\phi - \hat{\phi} = \frac{|H(-\alpha')|}{|H(\alpha')|} \sin(2\phi).$$

Another important source of error in phase estimation is due to the harmonic distortions in the fringe signal caused by the nonlinearities associated with the imaging device (CCD, CMOS) and multiple beam interference. In (8.10), we have assumed the ideal fringe signal to be consisting of only the zero and first order Fourier signal components. However, in practice, the presence of harmonics of

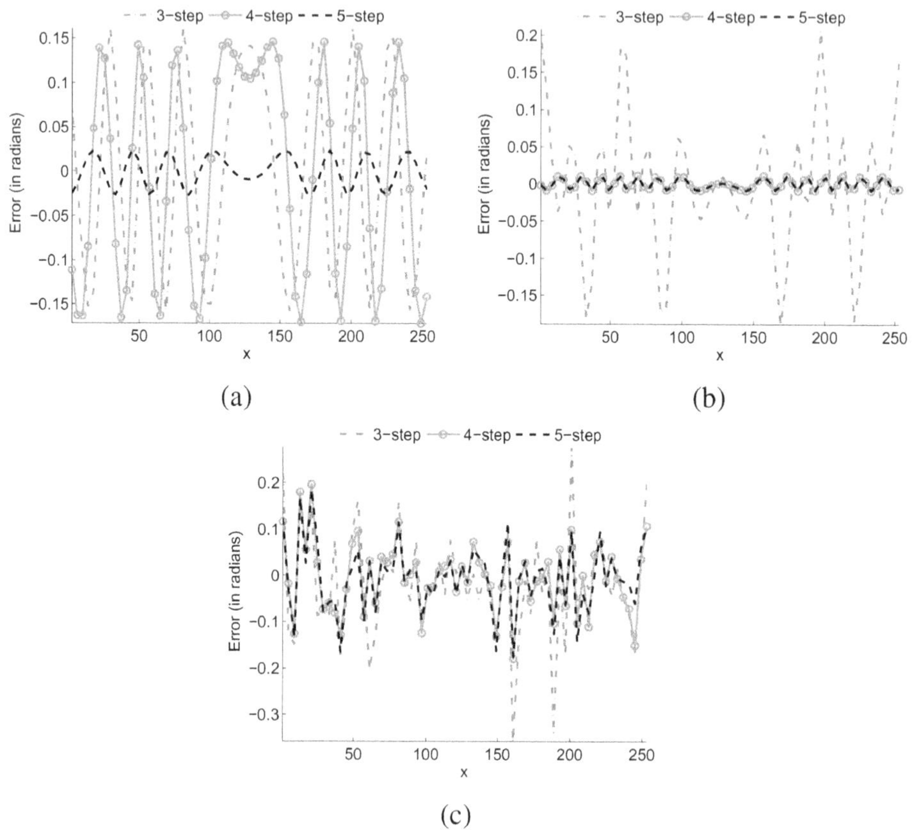

Figure 8.3. (a) Detuning error in phase estimation due to 20% phase step miscalibration (b) error in phase estimation in the presence of harmonic fringe distortion; (c) error in phase estimation in the presence of random additive white Gaussian noise. All values are in radians.

significant magnitude may distort the fringe signal from its pure sinusoidal nature. In line with (8.10), the harmonics distorted fringe signal can be represented as

$$\Gamma(n) = \sum_{l=-\infty}^{\infty} A_l \exp(i\phi l) \exp(i\alpha l n),$$

where all the harmonic components are taken into account. Since the magnitudes of the harmonic components decrease with the increase in their order, the harmonic rejection capability of these algorithms can be analyzed by assuming the presence of only the first few harmonics. The property that is desirable of any PSI algorithm resides in its ability to reject all the harmonic components excepting the fundamental signal for $l = 1$. The associated CP condition obtained from (8.11) is

$$P(\exp(i\alpha l)) = 0 \ \forall \, l \text{ except } l = 1.$$

Similarly the FTF condition follows as:

$$|H(\pm l\alpha)| = 0 \ \forall \, l \text{ except } l = 1.$$

Based on these conditions and on the roots of CP and FTF gain, associated with 3-step, 4-step and 5-step algorithms for $\alpha = \pi/2$, displayed in figure 8.2, it can be seen that the 3-step algorithm is capable of filtering $l = \{\ldots, -5, -4, -1, 0, 3, 4, 7, \ldots\}$ harmonic components. The 4-step and 5-step algorithms are capable of filtering comparatively more harmonic components $l = \{\ldots, -6, -5, -4, -2, -1, 0, 2, 3, 4, 6, 7\ldots\}$. The performance of these algorithms is evaluated with the harmonically distorted fringe signal example given in 8.1(a) with $l = 5$. The amplitude of the harmonics are set such that $A_l/A_{l+1} = 10$. The plot of errors in the estimation of phase in the presence of harmonic fringe distortion, as shown in figure 8.3(b), is indicative of the high robustness offered by the 4-step and 5-step algorithms in comparison to the 3-step algorithm.

Until this point, we have considered the fringe signal in the noiseless condition. However, the phase shifted interferograms are usually corrupted by random noise during the recording process [5, 11]. In general, this random noise is characterized as an additive white Gaussian noise. Upon the degree of robustness of the PSI algorithms against the noise will finally depend the accuracy of the estimated phase. Due to its whiteness, the spectral contribution of noise is distributed equally in the entire frequency plane. Accordingly, the noise robustness of these algorithms can be analyzed in function of the gain of the associated FTFs [12]. Ideally, in the presence of additive white Gaussian noise, the spectrum of the temporal fringe signal is shifted by a constant noise power. To be able to reject as much noise contribution as possible from the retrieved complex fringe signal, the desirable property for any PSI algorithm to own would be to have a minimum FTF gain outside the spectrum of the first order fundamental component of the fringe signal. Let us observe, once again, the plot of gains of the FTFs of the 3-step, 4-step and 5-step algorithms shown in figure 8.2. The area under the continuous spectrum gives us a fair idea about the noise rejection capability of these algorithms. The area of spectrum outside the spectral location of the desired complex fringe signal is maximum in the case of the 3-step algorithm, whereas it is minimum for the 5-step algorithm. It can therefore be deduced that the 5-step algorithm possesses higher noise robustness as compared to the 3-step algorithm. Indeed, this can be confirmed with the one-dimensional fringe signal example in the presence of noise. The errors in the phase estimates computed using the 3, 4 and 5-step algorithms are plotted in figure 8.3(c). The problem of noise in phase shifting interferometry has been addressed in detail in [11, 12]. The noise performance analysis in [12] indicates that the error in the estimated phase in the presence of noise is Gaussian in nature with zero mean and its variance is a function of the fringe amplitude and noise strength in the complex fringe signal. The phase error shown in figure 8.3(c) is in accordance with this observation.

A two-dimensional example of phase estimation is provided in figures 8.4 and 8.5. The phase shifted fringe patterns with the background intensity, $a(x, y)$, fringe amplitude, $b(x, y)$, phase step, $\alpha(x, y)$, and phase, $\phi(x, y)$ simulated at SNR = 20 dB are shown in figures 8.4(a)–(d), respectively. One of the phase shifted fringe patterns and the true wrapped phase are shown in figures 8.4(e) and (f), respectively. Figures 8.5(a), (b) and (c) show the estimated phase using the 3-step, 4-step and

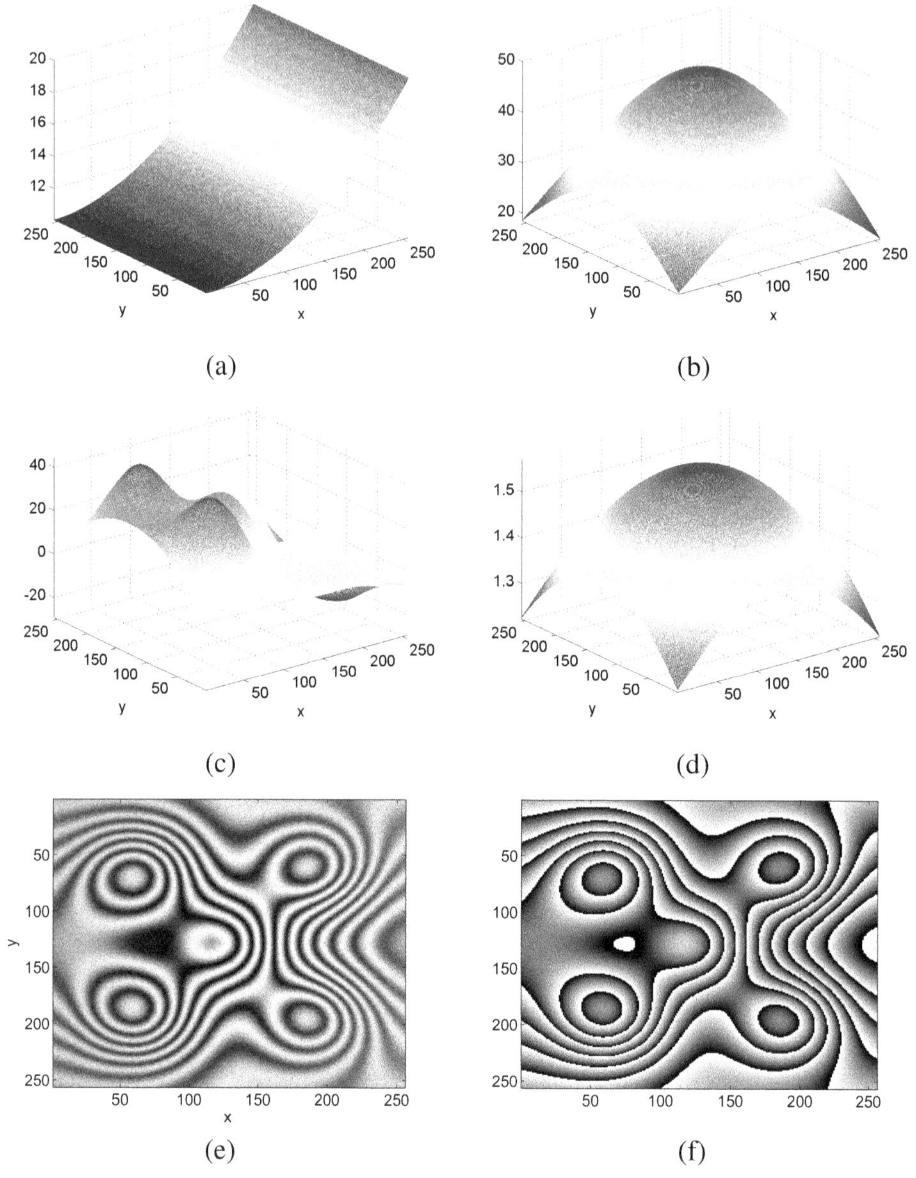

Figure 8.4. Simulated (a) background intensity, (b) fringe amplitude, (c) phase (in radians), (d) phase step (in radians) (e) one of the phase shifted fringe patterns at SNR = 20 dB, and (f) wrapped phase.

5-step PSI algorithms, respectively. The error maps in the phase estimations shown in figures 8.5(d), (e) and (f) indicate that the errors observed in the 3-step and 4-step algorithms exhibit twice the rate of variations as that of the true phase. On the other hand, the 5-step algorithm exhibits high robustness against the phase step miscalibration and noise.

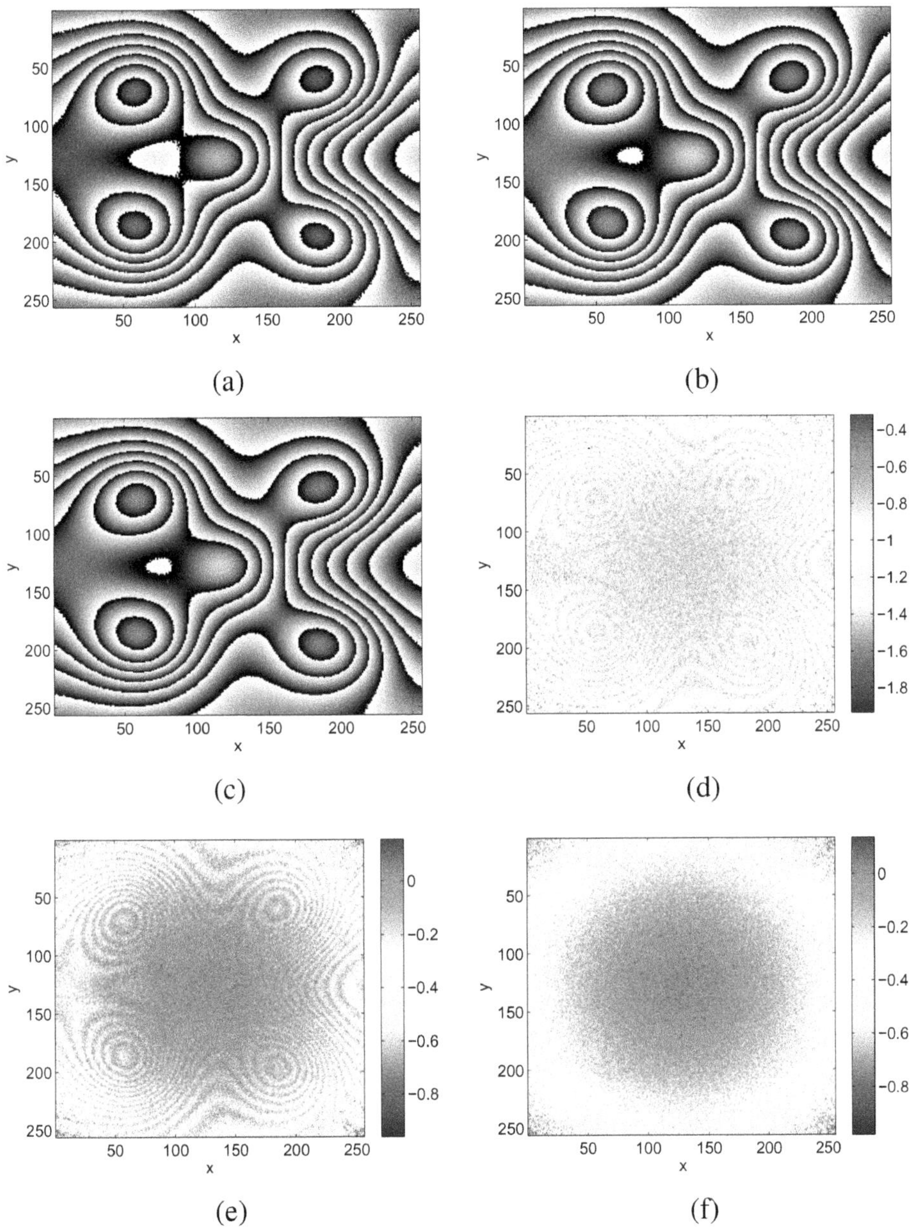

Figure 8.5. (a), (b) and (c) Estimated phase (in radians) and (d), (e) and (f) errors in phase estimation (in radians) using 3-step, 4-step and 5-step algorithms, respectively.

8.2 PSI algorithms with unknown and constant phase shifts

The PSI algorithms we have discussed so far offer the capability of estimating the phase in the presence of error sources that include phase step miscalibration, fringe harmonic distortion and random noise. However, the awkward part is that each of these algorithms require *a priori* knowledge of the phase steps. This section discusses those algorithms which have been developed with the aim of resolving such a problem by taking a two-step approach which consists of first estimating the phase step followed by estimating the phase using the estimated phase step. The additional step of phase step estimation is crucial for the estimation of the phase. These algorithms are discussed in the context of constant and unknown phase steps.

8.2.1 Carré algorithm

Perhaps the most popular algorithm designed in PSI with unknown phase steps is the 4-step Carré algorithm [1, 13]. This algorithm involves the process of nonlinear phase step estimation followed by linear phase estimation. The algorithm designed for $N_f = 4$, records phase shifted interferograms with phase steps of α. Accordingly, the phase step and phase estimate can be computed as

$$\tan(\hat{\alpha}/2) = \sqrt{\frac{3[\Gamma(1) - \Gamma(2)] - \Gamma(0) + \Gamma(3)}{\Gamma(0) + \Gamma(1) - \Gamma(2) - \Gamma(3)}}$$

$$\tan(\hat{\phi}) = \tan(\hat{\alpha}/2)\frac{\Gamma(0) + \Gamma(1) - \Gamma(2) - \Gamma(3)}{-\Gamma(0) + \Gamma(1) + \Gamma(2) - \Gamma(3)}.$$

It is interesting to note that for $\alpha = \pi/2$, the Carré algorithm boils down to the 4-step algorithm discussed in the previous section.

The Carré algorithm applied to the simulated fringe pattern in figure 8.4(e) provides the estimated phase as shown in figure 8.6(a). Since the algorithm proceeds by independently computing the phase step at each pixel, the phase is estimated accurately, as can be observed from the error map given in figure 8.6(b).

8.2.2 Generalized PSI algorithm based on subspace methods

Generalized PSI algorithms have been developed with a view to performing phase estimation in the case of unknown phase steps [14]. We have seen that a temporal fringe signal can be effectively modeled as a Fourier series up to a finite order, with order one in the ideal case. Thus, the problem of phase estimation can be treated from the perspective of parameter estimation, where the amplitudes, frequencies and phases of the Fourier components become the parameters to estimate, at least preliminarily.

In the context of holographic interferometry, although a typical PSI setup consisting of a single object illumination beam and a reference beam is sufficient for displaying, through fringe patterns, the deformation behaviour of an object submitted to load, it is only under particularly designed illumination systems that the setup becomes sensitive to either the in-plane or out-of-plane displacements. The use

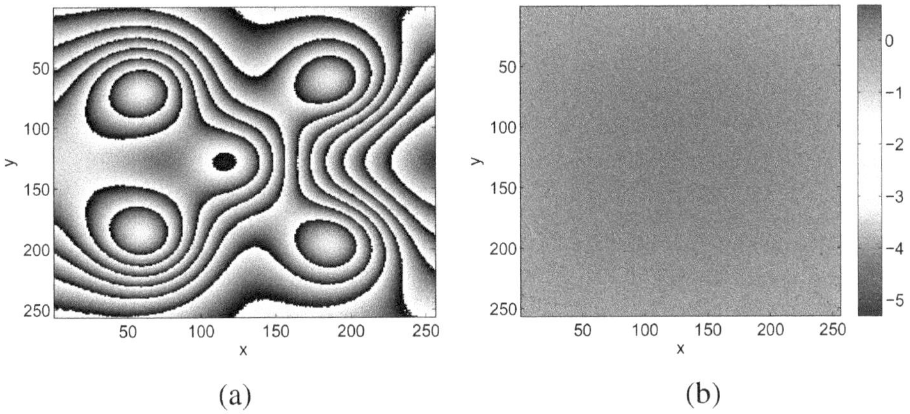

Figure 8.6. (a) Estimated phase using Carré algorithm; (b) error in the phase estimation (in radians).

of phase shifting techniques has been extended to embedding the capacity of multicomponent sensitivity to PSI setups [15, 16]. This is achieved by adding the feature of providing simultaneously multiple phase shifting to an interferometer. The setup consists of multiple object illumination beams placed in different planes around the object surface normal. Separate PZTs are placed along the path of each illumination beam. This configuration provides at its output a composite fringe system of the form moiré. The aim of the algorithm designed for this multi-beam setup is to extract the multiple phase information encoded in such a fringe system. We will now discuss the analysis of such a composite fringe pattern recorded in a multi-beam optical setup consisting of T number of object illumination beams. The Fourier series representation of the fringe signal can be expressed as

$$\Gamma(n) = a + \sum_{t=1}^{T} \sum_{l=-\infty, l\neq 0}^{\infty} A_t^{(l)} \exp(i\phi_t l) \exp(i\alpha_t l n),$$

where $A_t^{(l)}$ is the amplitude of the lth Fourier series component of the tth fringe signal; α_t and ϕ_t represent the phase step and phase associated with the tth fringe signal, respectively. For the simplicity of analysis, let us consider fringe signal representations to be perfectly sinusoidal ($|l| = 1$). Thus

$$\Gamma(n) = a + \sum_{t=1}^{T} A_t \exp\left(i\phi_t\right) \exp(i\alpha_t n) + \sum_{t=1}^{T} A_t \exp(-i\phi_t) \exp(-i\alpha_t n). \tag{8.18}$$

Equation (8.18) can be represented as

$$\Gamma(n) = a + \sum_{t=1}^{T} p_t q_t^n + p_t^*(q_t^*)^n, \tag{8.19}$$

where $p_t = A_t \exp(i\phi_t)$ and $q_t = \exp(i\alpha_t)$. In deformation analysis we are particularly interested in the simultaneous measurement of all three components of displacement. This requires the object to be illuminated with three object beams simultaneously, and the moiré fringe pattern is recorded. In this case, the temporal fringe signal is represented by $T = 3$. The three phases ϕ_1, ϕ_2 and ϕ_3 have to be estimated accurately in order to obtain the measurement of all the three displacement components. The fringe signal is a nonlinear function of the phase step and a linear function of the complex variable p_t carrying the information on the fringe amplitude and phase. The phase estimation can be obtained from the parameters p_t and α_t in (8.19). Once the phase steps α_t are estimated, the phases can be computed using the least squares method. As observed in (8.18), the temporal fringe signal can be modeled to be a linear combination of zero frequency signal (background intensity component) and complex sinusoidal signals with frequencies $\{-\alpha_T, \ldots, -\alpha_1, \alpha_1, \ldots, \alpha_T\}$. The phase shifted interferograms available for evaluating the phase being limited in number, the length of temporal fringe signal, N_f, is small. Periodogram is a commonly used method of frequency estimation [17]. However, since the periodogram based frequency estimation does not provide phase step estimation below its resolution limit of $1/N_f$, an increased number of algorithms taking advantage of the high resolution frequency estimation capability of subspace based methods have been reported in the literature [18]. We will describe under the following two subheadings the most popular subspace based methods of frequency estimation, restricting our discussion in the context of phase step estimation. Before moving on to the discussion on subspace based phase step estimation, let us now consider various notations of interest [17]:

- The fringe signal representation in (8.19) can be written as

$$\Gamma(n) = a + \sum_{t=1}^{2T} p_t q_t^n,$$

where

$$p_t = A_t \exp(i\phi_t),\ q_t = \exp(i\alpha_t) \qquad \text{for } t = 1, 3, \ldots, T$$
$$p_t = A_t \exp(-i\phi_t),\ q_t = \exp(-i\alpha_t) \qquad \text{for } t = 2, 4, \ldots, 2T.$$

- We consider the noisy temporal fringe signal as

$$\Gamma_\varepsilon(n) = \Gamma(n) + \varepsilon(n),$$

where the fringe signal is considered to be corrupted by the additive white Gaussian noise $\varepsilon(n)$ with zero mean and variance σ^2.

- The covariance of $\Gamma_\varepsilon(n)$ is defined as

$$r(n') = \mathbb{E}[\Gamma_\varepsilon(n)\Gamma_\varepsilon^*(n - n')] = \sum_{t=0}^{2T} A_t^2 \exp(i\alpha_t n') + \sigma^2\delta_{n',0},$$

where $\delta_{n',0}$ is the Kronecker delta ($\delta_{n',0} = 1$ for $n' = 0$, and $\delta_{n',0} = 0$ for $n' \neq 0$). It is important to note here that $\alpha_0 = 0$, $\alpha_2 = -\alpha_1$, ..., $\alpha_{2T} = -\alpha_{2T-1}$; $A_0 = a$, $A_2 = A_1$, ..., $A_{2T} = A_{2T-1}$ and $\phi_2 = -\phi_1$, ..., $\phi_{2T} = -\phi_{2T-1}$.

- The phase steps related vector and matrix notations are given as

$$\begin{aligned} \boldsymbol{d}(\alpha_t) &\triangleq [1 \exp(-i\alpha_t) \cdots \exp(-i(m-1)\alpha_t)]^T \\ \boldsymbol{D} &= [\boldsymbol{d}(\alpha_0)\, \boldsymbol{d}(\alpha_1) \cdots \boldsymbol{d}(\alpha_{2T})]. \end{aligned} \tag{8.20}$$

- Depending on the selection of the free parameter m we have,

$$rank(\boldsymbol{D}) = 2T + 1 \text{ if } m \geqslant 2T + 1 \text{ and if } \alpha_t \neq \alpha_{t'} \text{ for } t \neq t'. \tag{8.21}$$

- We consider the following definitions

$$\tilde{\boldsymbol{\Gamma}}_\varepsilon(n) \triangleq \begin{bmatrix} \Gamma_\varepsilon(n) \\ \Gamma_\varepsilon(n-1) \\ \vdots \\ \Gamma_\varepsilon(n-m+1) \end{bmatrix} = \boldsymbol{D}\tilde{\boldsymbol{\Gamma}}(n) + \tilde{\boldsymbol{\varepsilon}}(n)$$

$$\tilde{\boldsymbol{\Gamma}}(n) = [\Gamma_0(n) \cdots \Gamma_{2T}(n)]^T$$

$$\Gamma_t(n) = A_t \exp\big(i\alpha_t n + \phi_t\big)$$

$$\tilde{\boldsymbol{\varepsilon}}(n) = [\varepsilon(n) \cdots \varepsilon(n-m+1)]^T.$$

- The covariance matrix of $\tilde{\Gamma}_\varepsilon(n)$ is computed as

$$\boldsymbol{R} \triangleq \mathbb{E}[\tilde{\Gamma}_\varepsilon(n)\tilde{\Gamma}^*_\varepsilon(n)] = \boldsymbol{DAD}^* + \sigma^2\boldsymbol{I}_m, \tag{8.22}$$

  where

$$\boldsymbol{A} = \begin{bmatrix} A_0^2 & & 0 \\ & \ddots & \\ 0 & & A_{2T}^2 \end{bmatrix},$$

  and $\boldsymbol{I}_m$ is the identity matrix of size $m \times m$.
- Once the phase steps are estimated, the phase estimation is performed using the least squares method. We use the following notations

$$\alpha = [\alpha_0, \ldots, \alpha_{2T}]^T$$

$$\boldsymbol{\Psi} = \begin{bmatrix} 1 & 1 & \cdots & 1 \\ 1 & \exp(i\alpha_1) & \cdots & \exp(i\alpha_{2T}) \\ \vdots & & & \vdots \\ 1 & \exp(i\alpha_1(N_f-1)) & \cdots & \exp(i\alpha_{2T}(N_f-1)) \end{bmatrix} \tag{8.23}$$

$$\bar{\boldsymbol{\Gamma}}_\varepsilon = \big[\Gamma_\varepsilon(0), \ldots, \Gamma_\varepsilon(N_f-1)\big]^T \tag{8.24}$$

$$\boldsymbol{p} = [a, p_1, \ldots, p_{2T}]^T, \tag{8.25}$$

$$\tilde{\boldsymbol{p}} = (\boldsymbol{\Psi}^*\boldsymbol{\Psi})^{-1}\boldsymbol{\Psi}^*\bar{\boldsymbol{\Gamma}}_\varepsilon|_{\alpha=\hat{\alpha}}. \tag{8.26}$$

Multiple signal classification (MUSIC)

Multiple signal classification (MUSIC) based high resolution phase step estimation method exploits the properties of the eigenvalues and eigenvectors of the covariance matrix $\boldsymbol{R}$ computed with $m > 2T + 1$. Since $\boldsymbol{R}$ is a positive semidefinite matrix, its eigenvalues are nonnegative. The eigenvalues of $\boldsymbol{R}$ can be computed from its singular value decomposition as $\boldsymbol{R} = \boldsymbol{U}_R\boldsymbol{\Lambda}_R\boldsymbol{V}_R$. The eigenvalues appear on the diagonal of the singular value matrix $\boldsymbol{\Lambda}_R$ and they are arranged in a descending order as $\lambda_0 \geqslant \lambda_1 \geqslant \cdots \geqslant \lambda_{m-1}$. Let $\boldsymbol{U}_\Gamma = [\boldsymbol{u}_0, \boldsymbol{u}_1, \ldots, \boldsymbol{u}_{2T}]$ represent the eigenvectors associated with the eigenvalues $[\lambda_0, \lambda_1, \ldots, \lambda_{2T}]$, and $\boldsymbol{G}_\Gamma = [\boldsymbol{u}_{2T+1}, \boldsymbol{u}_{2T+2}, \ldots, \boldsymbol{u}_{m-1}]$ represent the eigenvectors associated with the eigenvalues $[\lambda_{2T+1}, \lambda_{2T+2}, \ldots, \lambda_{m-1}]$. Based on the definition of $\boldsymbol{D}$ and $\boldsymbol{A}$, we have,

$$rank(\boldsymbol{DAD}^*) = 2T + 1$$

and,

$$\lambda_t = \tilde{\lambda}_t + \sigma^2 \text{ for } t = 0, \ldots, 2T$$
$$\lambda_t = \sigma^2 \text{ for } t = 2T + 1, \ldots, m - 1,$$

where $\tilde{\lambda}_0, \ldots, \tilde{\lambda}_{2T}$ represent the eigenvalues of $\boldsymbol{DAD}^*$. This significant distinction between the two sets of eigenvalues results in the following relationship

$$\boldsymbol{RG}_\Gamma = \boldsymbol{G}_\Gamma\begin{bmatrix} \lambda_{2T+1} & 0 \\ & \ddots \\ 0 & \lambda_m \end{bmatrix} = \sigma^2\boldsymbol{G}_\Gamma \tag{8.27}$$

$$\boldsymbol{RG}_\Gamma = (\boldsymbol{DAD}^* + \sigma^2\boldsymbol{I}_m)\boldsymbol{G}_\Gamma. \tag{8.28}$$

From (8.28) and (8.27), we can deduce that

$$\boldsymbol{D}^*\boldsymbol{G}_\Gamma = 0.$$

The above relationship can be effectively used to obtain the estimation of the phase steps. The phase steps can be computed as the roots of the equation

$$\begin{aligned} \boldsymbol{d}^*(\alpha)\boldsymbol{G}_\Gamma G^*\boldsymbol{d}(\alpha) = 0 \\ \boldsymbol{d}^T(z^{-1})\boldsymbol{G}_\Gamma G^*\boldsymbol{d}(z) = 0, \end{aligned} \tag{8.29}$$

where $\boldsymbol{d}(z)$ is the vector equivalent to $\boldsymbol{d}(\alpha_t)$ given in (8.20) as $z = \exp(i\alpha)$. The angular positions of the roots of (8.29) lying near the unit circle in the complex z-plane provide the estimates of the phase steps. There are essentially T complex conjugate pairs of roots lying on the unit circle in the complex z-plane corresponding to the T fringe signal components.

Estimation of signal parameters by rotational invariance techniques (ESPRIT)
Let

$$\boldsymbol{D}_1 = [\boldsymbol{I}_{m-1}\,\boldsymbol{0}]\boldsymbol{D} \quad (m-1)\times(2T+1), \tag{8.30}$$

$$\boldsymbol{D}_2 = [\boldsymbol{0}\,\boldsymbol{I}_{m-1}]\boldsymbol{D} \quad (m-1)\times(2T+1), \tag{8.31}$$

where $\boldsymbol{I}_{m-1}$ is the identity matrix of dimension $(m-1)\times(m-1)$. It can be deduced from (8.30) and (8.31) that

$$\boldsymbol{D}_2 = \boldsymbol{D}_1\boldsymbol{C}, \tag{8.32}$$

where

$$\boldsymbol{C} = \begin{bmatrix} \exp(-i\alpha_0) & & 0 \\ & \ddots & \\ 0 & & \exp(-i\alpha_{2T}) \end{bmatrix}. \tag{8.33}$$

Note that the unitary matrix $\boldsymbol{C}$ carries the information on all of the phase steps. The transformation in (8.32) is termed rotation. The procedure of phase step estimation based on ESPRIT [17, 19] is described below:

- Similar to (8.30) and (8.31), matrices $\boldsymbol{U}_1$ and $\boldsymbol{U}_2$ are defined as

$$\boldsymbol{U}_1 = [\boldsymbol{I}_{m-1}\,\boldsymbol{0}]\boldsymbol{U}_\Gamma \qquad (m-1)\times(2T+1) \tag{8.34}$$

$$\boldsymbol{U}_2 = [\boldsymbol{0}\,\boldsymbol{I}_{m-1}]\boldsymbol{U}_\Gamma \qquad (m-1)\times(2T+1). \tag{8.35}$$

- From (8.22), we obtain,

$$\boldsymbol{R}\boldsymbol{U}_\Gamma = \boldsymbol{U}_\Gamma \begin{bmatrix} \lambda_0 & 0 \\ \ddots & \\ 0 & \lambda_{2T} \end{bmatrix} = \boldsymbol{D}\boldsymbol{A}\boldsymbol{D}^*\boldsymbol{U}_\Gamma + \sigma^2\boldsymbol{U}_\Gamma. \tag{8.36}$$

  It can be deduced from (8.36) that

$$\boldsymbol{U}_\Gamma = \boldsymbol{D}\boldsymbol{W}, \tag{8.37}$$

  where the nonsingular matrix $\boldsymbol{W}$ is defined as $\boldsymbol{W} = \boldsymbol{A}\boldsymbol{D}^*\boldsymbol{U}_\Gamma\tilde{\boldsymbol{\Lambda}}^{-1}$, with

$$\tilde{\boldsymbol{\Lambda}} = \begin{bmatrix} \lambda_0 - \sigma^2 & & 0 \\ & \ddots & \\ 0 & & \lambda_{2T} - \sigma^2 \end{bmatrix}. \tag{8.38}$$

- It immediately appears from (8.33)–(8.38) that

$$\boldsymbol{U}_2 = \boldsymbol{D}_2\boldsymbol{W} = \boldsymbol{D}_1\boldsymbol{C}\boldsymbol{W} = \boldsymbol{U}_1\boldsymbol{W}^{-1}\boldsymbol{C}\boldsymbol{W} = \boldsymbol{U}_1\overline{\boldsymbol{W}}.$$

- The estimate of $\overline{\boldsymbol{W}}$ can be computed as

$$\overline{\boldsymbol{W}} = \left(\boldsymbol{U}_1\boldsymbol{U}_1^*\right)^{-1}\boldsymbol{U}_1^*\boldsymbol{U}_2.$$

- The phase step estimates $\{\hat{\alpha}_1, \hat{\alpha}_3, \ldots, \hat{\alpha}_{2T-1}\}$ are computed as the arguments of $\{\upsilon_1, \upsilon_3, \ldots, \upsilon_{2T-1}\}$ where $\{\upsilon_{t=0}^{2T}\}$ are the complex eigenvalues of $\overline{\boldsymbol{W}}$.

Phase estimation

Once the phase steps are estimated, they are plugged into (8.23) to generate the Vandermonde matrix Ψ. Based on (8.24)–(8.26), the parameter vector p is estimated in a least squares sense. The phase estimates are subsequently computed as $\hat{\phi}_1 = \text{Arg}\{\tilde{p}(1)\}, \ldots, \hat{\phi}_T = \text{Arg}\{\tilde{p}(2T-1)\}$. The least squares approach of phase estimation enables the measurement of the background intensity and the fringe amplitudes as well. These quantities are obtained as $\hat{a} = |\tilde{p}(0)|$ and $\hat{b}_1 = 2|\tilde{p}(1)|, \ldots, \hat{b}_T = 2|\tilde{p}(2T-1)|$. Here, $b_1, \ldots, b_T$ represent the fringe amplitudes corresponding to the T signal components.

A PSI setup with a single object beam can be considered as a special case of the generalized PSI formulation considered in (8.19) with $T = 1$. Thus, the background intensity, the fringe amplitude and the phase estimates are computed as $\hat{a} = |\tilde{p}(0)|$, $\hat{b} = 2|\tilde{p}(1)|$ and $\hat{\phi} = \text{Arg}\{\tilde{p}(1)\}$. As an example, a simulation is performed with $N_f = 7$ at SNR = 20 dB, and the fringe parameters as given in figure 8.4. The estimates of background intensity, fringe amplitude ($b(x, y)/2$), phase step and the phase estimates computed using the ESPRIT based method ($m = 5$) are shown in figure 8.7(a)–(d), respectively. It is important to note that the subspace based method is capable of estimating the phase step at each pixel location. This feature offers an indispensable advantage in the phase estimation in the presence of divergent/convergent object illumination beams or, in other words, the spatially varying phase steps which such illumination provides. The errors in the phase estimation obtained using the ESPRIT and MUSIC methods are shown in figures 8.7(e) and 8.7(f), respectively. It is important to mention here that the accuracy of phase estimation depends on the value of m. In order to accurately compute the covariance matrix R and satisfy the rank condition described in (8.21), the value of m should be selected such that

$$2T + 1 < m < N_f.$$

It is suggested that m should not be set very close to N_f.

The parametric estimation of the phase step and phase using the subspace based methods also allows one to effectively handle the harmonic distorted fringe patterns. In the presence of harmonics, the parameter vector can be set to accommodate the parameters of harmonic components as its elements. If we consider the fringe distortion up to second order harmonic, then we can set $T = 2$ where $\alpha_2 = 2\alpha_1$. In essence, the Fourier series representation of the fringe signal is considered up to second order in this case. The higher order harmonic components can be treated in a similar manner. Under the assumption of the presence of harmonics, the required number of fringe patterns (N_f) also increases in order to accurately estimate the higher sized parameter vector. An example of harmonic distorted fringe pattern is shown in figure 8.8(a). Phase estimations are performed with $T = 2$, and $N_f = 7$ and 10. The errors in the phase estimation using ESPRIT in these two cases are shown in

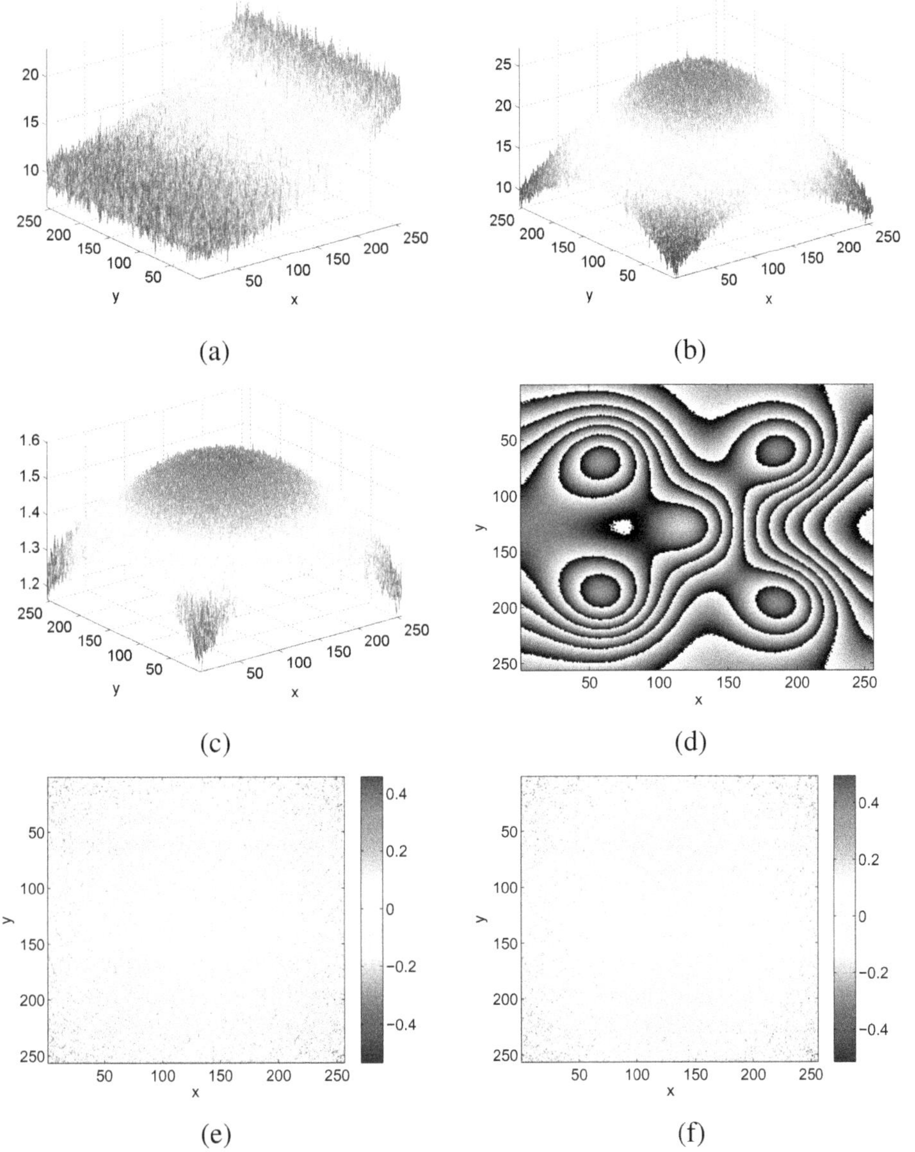

Figure 8.7. Estimated (a) background intensity (b) fringe amplitude ($b(x,y)/2$) (c) phase step (in radians) and (d) wrapped phase (in radians). Errors in the phase estimation using (e) ESPRIT and (f) MUSIC based methods.

figures 8.8(b) and (c), respectively. A higher number of fringe patterns helps in achieving improved robustness against the harmonic fringe distortion. Figure 8.8(d) shows the errors in the phase estimation using MUSIC.

A two-beam setup sensitive to the out-of-plane and in-plane components of displacement in the context of deformation analysis is reported in [15, 16]. Such a

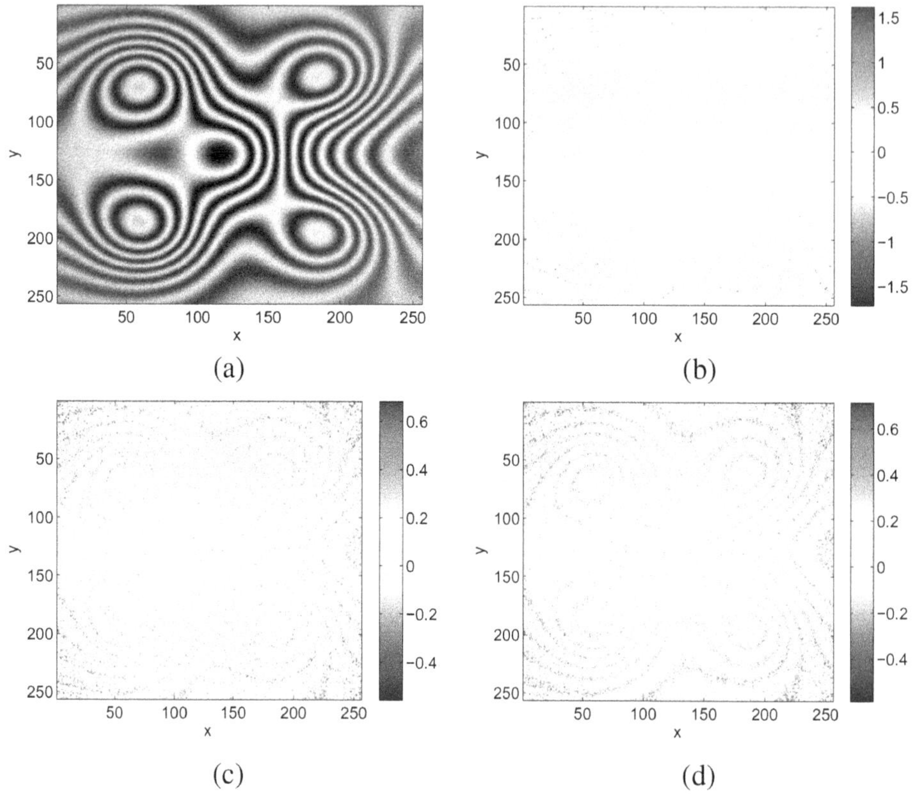

Figure 8.8. (a) Simulated fringe pattern containing second order harmonics. (b) Error in the phase estimation with $T = 2$, $N_f = 7$ (c) $N_f = 10$ using ESPRIT method. (d) Error in the phase estimation using the MUSIC method. All values are in radians.

setup, for example, can be extended to accommodate a higher number of object beams with the aim of encoding in the fringe system multiple physical quantities of interest. A setup with three object beams is sufficient to obtain all the three components of displacement. The fringe signal in this case can be represented as (8.19) with $T = 3$. As an example a moiré fringe pattern simulated at SNR = 20 dB is shown in figure 8.9. The fringe pattern is generated with $T = 3$, and phase steps of $\alpha_1 = \pi/2$, $\alpha_2 = \pi/3$ and $\alpha_3 = \pi/4$ are applied to the fringe system. It can be observed from the fringe pattern that it is very difficult to make a head or tail, even qualitatively of the fringe signal displayed in figure 8.9. The ESPRIT based fringe demodulation with $N_f = 20$ is employed to retrieve the three phase maps from the composite fringe pattern. The histogram of the accurately estimated phase steps is shown in figures 8.10(a)–(c). The retrieved phase maps are shown in figures 8.10(d)–(f). This example puts in the limelight the powerful capability of the subspace based methods in the simultaneous estimation of multiple phases corresponding to the multiple physical quantities embedded in the composite fringe signal.

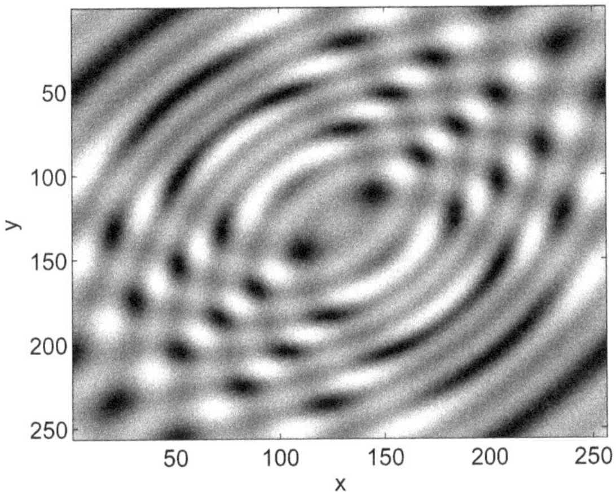

Figure 8.9. Simulated fringe pattern at SNR = 20 dB with $T = 3$.

8.3 PSI algorithms with random phase shifts

Although the generalized PSI algorithms offer a key advantage over their conventional counterparts in that they do not require *a prioi* knowledge on the phase step, they do, however, operate within the constraint of equal phase steps applied between two consecutive interferograms. This constraint restrains them from being applied to cases where the phase steps may vary between the consecutive interferograms. In such situations, the number of parameters that need to be estimated increases. The simplest approach is to estimate the distinct phase steps that are applied between each pair of consecutive interferograms. Several iterative PSI algorithms [20–22] are developed to handle the phase shifted interferograms with unknown and random phase steps. Due to their iterative nature, they require significant computation time to converge to a satisfactory solution. In order to facilitate an easier estimation of the unknown parameters, certain assumptions are imposed on their spatio-temporal variations. For example, in the generalized PSI algorithm proposed in [23], the background intensity, $a(x, y)$, and the fringe amplitude, $b(x, y)$, are approximated with two-dimensional polynomial functions of x and y with a pre-defined polynomial order. The number of polynomial coefficients that need to be estimated is much lower than the unknown intensity parameters. Temporal fringe demodulation techniques based on the recording of only two fringe patterns [24–26] have been reported to facilitate the measurement of dynamic events. The fringe patterns are normalized prior to phase step and phase estimation in order to reduce the number of unknown intensity parameters.

A powerful generalized PSI algorithm has been developed based on the concept of principal component analysis (PCA) [27, 28]. The main feature of the PCA based algorithm is its ability to accurately demodulate the randomly phase shifted interferograms in a computationally efficient manner. The computational efficiency is achieved on account of the non-iterative operation of the PCA based fringe demodulation. Of much interest in the image processing applications, the concept of

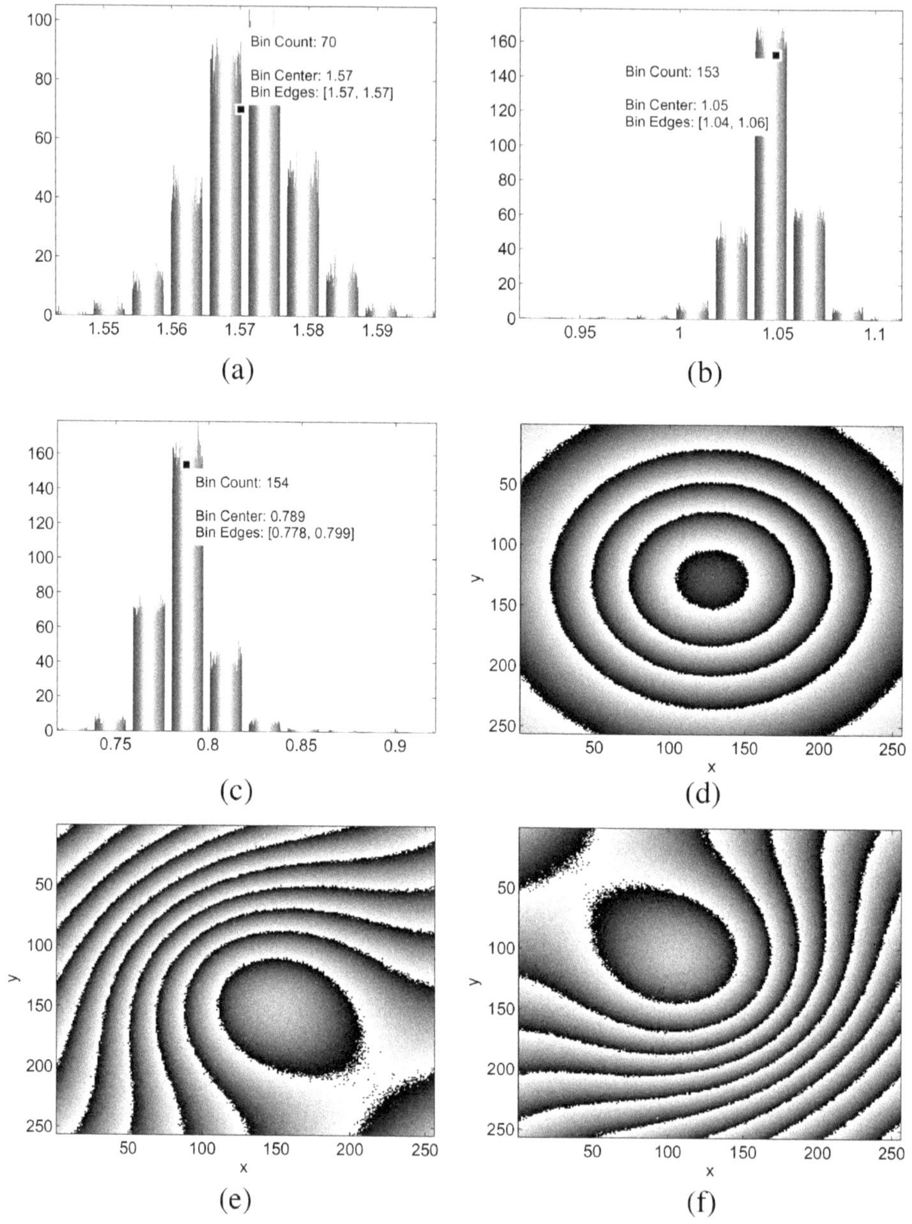

Figure 8.10. Histograms of the estimated phase steps α_1, α_2 and α_3 are shown in (a), (b) and (c), respectively. Corresponding estimated wrapped phases ϕ_1, ϕ_2 and ϕ_3 are shown in (d), (e) and (f), respectively.

the identification of the principal components using the PCA allows for extracting a fewer number of non-correlated images from a collection of a large number of correlated images. This small number of uncorrelated images, usually termed the **principal components**, represent most of the information in a least squares sense that is carried by a set of a large number of correlated images. In the generalized PSI with random phase steps, the temporal fringe intensity can be represented as

$$\Gamma(n) = a + b\cos(\phi + \alpha_n), \tag{8.39}$$

where, α_n represents the phase step imparted to the nth interferogram. Equation (8.39) can be expanded as

$$\begin{aligned}\Gamma(n) &= a + b\cos(\phi)\cos(\alpha_n) - b\sin(\phi)\sin(\alpha_n)\\ &= a + p_n\Gamma_c + q_n\Gamma_s,\end{aligned} \tag{8.40}$$

where $p_n = \cos(\alpha_n)$, $q_n = -\sin(\alpha_n)$, $\Gamma_c = b\cos(\phi)$ and $\Gamma_s = b\sin(\phi)$. Equation (8.40) represents the temporal fringe intensity as a linear combination of a, Γ_c and Γ_s. The weights p_n and q_n of this linear combination, known on account of the knowledge of α_n in the conventional PSI algorithms, allow for evaluating Γ_c and Γ_s by solving the linear equations of the form (8.40). On the other hand, these weights unknown in the generalized PSI with random phase steps, make the system of linear equations of the form (8.40) underdetermined, which is unsolvable. The PCA based PSI algorithms use the quadrature nature of the sought parameters, Γ_c and Γ_s, carrying the phase information. This fact can be expressed as

$$\sum_{x=0}^{N-1}\sum_{y=0}^{M-1}\Gamma_c(x, y)\Gamma_s(x, y) = 0.$$

For further analysis, the temporal fringe intensity, in (8.40), in the nth frame of the kth pixel is represented as

$$\Gamma_{kn} = a_k + p_n\Gamma_{kc} + q_n\Gamma_{ks},$$

where $k = [1, K]$ with $K = MN$; $\Gamma_{kc} = b_k\cos(\phi_k)$ and $\Gamma_{ks} = b_k\sin(\phi_k)$. The steps describing the procedure of PCA based temporal fringe demodulation are as follows:

1. A data matrix of size $N_f \times K$ is generated as

$$\mathbf{\Psi} = \begin{bmatrix} (a_1 + p_1\Gamma_{1c} + q_1\Gamma_{1s}) & \cdots & (a_K + p_1\Gamma_{Kc} + q_1\Gamma_{Ks}) \\ \vdots & \cdots & \vdots \\ (a_1 + p_{N_f}\Gamma_{1c} + q_{N_f}\Gamma_{1s}) & \cdots & (a_K + p_{N_f}\Gamma_{Kc} + q_{N_f}\Gamma_{Ks}) \end{bmatrix}.$$

2. In the PCA based fringe demodulation, the random phase steps are uniformly distributed in the range $(0, 2\pi)$. The background intensity filtered data matrix is computed as

$$\begin{aligned}\overline{\mathbf{\Psi}} &= \mathbf{\Psi} - \boldsymbol{m}_{\Psi}\\ &= \begin{bmatrix} (p_1\Gamma_{1c} + q_1\Gamma_{1s}) & \cdots & (p_1\Gamma_{Kc} + q_1\Gamma_{Ks}) \\ & \vdots\cdots\vdots & \\ (p_{N_f}\Gamma_{1c} + q_{N_f}\Gamma_{1s}) & \cdots & (p_{N_f}\Gamma_{Kc} + q_{N_f}\Gamma_{Ks}) \end{bmatrix},\end{aligned}$$

where

$$\boldsymbol{m}_\Psi = \begin{bmatrix} \left(\frac{1}{N_f}\sum_n \Gamma_{1n}\right) & \cdots & \left(\frac{1}{N_f}\sum_n \Gamma_{Kn}\right) \\ \vdots & \cdots & \vdots \\ \left(\frac{1}{N_f}\sum_n \Gamma_{1n}\right) & \cdots & \left(\frac{1}{N_f}\sum_n \Gamma_{Kn}\right) \end{bmatrix}.$$

3. The covariance matrix C_Ψ is computed as

$$\boldsymbol{C}_\Psi = \overline{\boldsymbol{\Psi}\boldsymbol{\Psi}^T}.$$

Here, the covariance matrix $\boldsymbol{C}_\Psi$ is a real and symmetric matrix of size $N_f \times N_f$. The (rs)th element of $\boldsymbol{C}_\Psi$ can be written as

$$\begin{aligned} C_{rs} &= \sum\nolimits_{k=1}^{K} (p_r \Gamma_{kc} + q_r \Gamma_{ks})(p_s \Gamma_{kc} + q_s \Gamma_{ks}) \\ &= \sum\nolimits_{k=1}^{K} p_r p_s \Gamma_{kc}\Gamma_{kc} + 2 p_r q_s \Gamma_{kc}\Gamma_{ks} + q_r q_s \Gamma_{ks}\Gamma_{ks}. \end{aligned} \tag{8.41}$$

Equation (8.41) and the definitions of p_n and q_n provide

$$\boldsymbol{C}_\Psi = \boldsymbol{\Gamma}_C \boldsymbol{P} + \boldsymbol{\Gamma}_S \boldsymbol{Q}, \tag{8.42}$$

where

$$\begin{aligned} \boldsymbol{P} &= \left[p_1 \cdots p_{N_f}\right]^T \left[p_1 \cdots p_{N_f}\right] \\ \boldsymbol{Q} &= \left[q_1 \cdots q_{N_f}\right]^T \left[q_1 \cdots q_{N_f}\right] \\ \boldsymbol{\Gamma}_C &= \sum\nolimits_{k=1}^{K} \Gamma_{kc}\Gamma_{kc} \\ \boldsymbol{\Gamma}_S &= \sum\nolimits_{k=1}^{K} \Gamma_{ks}\Gamma_{ks}. \end{aligned}$$

Since $\boldsymbol{P}$ and $\boldsymbol{Q}$ are the rank one matrices, we can deduce from (8.42) that C_Ψ is a rank two matrix with its eigenvectors given as

$$\boldsymbol{u}_P = \frac{\left[p_1, \ldots, p_{N_f}\right]}{\lambda_P}, \qquad \boldsymbol{u}_Q = \frac{\left[q_1, \ldots, q_{N_f}\right]}{\lambda_Q},$$

where

$$\lambda_P = \sum_{n=0}^{N_f} p_n^2, \ \lambda_Q = \sum_{n=0}^{N_f} q_n^2.$$

4. The diagonalization of the covariance matrix is generally performed using the singular value decomposition. The diagonal matrix D_Ψ can be computed as

$$\boldsymbol{D}_\Psi = \boldsymbol{U}\boldsymbol{C}_\Psi\boldsymbol{U}^T,$$

where the orthonormal transformation matrix $\boldsymbol{U}$ consists of $\boldsymbol{u}_P$ and $\boldsymbol{u}_Q$ as its first and second rows, respectively.

5. The principal components are computed using the hotelling transform as follows

$$y = \boldsymbol{U}\overline{\boldsymbol{\Psi}},$$

where the first and second rows of y represent the principal components as

$$y_{1k} = \{b_k \cos(\phi_k)\}, \qquad y_{2k} = -\{b_k \sin(\phi_k)\}.$$

6. The phase estimate is computed as

$$\phi_k = \arctan\left(\frac{y_{2k}}{y_{1k}}\right).$$

The PCA based temporal fringe analysis involves the computation of the SVD as its most important operation and which can be performed efficiently with the available software packages. Since the algorithm operates on the entire stack of fringe patterns simultaneously, the computation time is significantly low compared to the pixel-wise phase estimation methods.

The performance of the PCA based fringe demodulation depends on the number of recorded fringe patterns, N_f. That the accuracy is improved by the recording of a larger number of randomly phase shifted fringe patterns is shown with the simulation example given in figure 8.4. The error in the estimation of phase is shown in function of N_f in figures 8.11(a)–(c). On the other hand, the PCA based method also provides accurate phase estimation while operating on a set of linear phase shifted fringe patterns with comparatively lower values of N_f. Figure 8.11(d) shows the error in the estimation of phase computed for the case of linear phase shifts of $\alpha = \pi/3$, with $N_f = 5$.

Figures 8.12(a) and (b) show two of the $N_f = 10$ phase shifted fringe patterns, with $\alpha = \pi/3$, recorded in a classical holographic interferometry setup. The fringes correspond to the out-of-plane displacement of an aluminum plate subjected to deformation by a point load. For the purposes of the present discussion, the ESPRIT and PCA based methods are applied to this stack of fringe patterns to retrieve the corresponding phase map. Figures 8.13(a)–(d) show the estimated background intensity, the fringe amplitude, the phase step and the phase computed using ESPRIT. On the other hand, figure 8.13(e) shows the phase estimated using the PCA based method. Note that the apparent difference observed between the two estimated phase maps is only due to the constant offset phase.

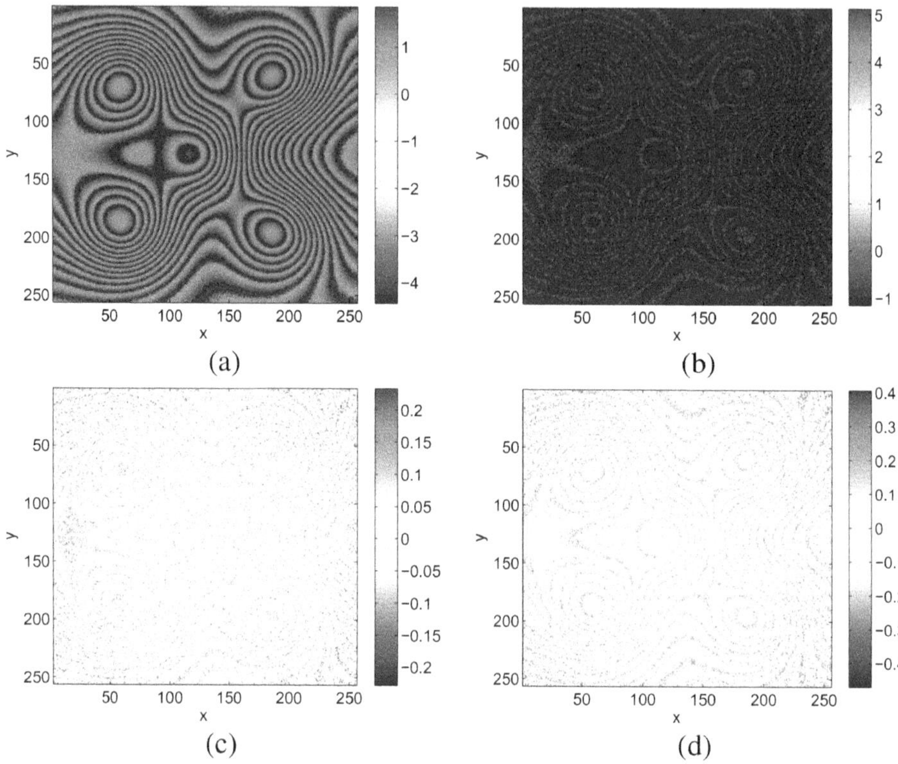

Figure 8.11. Errors in the estimation of phase using the PCA based demodulation of a set of fringe patterns with random phase steps in the range of $[0,2\pi]$ for (a) $N_f = 7$ (b) $N_f = 15$ and (c) $N_f = 20$. (d) Error in the estimation in the case of linear phase step of $\alpha = \pi/3$ with $N_f = 5$. All the error values are in radians.

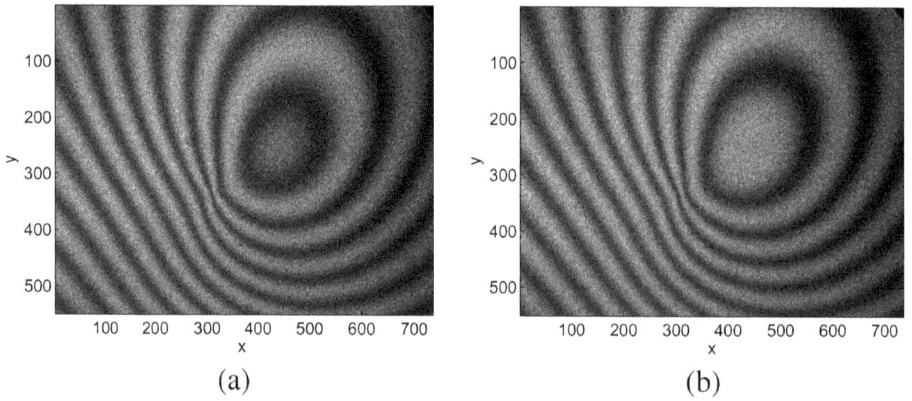

Figure 8.12. Two of the experimentally recorded phase shifted fringe patterns in a holographic interferometry setup.

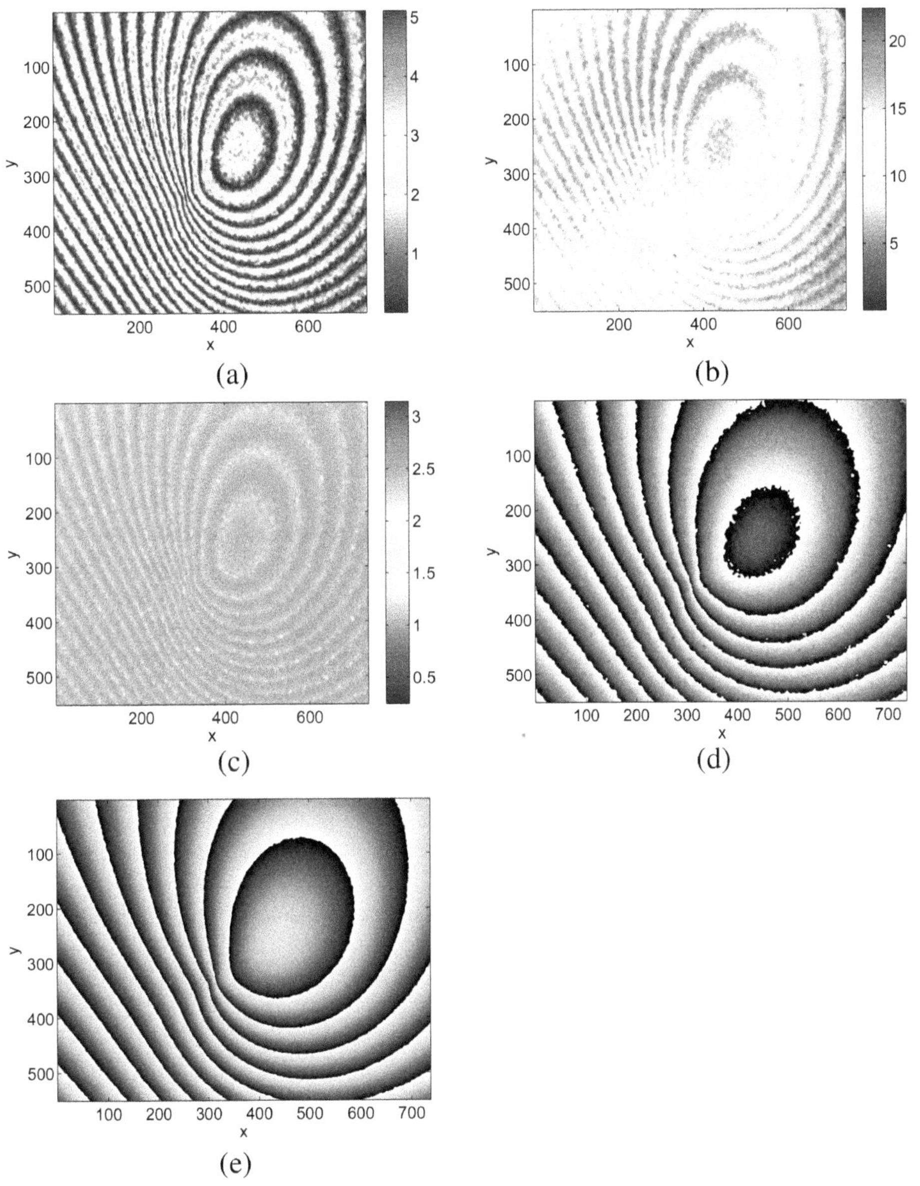

Figure 8.13. Estimated (a) background intensity (b) fringe amplitude (c) phase step (in radians) and (d) phase using ESPRIT. (e) Estimated phase using PCA.

References

[1] Creath K 1988 *V Phase-Measurement Interferometry Techniques in Progress in Optics* vol 26 (Amsterdam: Elsevier) pp 349–93

[2] Morgan C J 1982 Least-squares estimation in phase-measurement interferometry *Opt. Lett.* **7** 368–70

[3] Greivenkamp J E 1984 Generalized data reduction for heterodyne interferometry *Opt. Eng.* **23** 350–2
[4] Servin M, Quiroga J A and Padilla M 2014 *Fringe Pattern Analysis for Optical Metrology: Theory, Algorithms, and Applications* (Weinham: Wiley-VCH) pp 1–327
[5] Patil A and Rastogi P 2007 Moving ahead with phase *Opt. Lasers Eng.* **45** 253–7
[6] Surrel Y 1996 Design of algorithms for phase measurements by the use of phase stepping *Appl. Opt.* **35** 51–60
[7] Servin M, Estrada J C and Quiroga J A 2009 The general theory of phase shifting algorithms *Opt Exp.* **17** 21867–81
[8] Gonzalez A, Servin M, Estrada J C and Quiroga J A 2011 Design of phase-shifting algorithms by fine-tuning spectral shaping *Opt. Exp.* **19** 10692–7
[9] Hack E and Burke J 2011 Measurement uncertainty of linear phase-stepping algorithms *Rev. Sci. Instru.* **82** 061101
[10] Mosiño J F, Servin M, Estrada J C and Quiroga J A 2009 Phasorial analysis of detuning error in temporal phase shifting algorithms *Opt. Exp.* **17** 5618–28
[11] Rathjen C 1995 Statistical properties of phase-shift algorithms *J. Opt. Soc. Am.* **12** 1997–2008
[12] Servin M, Estrada J C, Quiroga J A, Mosiño J F and Cywiak M 2009 Noise in phase shifting interferometry *Opt. Exp.* **17** 8789–94
[13] Carré P 1966 Installation et utilisation du comparateur photoélectrique et interférentiel du bureau international des poids et mesures *Metrologia* **2** 13–23
[14] Patil A and Rastogi P 2005 Approaches in generalized phase shifting interferometry *Opt. Lasers. Eng.* **43** 475–90
[15] Rastogi P K 1992 Phase shifting applied to four-wave holographic interferometers *Appl. Opt.* **31** 1680–1
[16] Rastogi P K 1993 Phase-shifting holographic moiré: phase-shifter error-insensitive algorithms for the extraction of the difference and sum of phases in holographic moiré *Appl. Opt.* **32** 3669–75
[17] Moses R and Stoica P 1997 *Introduction to Spectral Analysis* (Upper Saddle River, NJ: Prentice Hall)
[18] Patil A, Langoju R, Rastogi P and Ramani S 2007 Statistical study and experimental verification of high-resolution methods in phase-shifting interferometry *J. Opt. Soc. Am.* **24** 794–813
[19] Roy R and Kailath T 1989 ESPRIT estimation of signal parameters via rotational invariance techniques *IEEE Trans. Acous. Speech. Sig. Proc.* **37** 984–95
[20] Cai L Z, Liu Q and Yang X L 2004 Generalized phase-shifting interferometry with arbitrary unknown phase steps for diffraction objects *Opt. Lett.* **29** 183–5
[21] Wang Z and Han B 2004 Advanced iterative algorithm for phase extraction of randomly phase-shifted interferograms *Opt. Lett.* **29** 1671–3
[22] Gao P, Yao B, Lindlein N, Mantel K, Harder I and Geist E 2009 Phase-shift extraction for generalized phase-shifting interferometry *Opt. Lett.* **34** 3553–5
[23] Juarez-Salazar R, Robledo-Sanchez C, Meneses-Fabian C, Guerrero-Sanchez F and Arvalo Aguilar L M 2013 Generalized phase-shifting interferometry by parameter estimation with the least squares method *Opt. Lasers. Eng.* **51** 626–32
[24] Vargas J, Quiroga J A, Sorzano C O S, Estrada J C and Carazo J M 2011 Two-step interferometry by a regularized optical flow algorithm *Opt. Lett.* **36** 3485–7

[25] Deng J, Wang H, Zhang F, Zhang D, Zhong L and Lu X 2012 Two-step phase demodulation algorithm based on the extreme value of interference *Opt. Lett.* **37** 4669–71
[26] Ma J, Wang Z and Pan T 2014 Two-dimensional continuous wavelet transform algorithm for phase extraction of two-step arbitrarily phase-shifted interferograms *Opt. Lasers Eng.* **55** 205–11
[27] Vargas J, Antonio Quiroga J and Belenguer T 2011 Phase-shifting interferometry based on principal component analysis *Opt. Lett.* **36** 1326–8
[28] Vargas J, Antonio Quiroga J and Belenguer T 2011 Analysis of the principal component algorithm in phase-shifting interferometry *Opt. Lett.* **36** 2215–7

www.ingramcontent.com/pod-product-compliance
Ingram Content Group UK Ltd.
Pitfield, Milton Keynes, MK11 3LW, UK
UKHW052039301026
9729IPUK00019B/306

9 780750 318877